KEY ENVIRONMENTS

General Editor: J. E. Treherne

SAHARA DESERT

The International Union for Conservation of Nature and Natural Resources (IUCN), founded in 1948, is the leading independent international organization concerned with conservation. It is a network of governments, non-governmental organizations, scientists and other specialists dedicated to the conservation and sustainable use of living resources.

The unique role of IUCN is based on its 502 member organizations in 114 countries. The membership includes 57 States, 121 government agencies and virtually all major national and international non-governmental conservation organizations.

Some 2000 experts support the work of IUCN's six Commissions: ecology; education; environmental planning; environmental policy, law and administration; national parks and protected areas; and the survival of species.

The IUCN Secretariat conducts or facilitates IUCN's major functions: monitoring the status of ecosystems and species around the world; developing plans (such as the World Conservation Strategy) for dealing with conservation problems, supporting action arising from these plans by governments or other appropriate organizations, and finding ways and means to implement them. The Secretariat co-ordinates the development, selection and management of the World Wildlife Fund's international conservation projects. IUCN provides the Secretariat for the Ramsar Convention (Convention on Wetlands of International Importance especially as Waterfowl Habitat). It services the CITES convention on trade in endangered species and the World Heritage Site programme of UNESCO.

IUCN, through its network of specialists, is collaborating in the Key Environments Series by providing information, advice on the selection of critical environments, and experts to discuss the relevant issues.

KEY ENVIRONMENTS
SAHARA DESERT

Edited by

J. L. CLOUDSLEY-THOMPSON

Foreword by

HRH THE DUKE OF EDINBURGH

Published in collaboration with the

INTERNATIONAL UNION FOR CONSERVATION OF
NATURE AND NATURAL RESOURCES

by

PERGAMON PRESS

OXFORD · NEW YORK · TORONTO · SYDNEY · PARIS · FRANKFURT

U.K.	Pergamon Press Ltd., Headington Hill Hall, Oxford OX3 0BW, England
U.S.A.	Pergamon Press Inc., Maxwell House, Fairview Park, Elmsford, New York 10523, U.S.A.
CANADA	Pergamon Press Canada Ltd., Suite 104, 150 Consumers Road, Willowdale, Ontario M2J 1P9, Canada
AUSTRALIA	Pergamon Press (Aust.) Pty. Ltd., P.O. Box 544, Potts Point, N.S.W. 2011, Australia
FRANCE	Pergamon Press SARL, 24 rue des Ecoles, 75240 Paris, Cedex 05, France
FEDERAL REPUBLIC OF GERMANY	Pergamon Press GmbH, Hammerweg 6, D-6242 Kronberg-Taunus, Federal Republic of Germany

First edition 1984

Library of Congress Cataloging in Publication Data
Main entry under title:
Sahara Desert.
(Key environments)
1. Desert biology—Sahara. 2. Sahara.
I. Cloudsley-Thompson, J. L. II. Series.
OH195.S3S24   1984   574.5'2652'0966   83-12136

British Library Cataloguing in Publication Data
Cloudsley-Thompson, J. L.
Sahara Desert. — (Key environments)
1. Sahara
I. Title II. Series
916.6   DT333

ISBN 0-08-028869-3

Printed in Great Britain by A. Wheaton & Co. Ltd., Exeter

The general problems of conservation are understood by most people who take an intelligent interest in the state of the natural environment. But if adequate measures are to be taken, there is an urgent need for the problems to be spelled out in accurate detail.

This series of volumes on "Key Environments" concentrates attention on those areas of the world of nature that are under the most severe threat of disturbance and destruction. The authors expose the stark reality of the situation without rhetoric or prejudice.

The value of this project is that it provides specialists, as well as those who have an interest in the conservation of nature as a whole, with the essential facts without which it is quite impossible to develop any practical and effective conservation action.

1984

Preface

The increasing rates of exploitation and pollution are producing unprecedented environmental changes in all parts of the world. In many cases it is not possible to predict the ultimate consequences of such changes, while in some, environmental destruction has already resulted in ecological disasters.

A major obstacle, which hinders the formulation of rational strategies of conservation and management, is the difficulty in obtaining reliable information. At the present time the results of scientific research in many threatened environments are scattered in various specialist journals, in the reports of expeditions and scientific commissions and in a variety of conference proceedings. It is, thus, frequently difficult even for professional biologists to locate important information. There is consequently an urgent need for scientifically accurate, concise and well-illustrated accounts of major environments which are now, or soon will be, under threat. It is this need which these volumes attempt to meet.

The series is produced in collaboration with the International Union for Conservation of Nature and Natural Resources (IUCN). It aims to identify environments of international ecological importance, to summarize the present knowledge of the flora and fauna, to relate this to recent environmental changes and to suggest, where possible, effective management and conservation strategies for the future. The selected environments will be re-examined in subsequent editions to indicate the extent and characteristics of significant changes.

The volume editors and authors are all acknowledged experts who have contributed significantly to the knowledge of their particular environments.

The volumes are aimed at a wide readership, including: academic biologists, environmentalists, conservationists, professional ecologists, some geographers, as well as graduate students and informed lay people.

John Treherne

Contents

CHAPTER 1

Introduction

J. L. CLOUDSLEY-THOMPSON

Department of Zoology, Birkbeck College, University of London

CONTENTS

1.1. FOREWORD

Named from the Arabic word meaning 'desert' or 'wilderness', the Sahara is the world's grandest and most awe-inspiring desert. Much of it probably man-made, most of recent origin, it has long been intimately related with mankind. It has witnessed bitter warfare throughout the ages and has invoked both the worst and the best in human nature. In his speech before the Battle of the Pyramids, in 1798, Napoleon Bonaparte said: 'Soldiers, consider that from the summit of these pyramids, forty centuries look down upon you.' He was not without a due sense of history. The scientist, too, must also be aware of history and evolution, as well as of animals and plants, soil and climate, day and night, summer and winter.

My own introduction to the Sahara, in 1941, was perhaps scarcely less dramatic than that of Napoleon's privileged soldiers. A fleeting vision, from the window of the troop train that bore us westwards to the war, of a gaunt camel beside a clump of palms silhouetted against a sea of silvery sand. The landscape, bathed in a flood of incandescent light from a crescent moon suspended in the velvet, star-spangled sky. A cool breeze giving little hint of the blazing heat of the days to follow, when Stuka dive-bombers would scream towards us out of the rising sun, their bombs hurtling earthwards like vicious black eggs to explode

1

with deafening crashes and clouds of yellow and black smoke. The rattle of Bren light machine-guns, their barrels almost red-hot as we replied with magazine after magazine—burning vehicles, crashed planes, the squat grey panzers of the Deutsche Afrika Korps firing high-velocity shells—and the sinister, square black screens of the deadly 88-mm anti-aircraft/anti-tank guns that wrought such havoc among our lightly armoured Crusader tanks. All these made an unforgettable, but by no means inappropriate, introduction to a pitiless environment in which human warfare, through the ages, has matched the severity of the incessant struggle for survival of every animal and plant.

In time of warfare, the signal for attack by the Shaiqiya of the northern Sudan would be a warbling cry, *lilli-lilli-loo*, given by one of the tribe's most beautiful girls, mounted on a gorgeously caparisoned camel. Then the cavalry would charge, uttering their sardonic war-cry *Salaam Aleikum*. Under the mistaken impression that bullets could not kill true believers, the tribe was almost annihilated when it attacked the Turks in this way at the Battle of Korti, in December 1820. Courage alone makes little impression against modern weapons. In addition to their swords and spears, Osman Digna's brave warriors, who fought the Battle of Sinkat in 1884, and were the only enemy soldiers ever to break a British square, were equipped with firearms captured from the ill-fated Hicks expedition of a year earlier. As well as possessing courage, fortitude and patience, however, many of the peoples of the Sahara are among the most hospitable and generous in the world.

In the blinding glare of the midday heat, when gazelles and camels huddle in the scanty shade of bare rock or sparse vegetation, and when the salty sweat stings one's eyes, the desert could easily qualify for one of Dante's infernal 'bowges' but, in the cool of evening, as a crimson sun tints the western sky with indigo and splashes the scattered clouds with violet rays, the bizarre and barren landscape is often strikingly beautiful. The Sahara exerts a strange fascination on every traveller, whatever his profession. There is something almost mystical in its eroded features, something that excites the imagination whilst stimulating academic interest. Everyone concerned with this volume must hope that it will not merely provide otherwise inaccessible scientific information, but that it will simultaneously evoke curiosity, thereby encouraging others to add to the store of human knowledge about this immense geographical region and its resilient inhabitants.

1.2. THE PHYSICAL ENVIRONMENT

From the Atlantic Ocean on the west, the Sahara stretches right across northern Africa to the Red Sea and the highlands of Ethiopia, an area of some $9,100,000$ km^2 (3,500,000 sq. miles). Comprising a major portion of the Great Palaearctic desert, it presents a variety of forms. Sometimes it is a rocky plateau, sometimes a pebbly plain. In some regions, there are huge depressions with dunes of shifting sand; in others, mountainous escarpments. As the observer approaches the central core, where rainfall is so infrequent as to be almost non-existent most years, he reaches an area where the landscape appears almost lunar because erosion by wind has for so long exceeded erosion by water. Even in such 'dead' spots as the central Sahara, however, the land is not completely lifeless. Particles of dried vegetation, carried by wind from the desert's edge, accumulate among dunes and in rocky depressions to support a small population of insects and the scorpions and lizards that prey on them.

The Sahara is a region of extreme aridity, high temperatures and violent winds. Most of the land surface is occupied by shifting sand dunes or *ergs* such as those in Algeria and Libya, wind-scoured plains strewn with gravel or boulders, known as *reg* or *serir* and *hammada* composed of denuded rock plateaux, smoothed and polished by wind abrasion and flattened by denudation. Libya's great *selima* sand sheet, which stretches for some 7800 km^2 (3000 sq. miles), consists of a thin layer of sand covering such eroded bedrock.

1.2.1. Rainfall

Desert rainfall is notoriously variable. The percentage departure from the normal (mean) annual rainfall exceeds 40% in the Sahara. At Tamanrasset, Algeria, 160 mm (6.3 in.) fell in one year; in another, only 6.4 mm (0.25 in.). During the three years following September 1933, only 2 mm, 3 mm and 5 mm, respectively, were recorded at Helwan, Egypt; but no less than 125 mm (5 in.) fell in the year 1945–46. Again, rain may fall heavily in one place and not be recorded in another, only a few km away. Thus, the monthly precipitation at Khartoum in July 1946 was 129.5 mm (5.1 in.) while Shambat, about 8 km distant, received only 38.7 mm (1.5 in.). Rainfall figures for the year in the two localities were 247.7 mm (9.7 in.) and 143.1 mm (5.6 in.) respectively. The previous year, however, the situation had been reversed, Khartoum receiving 90.6 mm (3.6 in.) and Shambat 224.8 mm (8.8 in.). The biological significance of average rainfall figures in the Sahara is, therefore, extremely limited. More important is the mean period between storms of sufficient magnitude for some water to remain long enough in favoured localities for seeds to germinate, grow into mature plants, and re-seed themselves. It is upon such plants that nomadic animals, and indirectly their owners, depend for food.

Removal of the vegetation by overgrazing greatly reduces the beneficial effects of rain, because most of the water quickly runs off the bare surface of the ground. Within a few minutes, dry *wadis* may become roaring torrents, removing any vestiges of top soil and eroding deep gulleys into the ravaged landscape. My wife and I experienced an example of desert rainfall when we were struck by a devastating thunder-storm one evening in northern Chad, just as we were beginning to prepare camp for the night. We barely had time to secure our possessions in the sudden gale before there was a torrential downpour. Within a few minutes, the entire landscape, brilliantly illuminated by the continual lightning, was inundated. Looking out from our vehicle, we seemed to be floating in a vast sea, the ripples from the wind giving the appearance of rapid currents. When the rain stopped abruptly after six hours, we literally had to shout to be heard above the fantastic chorus of croaking toads and stridulating insects. By sunrise all was quiet, and less than 20 km further on the ground was bone dry!

Although there is a tendency for rain to fall in sudden storms and at irregular intervals, the Sahara as a whole is characterized by extreme drought. The southern fringe receives rain in summer, the Mediterranean coast in winter.

1.2.2. Temperature

The Sahara is one of the hottest regions of the world with mean annual temperatures exceeding 30°C (85°F). The hottest months are June, July and August. A shade temperature of 58°C (136.4°F) was recorded at El Aziza, Libya, in September 1922. Like average figures for rainfall, mean temperatures are of little significance because ambient temperatures oscillate so greatly between day and night, summer and winter. It is the incidence of high temperatures liable to cause heat damage, and of low temperatures, which limit the growing season, that affects living organisms. An annual range of shade temperature from −2.0°C (28°F) to 52.5°C (126.5°F) has been recorded from Wadi Halfa, Sudan, and a daily range in summer of 29°C (52°F) at ln Salah in Algeria. The record fluctuation, however, appears to be from −0.5°C (31°F) to 37.5°C (99°F) within 24 hours at Bir Mighla, southern Tripolitania, in December (Cloudsley-Thompson, 1977).

1.2.3. Wind

Throughout large areas of the Sahara, the distribution of plants and animals is influenced greatly by the presence or absence of shelter from wind. In North Africa, the *khamsin*—so-called because it is said to blow for 50 days at a stretch—a suffocating dust-laden gale that cakes one with yellow mud, makes eating a problem and reduces visibility to zero. I remember one night, in Libya during the war, when I could not see a lighted torch from a distance of three paces. In Algeria, this wind is known as *sirocco*, in the central Sahara as *shahali* and elsewhere as *harmattan* or *simoom*. Whilst driving along the Route de Hoggar in the central Sahara in midsummer, 1969, my wife and I passed beneath a cloud of dust so dense that we had to use headlights although the day was still young and there was no sand in the air at ground level. The *haboobs* or dust storms of the northern Sudan can also reduce visibility to a few metres and bring about almost complete darkness at noon. Another familiar phenomenon, the dust devil or *ebliss* results from a sudden upward rush of heated air on a still day. Such whirlwinds often carry sand grains and other objects to great heights and many disperse or destroy small plants and animals in their way.

1.2.4. Soils

Shifting sands and bare rock occupy only about one-fifth of the Sahara. More than half of the area comprises soils known as *yermosols*, with shallow profiles over gravel or pebble beds. This cover, of Cainozoic age, has been forming for the past 50 million years and includes marine deposits in Libya, Egypt and Senegal, and continental deposits in the Chad basin. Mesozoic volcanic lavas also occur, especially in the north-east. Desert soils are produced almost entirely by mechanical and chemical weathering of rock. Chemical weathering is considerable despite the aridity because its effect is enhanced by the high temperatures. Although rain storms are rare, they exert a profound effect on bare rock and soil unprotected by vegetation. Furthermore, exposed rock experiences wide and rapid temperature variations. If moisture is present, it is broken up, perhaps by differential expansion or contraction of its mineral constituents, to form screes and aprons of rocky waste which gradually bury the bases of rocks and hills in the products of their own decay. Finally, rocks are abraded by wind-blown sand and ground away by torrential streams.

1.3. BIOLOGICAL ADAPTATIONS OF THE FLORA AND FAUNA

This brief and inevitably superficial introduction can do no more than set the scene for the detailed scientific chapters to follow, in which the physical, biotic and human environments of the Sahara will be subjected to expert scrutiny. The adaptations of plants and animals to the harsh climatic extremes of the desert are of tremendous interest to biologists, while the indigenous tribes of the Sahara, their history, culture and ways of life, are a fertile source of research material for the anthropologist. Finally, there is the eternal problem of desertification to which I have already alluded. Although the rise of technology has changed the face of the earth, it has led to the squandering of limited resources at an alarming rate. Even the peoples of Saharan countries—many of whom have never enjoyed, and probably never will enjoy, a full meal—are deeply affected by the world population explosion and consequent environmental degradation.

1.3.1. Water conservation

The desert environment is inhospitable to life for many reasons, but chief among these is shortage of water. Not only is lack of rain the primary cause of desert conditions, and the absence of clouds responsible for extremes of temperature, but low humidity itself has an adverse effect on plants and animals because the rate of evaporation is so great, especially at high temperatures. Desert plants are able to survive by virtue of complicated combinations of physiology, anatomy and life history. Annuals evade more extreme conditions by completing their life-cycles during the short rainy season, and passing the remainder of the year as fruits or seeds lying dormant in the soil. Many desert animals, likewise, withstand the dry season in diapause—a state of suspended development or growth, accompanied by greatly decreased metabolism. Thus, a vernal rain fauna appears at the time of inflorescence, when the desert is transformed by an abundance of plant and animal life. Flowers are visited by butterflies and moths, bees, wasps, hover-flies, bee-flies and other Diptera. The droppings of camels and goats are rolled away by dung beetles, and grass seeds are harvested by industrious ants. Termites extend their subterranean galleries to the soil surface and indulge in nuptial flights while predators, such as scorpions, camel-spiders (Solifugae), spiders, ant-lions, bugs, wasps, robber-flies and predatory beetles, glut themselves on an abundance of food.

The importance of a limited season of plant growth in arid environments is reflected in the rutting and birth of various species of gazelles which calve about one month after the onset of the rains when plenty of grazing is available. In contrast to most domesticated animals the camel, too, has a pronounced rutting season at the time of rainfall and its pregnancy lasts for 12 months. The fertility of jerboas, voles and other rodents is interrupted during dry weather and, at this time, the population level drops considerably.

Drought-evading ephemerals of the desert are not true xerophytes—they are really mesophytic because their activity takes place only when moisture is available. To some extent, the same may be true of animals, but usually even those that aestivate show some morphological and physiological adaptations to aridity. Truly xerophytic plants, and the animals of arid environments, show many striking similarities to one another in their responses to the rigours of drought and heat. Setae, hairs, scales and other structures that create a boundary layer reducing transpiration and the flow of heat from the environment are common to both. Plant stomata and insect spiracles are similar in the possession of a complex passage which resists the diffusion of water vapour. Cuticular transpiration in both plants and arthropods is reduced by the presence of waxes having high melting points, and both plants and animals may possess mechanisms for the excretion of surplus salt, the uptake of moisture from unsaturated air, the retention of metabolic water, and so on (Hadley, 1972).

Surface-to-volume relationships are an important factor in determining transpiration and heat flux relationships between plants, as well as animals, and their environments. Large succulents—euphorbias and cacti—heat up more slowly than small-leaved desert shrubs but the latter never reach such high temperatures because transpiration is less even from their relatively small, narrow leaves. Similarly, desert woodlice, such as *Hemilepistus* spp., are not only unusually large but are particularly resistant to water loss by transpiration through the cuticle when compared with species from more mesic environments. Cuticular permeability is exceptionally low in unfed desert ticks and the critical transition temperatures of their epicuticular waxes extremely high. Similarly, the permeability of insect cuticle is particularly low among desert species (Edney, 1979).

Plants and animals adapted to hot, dry, environments often show an enhanced ability to survive drought. Leaves of desert acacias may tolerate desiccation so extreme that they become brown and brittle, and yet are capable of recovery. The Saharan solifugid *Galeodes granti* can withstand a loss of two-thirds of its body weight, the gecko *Tarentola annularis* one of 35%, while the camel is unique among mammals in also being able to tolerate a loss of about one-third of its body weight.

1.3.2. Thermal adaptations

Membranes permeable to oxygen and carbon dioxide are also permeable to water vapour. The air leaving any photosynthetic or respiratory surface is normally saturated with water vapour and some loss of moisture is therefore inevitable. While the function of transpiration is to limit heat stress, the attainment of this objective conflicts with the need for water conservation. In the Sahara, where water shortage is acute, the balance between these two incompatible requirements almost invariably swings in favour of water conservation. The logical consequence of this is that all the smaller animals avoid excessive heat by their behavioural responses, while larger mammals such as camels, gazelles, eland, oryx and addax antelope, as well as the ostrich, tolerate hyperthermia during the day, storing excessive heat which is dissipated at night. In contrast, xerophytic plants are unable to avoid excessive insolation, but survive by the development of excessive rooting systems, which increase the uptake of water. In succulents, such roots spread laterally but are seldom more than 3–4 cm below the surface of the soil so that they exploit to the maximum every shower of rain even if water does not penetrate far into the ground. Non-succulent desert perennials, on the other hand, often have extraordinarily deep tap roots—those of *Acacia* spp. extending over 15 m to reach a water table far beneath the desert surface.

Just as plant roots are able to absorb water from damp soil, so many desert arachnids are able to absorb capillary moisture from damp sand. Some desert arthropods also demonstrate their natural superiority by taking up moisture from unsaturated air. These include mites and ticks (Acari), bristletails (Thysanura), book-lice (Psocoptera), fleas (Siphonaptera) and a few other wingless insects. Neither termites (Isoptera) nor adult desert beetles appear to possess this useful facility, but beetle larvae may do so. Even so, none of these animals can afford to utilize evaporative cooling for purposes of thermoregulation (Edney, 1979). Indeed, of those animals that are day-active, the majority seek shade when the sun is hottest or orient their bodies so that the least possible area is exposed to insolation. In this, locusts, lizards and camels show a thermal response in common with that of wilting plants whose leaves droop so that their flat surfaces are no longer at right angles to the rays of the sun.

Just as many desert plants and animals can tolerate extreme desiccation, so can they survive unusually high temperatures that would be lethal to their relatives from more humid regions. Not only do the larger homeotherms tolerate voluntary hyperthermia, as already mentioned, but many arthropods survive extremely high temperatures for 24 hours or more at very low humidities. Examples include *Galeodes granti* 50°C, the scorpion *Leiurus quinquestriatus* 47°C, and various tenebrionid beetles 43–46°C.

Smaller animals cannot afford to expend water for evaporative cooling but, in general, they do not need to do so because they escape from the midday heat by retiring into shady places or cool burrows. Desert rodents do not sweat, but they possess an emergency thermoregulatory mechanism and produce a copious flow of saliva in response to heat stress. This soaks the fur under the chin and throat, providing temporary relief when body temperatures approach lethal limits. Some reptiles, especially tortoises, also employ thermoregulatory salivation. In addition, tortoises discharge urine over the back legs when their temperatures are dangerously high. The function of the larger bladder of desert tortoises has long puzzled naturalists. We now know the answer: urine is stored not only as a defence against predatory enemies but also for emergency cooling (Cloudsley-Thompson, 1977).

Other morphological adaptations of animals to desert conditions include counter-current heat exchanges. Although the body temperatures of gazelles, for instance, may reach 46°C, the blood that supplies the brain is cooled by means of heat exchange in the carotid rete, a network of small blood vessels in the cavernous sinus. This sinus is filled with venous blood that drains from the nasal passage, where it has been cooled by evaporation from the moist mucous membranes. In a desert gazelle the brain temperature may thus be nearly 3°C lower than that of the blood in the central arteries. Counter-current heat shunts are important in reptilian thermoregulation. They help to maintain core temperatures at an optimum, both during hot periods of the day, and when cooling takes place at dusk, so that activity can be prolonged into the night (Louw and Seely, 1982).

Most so-called 'desert adaptations' of animals—such as webbed feet and toes fringed with elongated scales, the pointed rostrum and modified ears and eyes of desert lizards—are, in fact, specific modifications for living in sand. Adaptations of plants to life in deserts include sunken stomata, small, waxy leaves, and the water-storage tissues of euphorbias and cacti.

1.3.3. Animal defences

At the beginning of this article, I emphasized the severity of the struggle for survival imposed on desert plants and animals. Climatic extremes of arid environments are countered by behavioural means, by physiological adaptations that reduce water loss, conserve moisture or enhance the toleration of high temperature. Predation has also played a vital evolutionary rôle. Many desert plants are thorny as an adaptive response to intensive browsing, while desert animals are almost always either sand-coloured and cryptic, or black. The function of crypsis in deserts is nearly always one of defence: among spiders and insects it is invariably a reponse to predation. Black coloration, on the other hand, has an aposematic function—unless it occurs on black sand and lava flows where it may be cryptic. It may also be an evolutionary legacy; if so, it presumably had an adaptive function at one time. White may be cryptic on white sand or aposematic on darker substrates. The thermal significance of colour is comparatively slight (Cloudsley-Thompson, 1977).

1.4. THE PEOPLES

The first inhabitants of the Sahara were probably Negroes who retreated in the face of advancing Berbers, themselves afterwards pushed back by the Arabs. From a mixture of these emerged the three great ethnic groups of today: the Tuareg, Tibbu and Moors. Nomadic tribes are to be found all over the Sahara. Some are small and with little cultural organization. Desperately poor and owning little except the barest necessities of life, they sleep in the open and dress in skins and rags. Other tribes number several thousands each, live in comfortable tents, are rich in livestock and have elaborate political, social and economic systems (Briggs, 1960).

The Nemadi are non-pastoral nomads who lead solitary lives, avoiding the settlements and camps of others. Their tents are made of antelope skins and their clothing is extremely simple. They live by hunting addax, oryx, ostriches and bustards with the help of dogs which distract the prey until they can approach and kill it. In contrast, the Tuareg and Teda are huge tribes of pastoral nomads who, in the past, ranged over vast areas of the Sahara—the Tuareg from Libya to Timbukto, the Teda in the south-eastern regions. There are also numbers of Arab tribes such as 'Cha'amba' of the north-western central Sahara, who once enjoyed an unenviable reputation as bandits. Experts in desert warfare, they eagerly joined the French Saharan Camel Corps to fight their traditional enemies the Tuareg and Moors. It was one of their patrols that smashed for ever the military power of the Hoggar Tuareg at the Battle of Tit in southern Algeria, in May 1902.

Unlike food gathering, nomadism is not isolated from other forms of land use in the Sahara. Nomadic pastoralists exchange their produce—milk, meat and hides—with that of the cultivators who engage in dry farming or oasis agriculture. Some of the Bedouin of Libya, some Teda and the Gherib of Tunisia are semi-nomads who own plantations without actually cultivating them. In the old days, they used slaves to work for them, but today these tasks are performed by their descendants who are still linked economically and socially with their former masters (Briggs, 1960).

Pastoral nomadism can be a remarkably efficient adaptation to the vagaries of the desert environment. There is nothing random about the migrations of the nomads. In the dry season they move as far southward as they can go without entering the region of the tsetse fly. Here their cattle graze the stubble remaining from the crops of the sedentary farmers and, at the same time, manure the fields. In return, the nomads receive millet (*dura*) from the farmers. With the first rains, however, the grass springs up and the herds then move northward in search of fresh fodder. The migration continues until the northern edge of the Sahelian rain belt is reached, and then the return to the south begins anew, the cattle grazing the grass that grew up behind them on their northward journey. The traditional migration routes and the amount of time a herd may spend at any particular well are governed by rules worked out by the tribal chiefs. In this way, overgrazing and conflict are avoided.

1.5. RECENT EXPANSION OF THE SAHARA

Although there is no doubt about the reality of desert expansion or 'desertification'—the ominous process by which the world's existing deserts, including the Sahara, are extending their boundaries, while new patches of desert appear simultaneously in nearby arid and semi-arid regions—a dichotomy of opinions appears to exist as to its cause. Some scientists believe that the climate is inexorably changing, others that overgrazing by sheep and goats is to blame. In fact, both views are probably correct. When the desert expands as a result of overgrazing by domestic stock, the felling of trees for fuel, and bad agricultural processes, the 'albedo' or reflectivity of the land surface increases, and this may inhibit rainfall by enhancing atmospheric subsidence. At the same time, fine airborne dust can cause an inversion of temperature so that the air becomes hotter at higher altitudes preventing the formation of rain-bearing, cumulus cloud. No less than 60–200 million tonnes of soil were lost from the Sahara and blown into the Atlantic Ocean annually during the early 1970s (see Morales, 1978).

Does 'drought feed on drought', by setting up positive feedback loops? And does Man accelerate such feedbacks when he overstocks or overcultivates the desert margin? Hare (1977), who posed these questions, concluded that the tendency for desert surfaces to extend outwards from existing deserts, though real in some areas, is largely due to unwise response of human societies to the strains imposed by naturally recurring drought, and there is no convincing evidence that a lasting decrease of rainfall in the Sahara is taking place at the present time, although such desiccations have often occurred in the past. The extent to which Man is himself responsible for climatic change is not, however, relevant to the argument. As the Mole pointed out in *The Wind in the Willows*, when he cut his leg on Mr. Badger's doorscraper: 'Well, never mind what done it,' he said forgetting his grammar in his pain, 'It hurts just the same, whatever done it.'

1.5.1. Man-made desert

A traveller, moving towards the poles from the equatorial rainforest will cross several broad zones of savanna vegetation before he reaches the tropical and sub-tropical desert regions of the world. In Africa, this savanna has evolved contemporaneously with Man (Cloudsley-Thompson, 1969), and the destruction of the forest doubtless began when the use of fire was first acquired—probably well over 50,000 years ago when hearths were being built in eastern Africa in the era of the Acheulian hand-axe culture. Fire is an essential tool in shifting cultivation and, although fires occur naturally from time to time, deliberate firing by Man has a far greater effect upon the vegetation. This is because man-made fire covers the same ground more frequently than natural fires and, moreover, has a far greater effect upon the vegetation. Even in regions with high rainfall, the forest ecosystem never re-establishes itself once it has been destroyed over extensive areas. On the contrary, deforestation leads to rapid deterioration of soil conditions which eventually become so adverse that the land can only be used for grazing until eventually it has been reduced to semi-desert or desert.

It is not only through overgrazing by domestic animals that Man induces desert encroachment: at least half of all the timber cut in the world each day is used as fuel for cooking or as a source of warmth (see Glanz, 1977). As the trees thin out, the women have to spend more and more of their time collecting firewood. In West African cities it often costs more to heat a pot than to fill it, and dried animal dung is everywhere burned instead of being left to fertilize the soil.

Evidence of climatic change on the fringe of the Sahara within recent times is afforded by the studies of Wickens (1975) and others on the *qoz* dunes of the central Sudan. The orientation of these stabilized sand dunes indicates that the isohyets were 450 km to the south of their present position from 20000 to 15000 B.P.* and 200 km to the south from 7000 to 6000 B.P. Palaeobiological evidence suggests that there were northward shifts of 400 km between 12000 and 7000 B.P. and of 250 km between 6000 to 3000 B.P.—with corresponding parallel shifts of the vegetation. Rock paintings by the people of Tassili in the Sahara document the rise and fall of their culture. About 7000 B.P. they were hunters who pursued giraffe, antelopes and other animals of savanna. Then they took to the breeding of cattle, their murals of some 2000 years later depicting immense herds. The most recent pictures, illustrating camels, date from approximately 2000 to 3000 B.P., after which the culture was obliterated by military conquest.

1.5.2. Documentary evidence

Classical accounts provide overwhelming evidence for desert encroachment within the last two millennia. At the battle of Heraclea (280 B.C.), two Roman legions, under command of Valerius Lavinius, were defeated by the army of Pyrrhus, king of Epirus. This was the first occasion on which Roman infantry, equipped with short swords, came in conflict with a Macedonian phalanx armed with the dreaded barrier of long spears projecting from their front line. At the same time, Pyrrhus brought twenty elephants—strange beasts never before seen in Europe—with armed men in castles on their backs. He stationed them on the wings of his army to disconcert the Roman cavalry with their smell and monstrous trumpeting, for untrained horses will not face them (Cloudsley-Thompson, 1967). Pyrrhus could win battles but not wars, and some of his successes were so costly to his own forces that they became known as 'Pyrrhic victories'. Nevertheless, Hannibal is alleged to have rated Pyrrhus foremost of all generals in experience and ability, with Scipio second, and himself third (Scullard, 1974). In the battle of Trebia (218 B.C.), Hannibal employed the same tactics that Pyrrhus had used, but the elephants which the Carthaginians had brought across the Alps were the African species whereas Pyrrhus used Indian elephants. Ancient writers were unanimous in their opinion that Indian elephants were larger than Africans, a view reinforced by the historian Polybius in his description of the battle of Raphia (217 B.C.). In this, the only known encounter between elephants of different species, the 73 African elephants of Ptolemy IV of Egypt were terrified by the greater size of the 102 Indian elephants in the army of Seleucus Antiochus III, king of Syria, and at once turned tail before they got close. There has been much controversy about the claim of early authors that African elephants were inferior in size and strength to the Asiatic species—for it has commonly been assumed that the contrary is the case. There are, however, two main sub-species of African elephant: the larger bush elephant and the smaller forest elephant used by the Carthaginians and Egyptians. Ptolemy's animals were trapped in Abyssinia, Somalia and the eastern Sudan—regions that are now desert—and shipped to Egypt in a fleet of 'elephantoforoi' (Cloudsley-Thompson, 1967; Scullard, 1974).

In classical times, forest elephants—now restricted to the Congo basin and the forests of West Africa—were widely dispersed from the shores of the Mediterranean to the Cape of Good Hope. Hanno, the Carthaginian, who was sent out on an adventurous journey beyond the Pillars of Hercules some time before 480 B.C., reported that, after passing Cape Soloeis (modern Cape Cantin), 'we arrived at a lagoon full of high and thick-grown cane. This was haunted by elephants and multitudes of other grazing beasts.' This lagoon was probably the marshes of the river Tensift, north of Mogador and at the foot of the Atlas Mountains (Scullard, 1974).

---

* B.P.—before the present. For technical reasons B.P. dates (which are referred to A.D. 1950) are not directly convertible to the B.C./A.D. system.

FIG. 1.1. African bush elephant, Uganda. December 1963.

Elephants and lions were frequently represented in the Meroitic art of Nubia, and were almost certainly captured and tamed at Musawwarat es-Sofra, a day's camel ride from the Nile, south-east of Meroë less than 2000 years ago. Diodorus (1st century B.C.) described a tribe of elephant-eaters living in regions covered with thickets of trees growing close together, probably on the upper reaches of the Atbara river, while Strabo (*c.* 63 B.C.–A.D. 23) mentioned a hunting ground for elephants at Ptolemais, near the modern Aqiq. About A.D. 60, the emperor Nero sent an expedition, commanded by two centurions, or by a tribune and a centurion, to discover the source of the Nile.* According to Pliny and Seneca, they found the tracks of elephants and rhinos in forest around Meroë. During the 12th century A.D., the Arab traveller Idrisi saw elephants and giraffes near Dongola. Lions were widely distributed and common in Algeria until the time of the French occupation of that country. Cheetahs were plentiful also, and at one time were caught and trained for coursing game.

Most of the Saharan game was eliminated by indiscriminate hunting during the last century. Antelope, mouflon and ostriches have been exterminated over large areas and even gazelles are becoming scarce. However, it is difficult to imagine such animals surviving today in the barren desert that was able to support them until very recently. In the journal of his journey up the Nile during the years 1821 and 1822,

---

* This was probably a cover story. It seems more probable that the expedition would have been an intelligence mission to determine whether the country was worth conquering!

FIG. 1.2. Roman ruins: the Triumphal Arch of Diocletian at Sbeitla (Sufetula), Tunisia. April 1954.

FIG. 1.3. Bardia harbour, Libya; showing wind erosion of the bare rock. January 1942. The sewers of Rome, in the course of centuries, engulfed the wealth of North Africa.

Linant de Bellefonds commented on the woodedness of the countryside of the northern Sudan and mentioned that he heard a lion roaring at Ed Debba near old Dongola. Away from the river today, that region of the Sahara is almost complete desert, with practically no vegetation. In 1835 lions were plentiful around Shendi, 190 km north of Khartoum, and game was abundant at Kassala as late as 1883. Many other examples could be cited to show how much the flora and fauna of the Sahara have been impoverished in recent times (Cloudsley-Thompson, 1967).

The inhabitants of Leptis Magna in Tripolitania, Sufetula in Tunisia, and other Roman cities, were able to draw on the produce of hundreds of thousands of olive trees. The beds of the *wadis* were terraced, catchment walls along their sides prevented rain water from washing away the soil, and enough moisture was retained to make possible the cultivation of cereals, dates and even of vines. The Roman amphitheatre at El Djem, the third largest in the world and capable of seating 60000 people, contrasts with the present small Arab village and suggests how much standards of agriculture have been depressed in recent years. The land was not fertilized, however, in ancient times; it is said that the sewers of Rome, in the course of centuries, engulfed the prosperity of the Roman peasants and devoured the wealth of Sicily, Sardinia and the fertile lands of North Africa as well. Even today, the chief drawback to development in much of North Africa is not the climate so much as the monoculture of wheat, and other bad agricultural practices which, combined with overgrazing, result in destruction of the natural vegetation. Thus the increase of the Sahara in recent years is part of the man-made, large scale, shift of the vegetation belts (see Dregne, 1970).

1.5.3. The situation today

In trying to understand the present situation regarding desertification in the Sahel savanna bordering the Sahara, it may be significant to realize that much of the economic development and population expansion of the 1960s took place there during a brief period of abnormally high rainfall. What made the drought of 1969–1973 so devastating, however, was the increase in domestic animals and consequent overgrazing, rather than just a period of reduced rainfall—no greater fluctuation than is normal in semi-arid regions on the fringe of the desert. If the flocks and herds are allowed to build up again, there will inevitably be a worse disaster in a few years' time (Cloudsley-Thompson, 1977, 1978). Although pastoralism is a major cause of desert expansion, true nomadism, by its very nature, is transitory and so does not necessarily result in overgrazing.

The effects of overgrazing are more conspicuous in some parts of the Sahara than in others. For instance, many species of plants require only two-thirds as much rainfall on sandy soils as they do when growing on clay. The effects of overgrazing are therefore likely to be greater where clay predominates. Perennial grasses are the first element of the vegetation to suffer under heavy grazing. Their replacement by annual species is associated with a general reduction in the amount of plant cover. Bare ground generally absorbs less rain-water than it would in the presence of vegetation. Thus the amount of water entering the soil and becoming available for plants is reduced. Run-off is increased when the soil is compacted, and shallow-rooted plants die. Soil erosion then becomes steadily more serious.

In regions where the needs of the vegetation are delicately balanced by precarious rainfall, overgrazing and the utilization of firewood can have a disastrous and irreversible effect. Destruction of the acacia desert scrub around Khartoum and Omdurman has taken place to such an extent that charcoal now has to be brought from as far afield as Kosti, Wad Medani and El-Gueisi. It has been estimated that in the Sahel zone of Africa alone, nearly 3 tonnes of wood are consumed annually by each family—a total of some 50,000,000 tonnes a year. This state of affairs is exaggerated by the export of charcoal from Somalia and the Sudan to oil-rich Saudi Arabia, Kuwait and the Gulf States where charcoal barbecue cuisine has become fashionable! (Cloudsley-Thompson, 1978).

FIG. 1.4. Desertification in progress. Goats grazing in a *wadi* bed, Northern Sudan. December 1968.

Two contrary opinions are commonly expressed about goats. According to the first, they are a living testimony to the wisdom of Allah who created such a wonderful machine that it can transform even waste paper and other refuse into good milk. The second point of view stresses the fact that goats are a menace and there is little hope for rehabilitation of the Sahel zone unless they can be removed. Personally, I subscribe to the first view in that I do not believe that much of the desert can be rehabilitated without irrigation, whether goats are present or not. On the other hand, by causing even further degradation of the land, the goat is encouraging the desert to advance. Thus, the future of land that has not yet been reduced to desert is being mortgaged for milk and meat today (Cloudsley-Thompson, 1977, 1978). Although goats have a broader diversity of ecophysiological adaptations, are browsing animals, and require less water than sheep, the latter are more prone to follow the same tracks, thus engendering gulley erosion. Indeed, there is nowhere in the world where either sheep or goats may safely overgraze, but the dangers are greatest in arid lands.

The detrimental influence of man and grazing animals on the natural vegetation has been demonstrated experimentally on a number of occasions, and it has been emphasized that the conservation of soil and plant cover is imperative in arid regions where the unstable balance of nature can easily be tipped towards destruction by a slight disturbance of the ecosystem. The effects of overgrazing are manifold. Construction of additional wells and water-holes may enable grazing to be spread over a wider area, thus reducing the pressure around the original watering points. Unfortunately, such amelioration is usually short–lived, since the additional water made available merely enables the numbers of animals to increase still more, while the level of ground–water almost inevitably drops. There is no possibility that the number of domestic animals in the Sahara will be voluntarily reduced in the foreseeable future, for these represent the

Fig. 1.5. Goats scavenging refuse outside a Sudanese village on the banks of the Nile. April 1979.

sustenance of the nomads and an insurance for the poor farmer against the years of drought when his crops fail. He does not realize that these animals are actually endangering his very existence (Cloudsley-Thompson, 1977). Indeed, the opposite opinion is sometimes expressed: 'During the last drought we lost two-thirds of our cattle. If, therefore, we are able to build up the herds to three times their previous size, after the next drought we will be left with the same number that we originally possessed.'

The use of artesian wells and intensive farming of date palm in parts of North Africa have had a dramatic effect on the water balance. During the last hundred years the number of palms in the border region of the Tunisian Sahara has increased by more than 50%, reducing the level of subterranean water at a rate of 5 cm per annum. (A well at Azaouad in Mali, now over 100 m deep, has been deepened progressively over the centuries as the water table has fallen.) As many as fourteen cuttings of alfalfa (lucerne) can be harvested annually in certain Libyan irrigation schemes. Nevertheless, the water is a fossil resource which is not being recharged. At Al-Kufrah, where 50,000 ha are now irrigated with the most modern of sprinkler techniques, the water-table is sinking at a calculated rate of 35 m in 40 years.

Apart from overgrazing and wanton felling of trees for fuel, which no government has yet succeeded in preventing, cultivation in marginal areas is perhaps the major factor contributing to desert expansion. When drought follows years of good rainfall, cultivated soil that has been broken up by agricultural activities is readily blown away as fine silt. The Gedaref millet (*dura*) project in the Sudan, for example, is likely, in the long run, to prove a costly failure from which only locusts and grasshoppers will benefit. Small areas had previously been farmed traditionally with hoes and then left fallow for several years after harvesting. This did little harm to the environment, but the mechanized agriculture of today cannot fail to cause rapid soil erosion. Furthermore, monoculture is especially dangerous in the tropics, because pests can reproduce so rapidly at the prevailing high temperatures (Cloudsley-Thompson, 1978). Enough has been said to illustrate the part played by Man during the last few centuries in the expansion of the Sahara. One has only to travel with an observant eye to realize how little of the country remains in its pristine state (Cloudsley-Thompson, 1977).

FIG. 1.6. Nubian desert, Upper Egypt, September 1962. Although no rain is normally recorded in this region, seedlings were sprouting in the lee of the *jebels* (hills).

1.6. CONCLUSION

In these introductory pages, it has not been possible to mention more than a few of the intriguing topics that will be considered in detail by the authors of the various chapters that follow. Indeed, I have not really attempted to summarize the volume but rather, with the aid of oversimplified statements and a few subjective experiences, to convey some impression of the Sahara for the benefit of any reader who has not yet enjoyed this fascinating desert region at first hand. Those who know it well would neither miss much of value nor cause offence if they were to skip these preliminary remarks and turn immediately to the subjects that concern them.

For convenience, the material in each chapter has been treated as a self-contained unit. For example, many of the faunal chapters open with comments on climatic factors of especial significance to the group of animals under consideration. Some repetition is consequently inevitable but, in compensation, it should be possible to read through any chapter without requiring to refer to others in order to follow it properly. The Sahara desert is understood to extend as far south as the 100-mm/rainfall belt but, again, the interpretation has not been rigidly upheld. The Tibesti and Hoggar regions which, though semi-arid, are really only islands within the main arid area and are therefore not excluded from chapters in which discussion of them is appropriate.

Parts of this Introduction have been quoted from a chapter entitled 'The Future of the Desert' which appeared in *Resources, Environment and the Future* (Eds. W. B. Fisher and P. W. Kent, 1982), pp. 100–132. Thanks are due to the editors and publishers Deutscher Akademischer Austauchdienst for permission to reprint them.

REFERENCES

BRIGGS, L. C. (1960) *Tribes of the Sahara.* Harvard University Press, Cambridge, Mass.
CLOUDSLEY-THOMPSON, J. L. (1967) *Animal Twilight. Man and Game in Eastern Africa.* Foulis, London.
CLOUDSLEY-THOMPSON, J. L. (1969) *The Zoology of Tropical Africa.* Weidenfeld & Nicolson, London.
CLOUDSLEY-THOMPSON, J. L. (1977) *Man and the Biology of Arid Zones.* Arnold, London.
CLOUDSLEY-THOMPSON, J. L. (1978) Human activities and desert expansion. *Georg. J.* **144** (3), 316–323.
DREGNE, H. (Ed.) (1970) *Arid Lands in Transition.* American Association for the Advancement of Science, Washington, D.C.
EDNEY, E. B. (1979) *Water Balance in Land Arthropods.* Springer-Verlag, Berlin.
GLANTZ, M. H. (Ed.) (1977) *Desertification. Environmental Degradation in and around Arid Lands.* Westview Press, Boulder, Colorado.
HADLEY, N. F. (1972) Desert species and adaptation. *Amer. Sci.* **60**, 338–347.
HARE, F. K. (1977) Connections between climate and desertification. *Environmental Conservation,* **4**, 81–90.
LOUW, G. N. and SEELY, M. K. (1982) *Ecology of Desert Organisms.* Longman, London.
MORALES, C. (Ed.) (1979) *Saharan Dust. Mobiization, Transport, Deposition* (SCOPE 14). John Wiley, Chichester.
SCULLARD, H. H. (1974) *The Elephant in the Greek and Roman World.* Thames & Hudson, London.
WICKENS, G. E. (1975) Changes in the climate and vegetation of the Sudan since 20,000 B.P. *Boissiera,* **24**, 43–65.

CHAPTER 2

Climate

G. SMITH, MBE

Keble College, Oxford

CONTENTS

2.1. INTRODUCTION

The Sahara is the largest of the world's hot deserts. The area within which the mean annual precipitation is 100 mm or less is approximately as large as that of the continental United States. This 'dead heart' of the Sahara extends for about 3200 miles (5150 km) from the shores of the Atlantic Ocean to the Red Sea coast in Egypt and the Sudan; in some places in Libya and Egypt it reaches the southern shores of the Mediterranean Sea. Its extent from north to south varies between 600 and 800 miles (966 to 1287 km) and is narrowest on the Atlantic coast. Climatically the Sahara is the classic example of the world's hot deserts by reason of the combination of its maximum percentage of possible sunshine hours, high temperature, low humidity and lack of rainfall.

Climatologists normally make a three-fold division of the world's desert regions (Meigs, 1953). Hot. deserts in tropical and sub-tropical latitudes are distinguished from deserts in higher latitudes in continental interiors such as the Gobi and Kazakhstan deserts of central Asia. These continental deserts of mid-latitudes may be almost as arid as the Sahara, the Kalahari and central Australia and during the summer or high sun period, temperatures may rise very high; however, winters are cold or very cold. A third type of desert is distinguished; the 'cold water coast' deserts found on the west coasts of continents in tropical latitudes of which the Namib in South-west Africa and the Atacama desert in coastal Peru and northern Chile are the best examples. The Atlantic coastal fringe of the Sahara falls into this third type but inland it merges into the typical tropical continental hot desert. Cold-water coast deserts in tropical latitudes are distinguished from typical hot deserts of the Saharan type both in the character of their climate

17

and partly in cause. One controlling cause of both their character and extent is the presence offshore of cold ocean water, both in the form of an equatorward setting current and an upwelling of colder water from the ocean depths. The cold water offshore encourages the formation of cloud and sea fog, strengthens the sea breezes on the coast and brings about a marked reduction in temperature as compared with that experienced further inland. However, both the Namib and Atacama deserts are as intensely arid as much of the central and eastern Sahara. Only the Sahara desert combines within its limits of extreme aridity examples of the hot tropical interior and cold-water coast deserts.

Meteorologically and climatologically the Sahara is merely the central and western portion of an even larger region of extreme aridity which extends uninterruptedly across the Red Sea and Persian (Arabian?) Gulf to include the deserts of Arabia and the Makran coast before terminating in the arid region of Sind. High temperatures and low rainfall on the coasts of the Red Sea and the Persian Gulf match those found in the Sahara and in central Arabia.

The climate of the Sahara is essentially very simple and remarkably constant round the year or even from day to day. Few parts of the world experience less 'weather' in the sense this term is usually understood in temperate latitudes. For most of the time the weather is the same from day to day. It is a climate of almost constant sunny days and clear nights when the moon and stars have a remarkable clarity and brightness. From time to time occasional high cloud or scattered cumulus appear; on rare occasions cloud may develop sufficiently to produce light rain or a heavy downpour. On other occasions strong winds may raise enough dust or even sand to obscure the sky and reduce visibility drastically. Precipitation in the form of rain, hail and, in some higher regions, a very rare snowfall is so sporadic in time and place as to be a noteworthy or remarkable event. There is a hot season at the time of high sun from May to September and a relatively cooler season at the time of low sun. Even in the low-sun period the days are warm or even hot so that there is no winter in the sense it is understood in mid-latitudes. The nearest equivalent of winter comes in the night or early morning in the cool season when temperatures may fall to near or even below freezing point for an hour or so. Even in the hottest months the daily range of temperature is quite large so that the nights provide a welcome relief from the very high temperatures experienced during the day. The very low relative humidity and the frequent brisk to strong daytime wind are very important aspects of the weather and climate of the Sahara for they reduce the 'physiological temperature' so that a daytime temperature of 40°C in the Sahara may feel more comfortable than one in the range 30° to 35°C in the humidity of the wet tropics.

The heat and aridity of the Sahara have an important and often significant influence on the weather and climate of adjacent regions. Hot, dry winds blowing out of the desert, either seasonally or for a few days at a time, 'export' Saharan heat and dryness to the Mediterranean lands, Atlantic Islands, such as the Canaries, Madeira and the Cape Verdes, and the Guinea lands on the southern margin of the desert. Hot winds such as the Leveche, the Sirocco, the Ghibli and Khamsin are examples of such weather phenomena experienced in southern Spain and Italy and all along the southern shores of the Mediterranean (Met. Office, 1962). The Harmattan is the local name for the Northeast Trade Winds in the West African countries south of the Sahara. It blows for most of the year and brings Saharan dust, high temperatures and low humidity (Walker, 1958). Saharan dust may fall hundreds of miles out in the Atlantic and more rarely, when the upper winds are favourable, Saharan dust has fallen as far north as Britain or Germany, either as 'red rain' when it has been washed out of clouds, or as a light cover of fine red dust (Tullet, 1978).

2.2. THE UNDERLYING CAUSE OF THE CLIMATE OF THE SAHARA

The Sahara is almost exactly bisected by the Tropic of Cancer; other hot desert regions are similarly located athwart the northern and southern tropics. This simple fact of coincidence goes some way towards explaining the causes of the nature and extent of the Saharan climate but in detail the causes are rather more complex. Hot deserts are located in tropical latitudes because the pattern of world pressure and wind systems, which meteorologists call the General Circulation of the Atmosphere, produces a particular combination of atmospheric conditions in these latitudes highly unfavourable to the formation of cloud and rain. The great extent and hyper-aridity of much of the Sahara, as compared with other hot deserts, is a consequence of a number of regional factors peculiar to North Africa. The African continent is widest in the latitude of the Tropic of Cancer. It is separated from the continents of Europe and Asia by relatively narrow and largely landlocked seas, the Mediterranean and the Red Sea, whose surface waters behave very differently in terms of surface temperature from those of the open oceans which greatly modify the temperature and humidity of air masses which traverse them. Much of the Sahara is thus far removed from oceanic influences; the Sahara is the most continental of hot deserts (Brooks, 1932). Consequently global and regional factors unite to explain the nature of the climate of the Sahara.

Like most models or generalized representations of very complex atmospheric processes the General Circulation of the Atmosphere is a highly simplified concept (Chandler and Musk, 1976). Fortunately it is simpler to understand, and fits observed facts better and more clearly, within the tropics and sub-tropics where the earth's wind and pressure systems are relatively persistent and not obscured by the frequent and irregular day to day changes such as occur in higher latitudes. Within the approximate latitudes of 30° North and South of the Equator the General Circulation can be regarded as a very simple form of heat engine. It is represented in plan and in section in Fig. 2.1. Within a few degrees either side of the Equator the lower and middle levels of the earth's atmosphere are heated by a combination of the intense solar radiation resulting from the high angle of the sun and the latent heat of condensation released into the atmosphere by the great masses of cloud which form in this rainy zone of converging Trade Winds and Monsoons. The heated air rises and at high levels in the upper troposphere moves polewards in both hemispheres. As it does so it comes more and more under the influence of the Coriolis Force, an apparent force resulting from the rotation of the earth from west to east (Riehl, 1978).

Thus the upper winds within the tropics become increasingly westerly and increase in speed so that at about latitudes 30° North and South there is a covergence of air into a belt of strong westerly winds called the Sub-Tropical Jetstream. Beneath the jetstream air subsides thus producing a belt of relatively high surface atmospheric pressure in these latitudes, manifested on surface charts of barometric pressure as a series of anticyclones, called the sub-tropical anticyclones. From these anticyclones air moves polewards as the mid-latitude westerly winds and equatorwinds as the Northeast and Southeast Trade winds in the

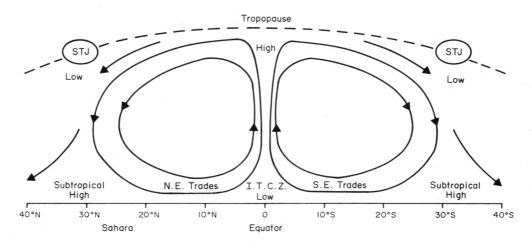

FIG. 2.1. The location of the Sahara desert in relation to the Hadley Cells within the tropics.

northern and southern hemispheres respectively. To summarize, there is a region of surface convergence and upper divergence of airstreams in equatorial latitudes and conversely a region of upper convergence and surface divergence in the sub-tropics. Seen in three dimensions there are two separate cells on either side of Equator with ascending air in equatorial latitudes and descending air in sub-tropical latitudes. These cells are known as Hadley Cells after the English scientist Hadley who first suggested this scheme in the 18th century (Frisinger, 1977). In the descending limb of the Hadley Cell the air is warmed adiabatically as a consequence of the increasing air pressure and its relative humidity is thus decreased since the air can gain no moisture until it comes into contact with the surface of the earth. The combination of circumstances beneath the Sub-Tropical Jetstream and for some distance on its equatorward side is thus highly unfavourable to the formation of cloud and rain. The surface sub-tropical anticyclones continually feed warm dry air into the Trade Winds which can only gain moisture and lose their inherent stability as they cross the ocean and approach the equatorial zone of surface convergence. The sub-tropical anticyclones are continually replenished by dry, warm and stable air from aloft. By contrast within the belt where the Trade Winds converge near the Equator conditions are highly favourable to cloud and rain as moist warm air ascends in what is termed the Inter-Tropical Convergence Zone (ITCZ).

This whole system of surface and upper winds and pressure belts shifts some 5° to 15° north and south of the Equator seasonally with the apparent movement of the overhead sun. Only those parts of the world which remain throughout the year under the influence of the descending limb of the Hadley Cell are typical hot deserts. Between the hot desert belt and the zone affected throughout the year by the Inter-Tropical Convergence is an intermediate zone, the seasonally wet and dry tropics, alternately under the influence of the migrating belt of equatorial rains and the dry trade winds. In the case of the Sahara both the great east to west extent of the African continent and the proximity of the equally arid regions of south-west Asia intensify the aridity of the Trade Winds and thus increase the area where extreme drought prevails throughout the year.

On the southern margins of the Sahara, where the zone or aridity merges into the zone briefly affected in the northern hemisphere summer by the migration of the Inter Tropical Convergence, occasional summer rainfall occurs. Precipitation is unreliable and highly variable in time and place so that the boundary of the desert in effect shifts from year to year or from one period to another. On the northern margins of the Sahara there is a similar transitional zone of low and unreliable rainfall, but here precipitation occurs in the winter season when the area comes under the influence of the surface westerly winds of mid-latitudes with their migratory frontal disturbances and associated disturbed weather, cloud and rain. This northern transitional zone lies to the south of the Atlas Mountains in Morocco and Algeria, through southern Tunisia and in a very narrow coastal strip along the Gulf of Sirte and from about Tobruk in Cyrenaica to the Nile delta. The hilly regions of Cyrenaica and the Atlas Mountains and coastal districts of Morocco, Algeria and Tunisia receive sufficient winter precipitation to be classified as having a sub-humid or even humid Mediterranean type of climate. In this region the summers virtually share the characteristics of the Sahara. Conversely in the winter season large areas of the countries to the south of the Sahara have weather which is almost indistinguishable from that in the deserts to the north; intense drought, abundant sunshine and persistent dry and often dust-laden north-easterly winds, the Harmattan or North-east Trades.

2.3. SURFACE PRESSURE, WINDS AND AIR MASSES OVER THE SAHARA

The Sahara owes its aridity directly to its position throughout the year on the southern side, or directly under, the high-level Sub-Tropical Jetstream. Like the surface pressure and wind systems this feature of the upper air circulation migrates from a mean winter position at about latitude 30°N to a mean summer position above the coast of north-west Africa, across the eastern Mediterranean and then further north across Turkey and Iran. This position further to the north over the Middle East is related to the great distortion of the upper level pressure and wind systems over the Himalayan–Tibetan region which is an important contributory factor in the mechanism of the Indian Monsoon.

Mean surface pressure and winds over North Africa and the Mediterranean–Middle eastern region in January and July are illustrated in Figs. 2.2 and 2.3. In January surface pressure over the Sahara is high compared with the coastal regions of West Africa, where the ITCZ lies near the coast, and the countries around the Mediterranean, affected at this season by a series of cyclonic depressions migrating from west to east. In July the ITCZ and the equatorial low-pressure trough have moved north to a mean position at about 18°N over the Sahel region to the south of the Sahara. The central and eastern Sahara is a region of relatively low surface pressure with persistent north to north-easterly winds blowing at an angle to the pressure gradient between the surface sub-tropical anticyclone, now only evident over the North Atlantic, and the two separate centres of low pressure over south-west Asia and the south central Sahara. Both these surface low-pressure areas are 'heat lows' and are shallow features in the lower atmosphere consequent upon the intense heating of these almost cloud-free regions at this season. Upper air charts of pressure show that at a pressure level of 850 mb and above (i.e. at heights above 1500 m) low pressure is replaced by a region of high pressure with dry subsiding air and diverging air flow. This is a situation quite unfavourable to the formation of cloud and precipitation and indicates that the sub-tropical anticyclone has merely been lifted off the surface; or rather masked by the heat of the lowest layers of the atmosphere. As a controlling feature of the weather and climate of the Sahara it is still as powerful as at other seasons.

The dryness and stability of the North-east Trade Winds blowing across the Sahara are reinforced by the extent of the land masses of North Africa and Eurasia. In summer, when these winds appear from a surface chart to originate much further to the north over the Mediterranean and Middle East, the air involved has little opportunity to pick up moisture as it would if it had had an oceanic track. Furthermore, much of the air involved has subsided from higher levels in the descending limb of the Hadley Cell beneath the STJ then located in its most northerly position. At this season the coastlands of North Africa and much of the Middle East share the complete drought of the Sahara. In winter when the STJ is further to the south above the northern Sahara these northern borderlands of the Sahara lie polewards of the STJ and consequently experience from time to time the more disturbed and changeable weather with cyclonic storms characteristic of the westerly wind belt of temperate latitudes.

For much of the year the Sahara is dominated by warm dry air of a type called by meteorologists, Tropical Continental. This air mass is hot and dry because it has acquired these properties in subsiding from higher levels and from the nature of the surface of the Sahara desert. Colder and more humid air of Polar and Arctic origin which has travelled south across the North Atlantic and Mediterranean does reach the Sahara from time to time particularly in winter and spring. However, the lower layers of such air masses are warmed on crossing the Mediterranean or the Atlantic; although they bring cloud and rain to the coast of North Africa they are of very little significance in making for disturbed weather in the Sahara. The presence of such air with a northerly origin can sometimes be observed by an increase of upper cloud, a slight fall of temperature, and an increase of wind speed; but more rarely by the growth of shower clouds and an occasional fall of rain.

During the summer months a very different type of air may penetrate into the central Sahara from the south. Warm, humid Tropical Maritime air of South Atlantic origin moves north as a south-westerly airstream behind the ITCZ and dominates the countries with a coastline on the Gulf of Guinea. These winds, which are really deflected South-east Trades, are often called the Guinea Monsoon. At and near its northern limit it is a very shallow current of air and is overlain by the hotter and very dry air of the Harmattan or North-east Trades. This forms a temperature inversion, a form of virtual lid, which inhibits rising air and thus condensation, cloud and rain for some considerable distance south of the ITCZ where the two opposing airmasses converge at the surface (Walker, 1958). The sporadic and unreliable rains in summer on the southern margin of the Sahara only occur when the ITCZ moves particularly far to the north or when a tropical weather disturbance, of the type which brings much precipitation to the seasonally wet countries of West Africa, produces an unstable situation in the upper atmosphere by

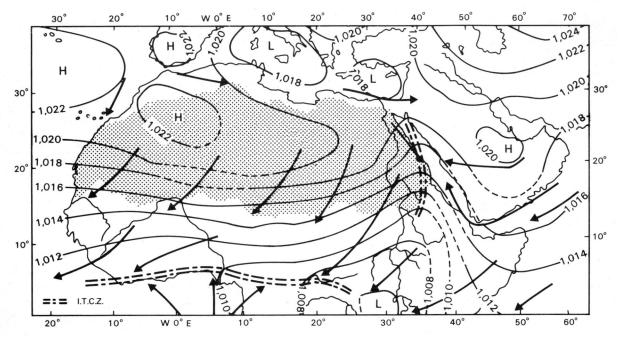

FIG. 2.2. Surface winds and pressure over North Africa in January; based on mean values. The stippled area shows the approximate extent of the Sahara as defined in this book; less than 100 mm annual rainfall.

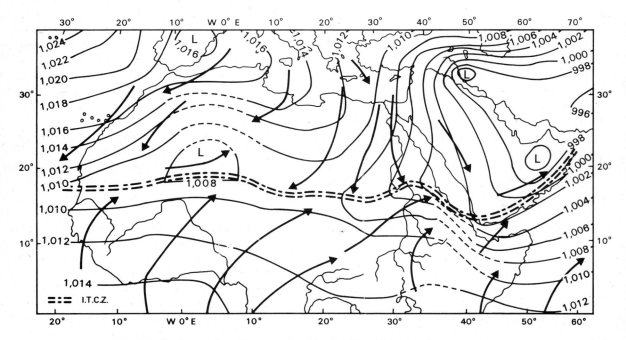

FIG. 2.3. Surface winds and pressure over North Africa in July; based on mean values.

destroying the inversion layer between the lower south-west current and the upper north-easterly flow (Eldridge, 1957).

The arid Sahara is thus a form of climatic divide or a broad frontier zone between the tropical summer rain region of the West African Sahel and the winter, or cool season, rainfall regime of the semi-arid shores of the southern Mediterranean. The consistently fine, hot and dry weather of the central Sahara is only very rarely broken by the influence of atmospheric disturbances from either the north or the south which may bring the rare and scanty rainfall to this otherwise extremely arid region. Such isolated and rare rainfall may occur in any month of the year but north of about latitude 20° to 22°N this is more likely to occur in the period October to February while to the south of this latitude such falls are more likely in the period July to September.

In the following sections of this chapter the individual elements of the climate of the Sahara are briefly described and illustrated with particular reference to climatic data available from five representative places with relatively long period meteorological records. These are Dakhla (Egypt) and Tamanrasset (Algeria) in the central Sahara; Bilma (Niger) on the southern margin of the desert; Salum (Egypt) on the Mediterranean coast; and Villa Cisneros (Western Sahara) on the Atlantic coast (Griffiths and Soliman, 1972; Meteorological Office, 1978; Rudloff, 1981).

2.4. GLOBAL RADIATION AND SUNSHINE

The Sahara is the most extensive region of the world with an average of over 10 hours of sunshine per day throughout the year and also with an annual average of global radiation exceeding 550 Langleys per day (one Langley is equal to one gram-calorie per sq. cm). These very high values around the year are a consequence of the absence of cloud. Maps of both these elements can be found in Griffiths and Soliman (1972) and Thompson (1964): the latter source contains reasonably large-scale maps of both these elements for the African continent for each month and for the year as a whole. Measured values of global radiation are only available for two or three places in the Sahara but can be calculated with a reasonable degree of certainty from the thirty or more places for which daily values of sunshine hours are available.

Mean monthly and annual values of global radiation and hours of sunshine for Tamanrasset in the central Sahara are given in Table 2.1, together with values of sunshine hours for Kufra, Bilma, Salum and Villa Cisneros. Kufra has been selected because the values are probably similar to those of Dakhla. For Tamanrasset and Villa Cisneros the sunshine hours have also been expressed as a percentage of the maximum possible duration assuming a completely clear sky.

Comparison of the values of global radiation with those of sunshine hours at Tamanrasset shows that radiation values in June are about 50% greater than those in December a much greater difference than that for sunshine hours. This difference reflects both the greater day length and the higher angle of the sun in the northern hemisphere summer. This should be borne in mind when interpreting the monthly values of sunshine hours at the other places. The factor of increasing day length is also relevant to an understanding of the significance of daily hours of sunshine.

Sunshine hours per day expressed as a percentage of the maximum possible at Tamanrasset indicate that the months of May and June and August and September are marginally more cloudy than the rest. This is the reverse of the situation on the northern margin of the Sahara where data for Tripoli, Bengazi and Cairo show that the months November to April all have a lower percentage of possible sunshine duration than the months May to October. Cairo falls within the limits of the Sahara as defined in this book. The smaller number of sunshine hours in all months at Villa Cisneros on the Atlantic coast shows the influence of the cold waters of the Canaries Current in increasing cloud and bringing fog to these shores. Sunshine hours are most reduced during the months July to October at the time when the contrast between land and sea temperatures is greatest.

TABLE 2.1. Radiation and Sunshine

	J	F	M	A	M	J	J	A	S	O	N	D	Year
TAMANRASSET													
Mean global radiation (langley/day)	437	508	589	639	656	636	654	606	517	480	427	394	545
Sunshine hours/day	9.6	9.5	10.4	10.9	10.6	10.2	11.5	10.7	9.6	10.0	9.5	8.8	10.1
% Possible	88	83	87	85	80	76	86	83	77	86	85	80	83%
lat. 22°42′N													
VILLA CISNEROS													
Sunshine hours/day	8.1	8.7	8.8	8.8	9.6	9.5	8.4	8.5	8.3	7.9	7.9	7.6	8.5
% Possible	75	77	73	69	72	70	62	66	67	69	72	71	70%
lat. 23°42′N													
KUFRA													
Sunshine hours/day	8.9	9.1	9.5	9.5	11.2	11.4	12.2	12.1	10.3	9.9	9.7	8.4	10.2
lat. 24°13′N													
BILMA													
Sunshine hours/day	9.4	9.9	9.8	9.9	10.4	10.6	10.8	10.3	10.2	10.3	10.2	9.4	10.8
lat. 18°39′N													
SALUM													
Sunshine hours/day	7.0	7.5	9.0	9.9	9.9	12.3	12.7	12.2	10.9	9.7	7.3	6.8	9.6
lat. 31°53′N													

Mean number of sunshine hours per year:

Tamanrasset	3686 hours
Villa Cisneros	3103 hours
Kufra	3723 hours
Bilma	3686 hours
Salum	3504 hours

2.5. AIR TEMPERATURE IN THE SAHARA

From the limited number of long period meteorological records available, as well as from the occasional spot readings of travellers, it is clear that large areas of the Sahara, as well as places on its semi-arid fringes, have experienced extreme maximum air temperatures of over 50°C. In accordance with standard meteorological practice such readings are 'shade' temperatures in an instrument screen. There is rarely any shade in the Sahara and therefore these readings are less representative of actual temperatures experienced during the heat of the day than they might be elsewhere. Similarly, owing to the absence of vegetation and soil moisture, the ground temperature will be very much higher.

At the other extreme most places in the Sahara experience air temperatures below freezing point at least once a year. Table 2.2 shows the mean daily maximum and a mean daily minimum air temperature for each month for five representative places; and also the highest and lowest temperatures ever recorded in each month. Almost all parts of the Sahara have a clear seasonal rhythm of temperature with the highest temperatures during the months June to August and the lowest during the months December to February. A notable exception is the Atlantic coastal fringe in Morocco, West Sahara and Mauritania where both the highest mean daily temperatures and the extreme maximum values are delayed until the months August to October. This is a typical maritime feature, but exaggeratd here by the combined effect of the cold ocean current in the midsummer months and the greater frequency of hot winds blowing off the land in autumn. This is illustrated by the temperatures for Villa Cisneros in Table 2.2.

TABLE 2.2. Mean Daily Maximum and Minimum Temperatures and Extreme Values

Temperature°	J	F	M	A	M	J	J	A	S	O	N	D	Year
TAMANRASSET (Alt. 1405 m) Lat. 22°42′N													
Mean daily max.	19	22	26	30	33	35	34	34	33	29	26	21	28
Mean daily min.	4	6	9	13	17	21	22	21	19	15	11	6	13
Absolute max.	26	28	32	36	38	38	39	38	37	34	30	27	39
Absolute min.	−7	−4	0	3	4	15	17	17	14	8	−1	−3	−7
DAKHLA (Alt. 110 m) Lat. 25°29′N													
Mean daily max.	22	24	28	33	37	38	38	39	36	33	28	23	32
Mean daily min.	4	5	9	14	19	23	23	23	20	17	11	6	14
Absolute max.	36	39	43	47	48	50	49	46	44	44	42	35	50
Absolute min.	−3	−4	0	2	7	13	16	16	11	8	2	−3	−4
BILMA (Alt. 355 m) Lat. 18°39′N													
Mean daily max.	27	30	35	38	42	43	42	40	41	38	33	27	36
Mean daily min.	7	9	13	17	21	23	23	23	21	16	11	8	16
Absolute max.	37	40	45	46	47	49	47	48	45	43	41	37	49
Absolute min.	−3	0	1	8	13	14	15	16	12	7	3	−1	−3
SALUM (Alt. 170 m) Lat. 31°53′N													
Mean daily max.	19	20	21	24	26	30	31	30	29	27	25	20	25
Mean daily min.	9	10	11	13	17	20	21	22	20	18	15	11	16
Absolute max.	27	33	41	42	43	46	44	43	42	40	36	32	46
Absolute min.	3	3	4	6	9	14	16	16	14	12	7	5	3
VILLA CISNEROS (Alt. 11 m) Lat. 23°42′N													
Mean daily max.	22	23	23	23	24	25	26	27	27	27	25	22	24
Mean daily min.	13	14	15	16	16	17	18	19	19	18	17	14	17
Absolute max.	31	34	36	37	35	36	38	39	42	39	37	29	42
Absolute min.	9	9	10	12	12	12	12	13	12	13	12	9	9

The moderating effect of altitude is also shown by the lower mean daily maximum and extreme maximum values during the warmer months at Tamanrasset on the south-western slopes of the Ahaggar mountains in the west central Sahara. These values are much lower than those at either Bilma or Dakhla. There is a much smaller difference between Tamanrasset and these places at a lower altitude in terms of the mean daily minimum and extreme minimum temperatures. The moderating influence of the sea on conditions in those parts of the Sahara which reach the shores of the Mediterranean is illustrated by the temperature values for Salum in Egypt. The daily range of temperature is lower in all months as compared with places away from the coast and air frosts have not been recorded here. Such conditions only apply to a very narrow coastal strip. On this coast the absolute maximum temperatures occasionally rise very high; a maximum temperature of 40°C or more has been recorded at Salum in every month from March to October. This occurs when the hot 'Khamsin' wind blows and draws dry heated air from the central Sahara towards the Mediterranean.

2.6. ATMOSPHERIC HUMIDITY AND EVAPORATION

Except on the shores of the Atlantic and the Mediterranean, both relative and absolute humidity are low in the Sahara particularly in view of the high temperatures which occur so often. Away from the coast a relative humidity of 100% is rarely observed except in those areas where there is a local source of moisture, as there is in an oasis or in the Nile valley and delta. In such areas morning mist or fog may occur in winter indicating that locally the relative humidity is 100% or near. Relative humidities below 20% are regularly

TABLE 2.3. Mean Monthly Relative Humidity

Relative humidity %	J	F	M	A	M	J	J	A	S	O	N	D
TAMANRASSET												
0700 hrs	37	34	31	35	34	27	25	29	29	37	40	40
1300 hrs	21	27	20	18	23	20	17	20	23	21	24	21
DAKHLA												
0800 hrs	56	52	44	38	35	34	33	36	42	47	50	57
1400 hrs	38	35	29	27	24	22	22	23	26	31	33	38
BILMA												
0700 hrs	40	33	28	22	29	30	39	56	38	33	39	43
1300 hrs	17	13	12	9	14	14	18	29	19	16	18	19
SALUM												
0730 hrs	75	75	71	65	65	63	70	77	74	74	75	75
1330 hrs	52	50	50	51	58	56	59	62	62	56	56	52
VILLA CISNEROS												
0500 hrs	75	79	82	83	85	86	88	88	89	87	84	75
1100 hrs	51	51	53	55	58	61	63	63	63	61	56	46

attained during the warmest months in the heat of the day and readings as low as 5% are not infrequent. Mean monthly value of relative humidity in the early morning and near midday for the five representative meteorological stations in the Sahara are given in Table 2.3.

The higher midday values at Dakhla as compared with those at Tamanrasset and Bilma probably reflect the availability of a local moisture source in the irrigated land. The much higher values of relative humidity at Villa Cisneros and Salum show the double influence of the sea; both in providing a source of atmospheric moisture and in reducing the temperature. Relative humidity is both a function of air temperature and of the actual water-vapour content of the air so that normally values of relative humidity are inversely correlated with air temperature. Thus on very hot days in the Sahara when the relative humidity may be well below 20% the absolute humidity of the air, which can be expressed as the dew point or as vapour pressure, may be as great or greater than on an average wet day in summer in Britain.

The combination of high relative humidity, high temperature, maximum solar radiation and frequent moderate to strong surface wind creates conditions highly favourable for rapid evaporation if a source of surface moisture is available. The evaporating power of the air, often termed potential evaporation or potential evapotranspiration, is a difficult and rather theoretical concept in the absence of a permanent moisture source. Most evaporation measurements in the Sahara have been made with Piche-type instruments mounted in an instrument screen but some measurements from open water evaporating pans have been made. Estimates of potential evaporation can also be made from meteorological data using empirical formulae such as those devised by Thornthwaite and Mather (1957) and Penman (1948). Such measurements and calculations are of considerable practical value; for example, they are very relevant in estimating the water requirements for irrigating crops and for calculating the evaporation loss from the surface of Lake Nasser, which is the largest water body in the Sahara. Because the air layer immediately above a wetted surface quickly takes up moisture the actual evaporation over such a surface will be much lower than that theoretically calculated over bare dry ground. This is known as the 'oasis effect'. A detailed but readable account of this complex problem can be found in Ward (1975). Estimates of daily evaporation in the arid Sahara vary between a mean value of 5 mm per day and as much as 8 mm per day, or between 1.8 and 2.9 m per year. For the higher estimate the seasonal variation is from about 4 mm per day in the cooler months to 11 mm per day in the hotter months. During weather conditions most suitable for high evaporation, with a strong wind, high temperature and low relative humidity, daily values may well exceed 20 mm. This high value is estimated from Piche measurements taken by the Egyptian Meteorological Department and quoted by Griffiths and Soliman (1972).

2.7. CLOUD AND PRECIPITATION

As already described and explained the air over the Sahara is very dry for most of the year. For this reason the sky is usually cloudless save for some high cirrus-type cloud often associated with the Sub-Tropical Jetstream and visible on cloud satellite photographs. Over most of the central Sahara cloud cover averages one-eighth or less of the sky, but on the northern and southern margins may increase to an average of one-quarter to three-eighths during the months when sporadic rain is more likely. The Atlantic coast with its more humid atmosphere is an exception; here cloud cover average three-eighths to one-half round the year.

Although precipitation in the Sahara is rare and very light it forms one of the few manifestations of weather as distinct from climate. Days with rain, and the occasional desert downpour, are of considerable meteorological interest as well as being an occasional blessing or a disaster to the inhabitants. Mean monthly or annual values of precipitation are almost meaningless in such an arid region. Kendrew (1961) notes that at In Salah in the central Algerian Sahara a significant shower may occur on average once every 10 years. Similar probabilities of a heavy shower apply to most of the Libyan and Egyptian desert, which appear to be even more intensely arid than the western Sahara. Some years ago, when the author of this chapter visited Kharga Oasis in Egypt, he was told that there had been unprecedented rain earlier that year and that perhaps 25 mm had fallen in two or three downpours. It was alleged that children aged 7 in Kharga had never seen rain before. Such statements probably ignore the occasional very light falls of rain or drizzle; but they are as expressive of the aridity of the Sahara as is the complete absence of *wadis*, or the courses of ephemeral streams, which are such a feature of desert regions which regularly receive some rain each year.

Rare but heavy showers of rain are often accompanied by thunder and brief squalls of wind with gusts reaching gale force (Pedgley, 1974). The rapid run-off of surface water after such storms can cause extensive local flooding, damage to dwellings and crops in an oasis, and on occasions considerable loss of life. A flooded wadi can create devastation many miles from the area where the downpour actually occurs and for this reason it is unwise to set up camp for the night in a wadi or depression if distant thunder, lightning or massive cloud are observed.

Table 2.4 gives the mean monthly rainfall and the maximum fall experienced in 24 hours at Tamanrasset, Bilma and Salum. At all three places the mean annual rainfall has been exceeded by a single

TABLE 2.4. Mean Monthly Precipitation and Maximum Fall in 24 hours

	J	F	M	A	M	J	J	A	S	O	N	D	Year
TAMANRASSET													
Mean monthly fall, mm.	4	1	1	2	6	4	3	10	7	2	2	2	44
Max. in 24 hours	7	3	5	20	48	19	5	35	48	17	19	21	48
BILMA													
Mean monthly	<1	0	0	<1	1	1	3	10	5	2	0	0	22
Max. in 24 hours	4	0	0	1	8	10	10	30	33	49	0	0	49
SALUM													
Mean monthly fall, mm.	12	12	12	1	3	0	0	0	1	5	28	21	95
Max in 24 hours	38	17	30	8	16	1	0	0	6	58	121	33	121

day's rainfall. Similar extremes and great variability are typical of any long-period rainfall record in the Sahara and even of places on its semi-arid margins with annual rainfalls of up to 250 mm.

The mountainous regions of the Ahaggar and Tibesti, which rise to heights of over 2900 m, are surrounded by a network of wadis which indicate that precipitation on these higher areas is both greater and more reliable. Owing to the absence of rain gauges precipitation here can only be estimated but it may be as great as 250 mm per year or more. Their summits are high enough to rise above the trade wind inversion and to trigger off uplift of moister and more unstable air when atmospheric conditions are favourable. It seems that this can happen when the ITCZ is furthest north in the summer months as well as in winter when colder air is introduced from higher latitudes under the influence of disturbances in the mid-latitude westerlies. Precipitation in winter here may be in the form of snow, and snow also occurs occasionally on the northern Red Sea Hills in Egypt. Snow is very rare at or near sea level on the Mediterranean coast of the Sahara, but is more frequent in the Jebel Akhdar region of Cyrenaica.

Throughout the Sahara it is not unusual for cloud to build up and for rain to be observed to fall from the high cloud base and then to evaporate in the very dry air beneath the cloud. Such instances are more frequently observed on the northern and southern margins of the desert when atmospheric conditions are marginal for the formation of precipitation.

2.8. WIND AND VISIBILITY

These two climatic elements are very closely related in the Sahara. Moderate or strong winds raise dust and sand grains and drastically reduce the range of visibility. Dust is raised from the loose dry surface of the desert and may be carried up to considerable heights. The extent of this wind deflation or erosion depends as much upon the nature of the surface as upon the actual wind speed. Only the stronger winds, for example with a mean speed of about 25 knots or more (about 50 metres per second), can raise the larger particles or sand grains. As distinct from dust particles, sand grains are carried along in the lowest 2 or 3 metres of the air in a series of jumps or hops (Bagnold, 1941). Dust particles can be carried up to great heights depending on the stability of the atmosphere and the precise upper air conditions. Dust is usually trapped below the trade wind inversion at about 2000 to 3000 m. The contrast between the hazy, dusty atmosphere below this level and the brilliant blue sky above is very obvious when ascending through this layer in an aeroplane on most days in the Sahara. The dust haze in the lower layers of the atmosphere is one reason for the brilliant red sunsets in the desert.

Winds of gale force and above (34 knots) are infrequent in the Sahara but they may occasionally occur anywhere. They result in severe dust or sandstorms which are a most unpleasant, and often dangerous, experience particularly if they are associated with high temperatures. The dust and sand particles, swept up from a hot desert surface, help to raise the air temperature even higher. The dust and sand particles can penetrate clothing, luggage and tents in an astonishing manner. This adds to the misery quite apart from the direct and harmful effects on skin, eyes and respiratory tracts.

Dust and sandstorms in the Sahara fall into three main types:

(a) On the northern margins of the Sahara strong winds in winter and spring are often associated with a shallow barometric low which develops over the northern Sahara or with a deeper depression in the Mediterranean. These disturbances move from west to east and bring hot, dusty air from the desert as far as the Nile Delta lands and the Mediterranean coast. Dust or even sandstorms are usually associated with these hot 'Khamsin' winds. Oliver (1947) gives an interesting account of how there was a dramatic increase in the frequency of duststorms in the Egyptian desert west of the Nile Delta as a direct consequence of military operations between 1940 and 1943. Tanks, and other military vehicles, broke up the hitherto undisturbed hardpan of the desert surface and thus released many more fine dust particles so that relatively lighter winds caused duststorms.

(b) On the southern margins of the Sahara in the northern Sudan and in the desert regions of Chad, Niger and Mali strong wind squalls lasting for an hour or more are often associated with small-scale tropical weather disturbances moving from east to west. These disturbances are called Haboobs in the Sudan and Tornados or Disturbance Lines in West Africa. When no rain falls, or when the rain does not reach the ground, the squall advances as a thick wall of dust which merges with the heavy clouds from which rain may eventually fall and thus abruptly end the duststorm (Bhalotra, 1963 and Eldridge, 1957).

(c) More extensive duststorms may occur anywhere in the Sahara if the barometric pressure gradient is sufficiently steep as to strengthen the normal Trade Wind or Harmattan above the critical level required to raise dust from the surface.

2.9. CLIMATIC CHANGE IN THE SAHARA

The question of changes of climate on both the northern and southern margins of the Sahara has been a subject of much discussion and controversy for many years. Some have even argued that the central Sahara was sufficiently moist during some periods of the Pleistocene as to permit grazing and hunting. Historical, archaeological and geological records provide abundant evidence that, at least prior to 4000 year B.P., the margins of the Sahara experienced more than one period when conditions were wetter than at present (Butzer, 1971). However, there is more doubt as to whether the very arid central Sahara was very different from the present throughout much of the Pleistocene. A recent review of the evidence for climatic change and fluctuation in the Sahara during the last two millennia (Nicholson, 1980) suggests that relatively minor but short-lived changes in the amount and regime of rainfall have occurred on both the northern and southern margins of the Sahara.

Suggestions have often been made that man might change the climate of the Sahara and induce rainfall by planting trees, creating more lakes but larger than Lakes Chad and Nasser, seeding the rare clouds with silver iodide, or even putting down extensive strips of a black substance such as bitumen or soot on the desert. All or some of these it is argued might induce rain. Both the evidence for such attempts elsewhere, and our knowledge of the meteorological processes operating in the atmosphere above the Sahara, suggest that such ideas are more appropriate to science fiction. Their effect would be either too local or too insignificant to overcome the very powerful atmospheric processes which make the Sahara the world's largest hot desert.

REFERENCES

BAGNOLD, R. A. (1941) *The Physics of Blown Sand and Desert Dunes*. Methuen, London.

BHALOTRA, Y. P. R. (1963) Meteorology of the Sudan; *Sudan Meteorological Service, Memoir no. 6*, pp. 64–72. Khartoum.

BROOKS, C. E. P. (1932) Le climat du Sahara et de l'Arabie. In *Le Sahara*, edited by M. HACHISUKA, pp. 27–105. Societe d'Editions Geographiques, Maritimes et Coloniales, Paris.

BUTZER, K. W. (1971) *Environment and Archaeology*. Aldine Publishing Company, Chicago.

CHANDLER, J. F. and MUSK, L. F. (1976) The atmosphere in perpetual motion. *Geog. Mag.* **49**(2), 93–102. London.

ELDRIDGE, R. H. (1957) A synoptic study of west African disturbance lines. *Quart. J. Roy. Meteorl. S.* **83**(357), 303–314.

FRISINGER, J. H. (1977) *The History of Meteorology to 1800*, pp. 125–128. Science History Publications, New York.

GRIFFITHS, K. F. and SOLIMAN, K. H. (1972) The northern desert (Sahara). In *Climate of Africa*, edited by J. F. GRIFFITHS, pp. 75–131. Vol. 10 of *World Survey of Climatology*. Elsevier, Amsterdam.

KENDREW, W. G. (1961) *The Climates of the Continents*, p. 45. Clarendon Press, Oxford.

MEIGS, P. (1953) World distribution of arid and semi-arid homoclimes. In *Arid Zone Hydrology*, pp. 203–210. UNESCO, Paris.

METEOROLOGICAL OFFICE (1962) *Weather in the Mediterranean*, Vol. 1, pp. 82–87. H.M.S.O., London.

METEOROLOGICAL OFFICE (1978) *Tables of Temperature, Relative Humidity and Precipitation for the World, Part IV, Africa, the Atlantic Ocean south of 35°N and the Indian Ocean*. H.M.S.O., London.

NICHOLSON, S. E. (1980) Saharan climates in historic times. In *The Sahara and the Nile*, edited by A. J. MARTIN WILLIAMS and HUGHES FAURE. Balkema, Rotterdam.

OLIVER, F. W. (1945) Dust storms in Egypt and their relation to the war period as noted at Maryut. *Geog. Jr.* **106**, 26–49.

PEDGLEY, D. C. (1974) An exceptional desert rainstorm at Kufra, Libya. *Weather*, **29**(2), pp. 64–70.

PENMAN, H. L. (1948) Natural evaporation from open water, bare soil and grass. *Proc. Roy. Soc. A*, **193**, pp. 120–145.

RIEHL, H. (1978) *Introduction to the Atmosphere*, pp. 143–149. McGraw-Hill, Tokyo.

RUDLOFF, W. (1981) *World Climates, with Tables of Climatic Data and Practical Suggestions*. Wissenschaftliche Verlagsgesellschaft, mbh., Stuttgart.

THOMPSON, B. W. (1964) *The Climate of Africa*, Maps 2–27. Oxford University Press, Nairobi.

THORNWAITE, C. W. and MATHER, J. R. (1957) Instructions and Tables for Computing Potential. *Evapotranspiration and the Water Balance, Publications in Climatology*, Vol. 10, pp. 185–311.

TULLET, M. T. (1978) A dust fall on 6 March 1977. *Weather*, **33**(2), 48–52.

WALKER, H. O. (1958) The monsoon in West Africa. *Ghana Meteor. Department, Note No. 9*. Accra.

WARD, R. C. (1975) *Principles of Hydrology*, pages 71–131. McGraw-Hill (UK), Maidenhead.

CHAPTER 3

Geology

M. WILLIAMS

School of Earth Sciences, Macquarie University, Australia

CONTENTS

3.1. INTRODUCTION

Apart from being our largest desert, the Sahara is also a superb geological museum. Rocks of every age and composition crop out in one part of the desert or another, and certain major geological structures like the West African craton have been stable for over 1500 million years. Wind-blown sands only blanket about a fifth of the total area and, except in a few favoured localities, the plant cover is sparse or absent. As a result, landforms and geological structures are impressively stark and well exposed.

In addition to the clearly visible older geological formations, more recent biological and archaeological remains are often excellently preserved. For instance, near Adrar Bous, an isolated granite mountain in the northern Ténéré desert of Niger (Fig. 3.1), Professor Desmond Clark and co-workers recovered a barbed bone point used by some Late Stone Age harpoonist over 7000 years ago Clark *et al.*, 1973). From the foot of the same mountain came the nearly complete skeleton of a domesticated Neolithic cow, *Bos brachyceros*, radiocarbon-dated to about 5000 years ago.

When confronted with such a rich prehistoric fossil legacy, the first question that springs to mind is when did the Sahara as we know it today first come into being. The birth of the Sahara was not a sudden or catastrophic event, but proceeded by fits and starts. Progressive desiccation was not caused by prehistoric peoples, but they did contribute to local destruction of the plant cover, a process that continues to this day.

The aim of this chapter is to trace the gradual evolution of the Sahara landscape, starting with its origins far back in the Precambrian, and concluding with the climatic vicissitudes of more recent years.

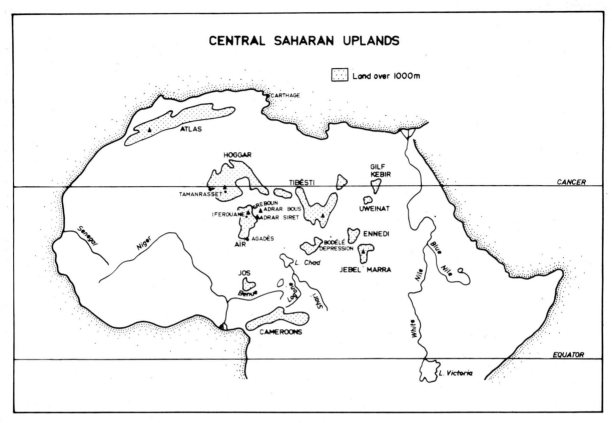

FIG. 3.1. Central Saharan Uplands.

3.2. TECTONIC SETTING

The oldest rocks to be found in the Sahara today are the various Precambrian formations. The youngest of these (designated Precambrian cover rocks in Fig. 3.2) are relatively undeformed and unaltered. They are separated by a marked erosional unconformity from the underlying Archaean and other Precambrian basement rocks which were strongly folded and metamorphosed on at least several occasions before being planed off by erosion.

Because much of the Precambrian landscape is concealed beneath an often considerable thickness—in some cases up to 10,000 metres—of younger rock formations, these ancient rocks are only exposed over limited portions of the desert, as is clear from Fig. 3.2.

Nevertheless, the influence of these ancient basement rocks upon the later geological history of North Africa is out of all proportion to their limited present-day surface distribution. Indeed, the pattern of subsequent faulting, vulcanism and other igneous activity reflects the influence of geological structures which developed some 550 million years ago, during the Pan-African orogenic event (Clifford, 1970), and in many instances during the even earlier phases of mountain building shown on Fig. 3.3.

Attempts to date the various orogenic events which created the structures that controlled the pattern of Saharan sedimentation during the past 570 million years of Palaeozoic and later times have often proved frustrating (Brock, 1981). Overprinting by later orogenic events often means that ages obtained by

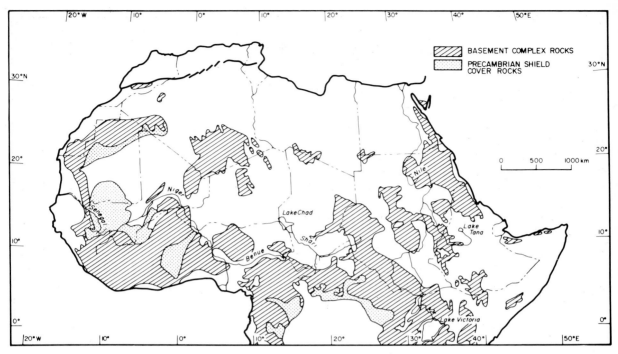

FIG. 3.2. Basement Complex rocks in and around the Sahara. Distribution of Basement Complex rocks and Precambrian shield ancient cover rocks generalized from 1:15 M Tectonic Map of Africa (Unesco, 1968). Political boundaries on this and later figures from Times Atlas (1972).

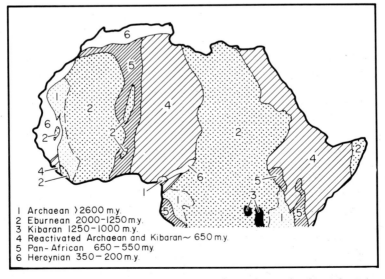

FIG. 3.3. Map of major orogenic events and associated structural zones in North Africa (after Fabre, 1974, Fig. 2).

radiometric dating or from palaeomagnetic assays are minimum ages only, in effect telling us when the radiometric clock was last reset to zero. This is the main reason why large areas of igneous and metamorphic rocks in North Africa, spanning a time range of 3000 million years, are lumped together as Basement Complex rocks (Fig. 3.2).

A glance at the diagonally hatched areas in Fig. 3.3 shows that about half of North Africa was affected (or reactivated) by the Pan-African earth movements which began about 650 million years ago. On the other hand, except for local movements, the dotted areas shown on Fig. 3.3 (which together comprise nearly 40% of the Sahara) have been stable for at least 1250 million years. Prolonged stability of these ancient cratonic areas has allowed denudation to proceed to the point where much of the western Sahara is now underlain at shallow depth by the worn down stumps of Precambrian basement structures.

By the start of Phanerozoic or post-Precambrian time, some 570 million years ago, much of the Sahara consisted of vast erosional plains, with an overall regional slope from south to north. Flow directions inferred from sediments laid down by Cambrian and early Palaeozoic rivers reveal a landscape devoid of discrete sedimentary basins, with gentle slopes and low relief (Beuf et al., 1971).

With the pre-Mesozoic or Hercynian earth movements of 350–200 million years ago, the Palaeozoic cover was folded, and a characteristic basin and swell topography developed. Later updoming of the Ethiopian–Arabian swell, which began during the late Cretaceous, was the forerunner to the Tertiary uplift, volcanism and rifting of East Africa. Other Mesozoic and Tertiary earth movements saw the uplift of the Hoggar, the Aïr, Tibesti and Jebel Marra, with accompanying upward flexure of the underlying Precambrian basement rocks. Uplift, faulting and vulcanism continued, and even in some instances accelerated, during the last two million years, but in most cases the major elements of the Saharan landscape were well established by the end of the Tertiary, before the continuing desiccation and rapid climatic changes that characterized the Quaternary.

3.3. STRATIGRAPHY AND SEDIMENTATION

As a result of the very low relief and low elevation of the early Palaeozoic Sahara, major changes in sedimentation often reflected comparatively slight earth movements. Towards 450 million years ago a further wave of uplift ushered in the late Ordovician glaciations of the Sahara (Beuf et al., 1971). Still preserved in the rocks surrounding the present Hoggar mountains are the glacial striations, erratics and moraines of this dramatic incident in the geological history of North Africa. A succession of marine incursions and withdrawals followed the melting of the Saharan ice caps. The rapid early Silurian rise in sea level was a world-wide event, comparable to the post-glacial sea-level rise of the last 17,000 years. Marine incursions triggered by tectonic downwarping were more localized in space, and took far longer. The Silurian seas withdrew once more, and during Devonian times much of the Sahara was again dry land. Marine and continental sediments continued to accumulate during the Carboniferous, mantling North Africa in limestones, mudstones and sandstones.

The Hercynian earth movements which were especially vigorous in the western Sahara and Atlas were associated with opening of the North Atlantic, and creation of the present continental margin of the western Sahara. A further outcome was folding and later erosion of the hitherto more or less horizontal Palaeozoic sedimentary rocks, the debris from which was spread northwards by Triassic and younger rivers. Silicified tree-stumps preserved within these Mesozoic sandstones and mudstones, often in positions of growth, attest to rapid burial of the North Saharan woodlands by actively aggrading rivers from the south and east.

Opening of the South Atlantic and separation of the African and South American plates was followed later in the Mesozoic by several marine incursions. During one of these much of the central Sahara was submerged beneath a shallow equatorial sea from which protruded the mountainous Hoggar, and the

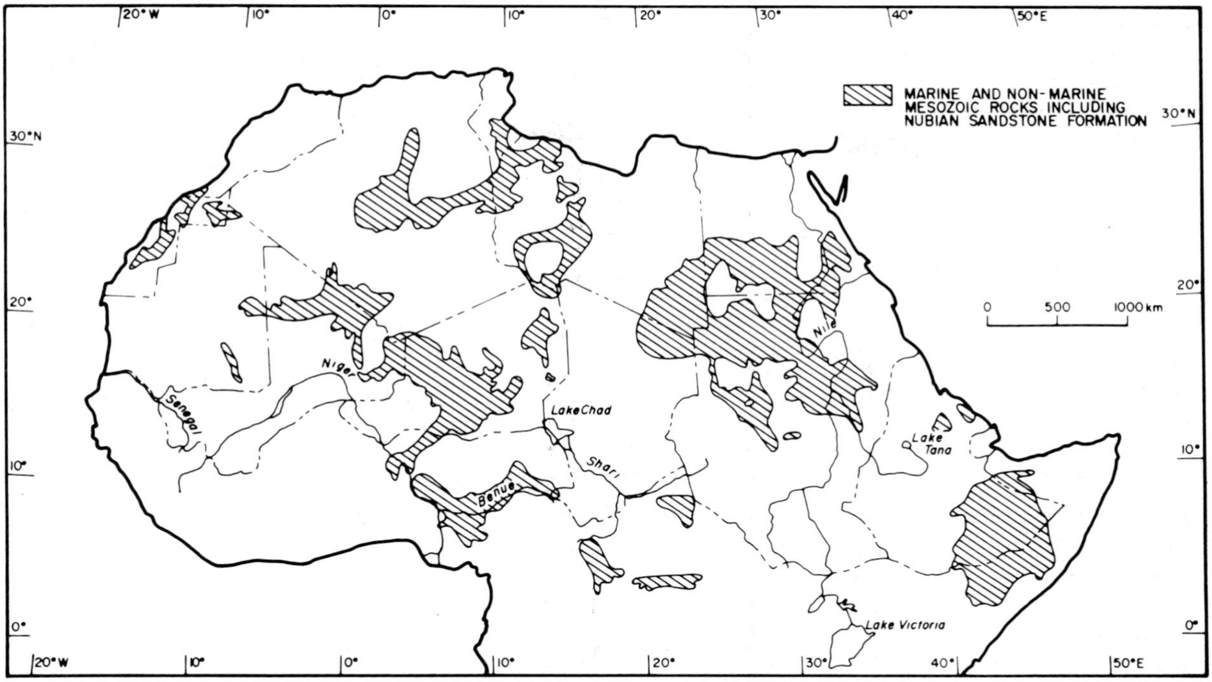

FIG. 3.4. Mesozoic sedimentary formations of the Sahara. Distribution of marine and non-marine Mesozoic rocks, including the Nubian Sandstone Formation, generalized from 1:10 M preliminary geological map of Africa (Unesco, 1967) and from Vail (1978, Fig. 7).

stable West African craton. Today, the horizontal sandstones and limestones of these marine and non-marine Mesozoic formations form vast gravel-strewn plateau surfaces or *hamada*, comparable alike in age and appearance to the stony tablelands of the present Australian desert. Figure 3.4 shows the present-day distribution of Mesozoic rock outcrops in the Sahara.

During the early Tertiary there was a long interval of intense weathering with very little erosion, at least in the tectonically-stable lowlands of the southern Sahara (Faure, 1962). Near-surface mobilization and redeposition of iron and silica at this time is reflected in the resistant layers of ironstone or silicified rock which today cap many of the Tertiary and Mesozoic plateau surfaces.

Middle Tertiary uplift initiated a further phase of widespread erosion in and around the major uplands. Much of the deeply weathered mantle was stripped from the hills of the southern Sahara. Sandy detritus from mountains in and around the Sahara was ferried by major rivers into the closed sedimentary basins which developed as a result of the Tertiary earth movements. Towards the close of the Tertiary wind-blown desert sands make their first appearance in the Chad basin (Servant, 1973). Elsewhere in the Sahara the late Tertiary vegetation was already well adapted to aridity (Maley, 1980). The Sahara was becoming a desert.

3.4. VULCANISM AND OTHER IGNEOUS ACTIVITY

Massive outpourings of lava accompanied the Tertiary and Quaternary uplift of what are now major Saharan massifs like the Hoggar, Tibesti and Jebel Marra. Before dealing with these Cainozoic events of the past 70 million years, it is appropriate to go back in time, and to compare Fig. 3.3 with Fig. 3.5.

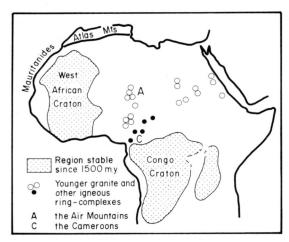

FIG. 3.5. Distribution of Palaeozoic, Mesozoic and younger igneous complexes in relation to stable cratons. Black circles denote Tertiary granites; open circles denote Younger Granites of Nigeria and Niger (300–150 m.y.), and ring complexes of Sudan, Libya and northern Ethiopia (550–45 m.y.). After Black and Girod (1970, Fig. 3) and Vail (1978, Fig. 9).

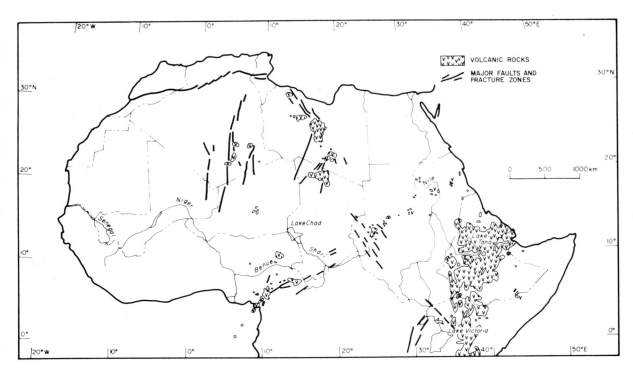

FIG. 3.6. Cainozoic volcanic rocks and associated lineaments. Distribution of volcanic rocks generalized from 1:15 M Tectonic Map of Africa (Unesco, 1968) and from Vail (1972, Fig. 4). Major lineaments and fractures after 1:10 M preliminary geological map of Africa (Unesco, 1967) and Vail (1972, Fig. 4).

Figure 3.5. shows in diagrammatic form the distribution of various igneous intrusions, all of which are flanked by two very ancient and stable cratons. Radiometric ages obtained on the ring-complexes or Younger Granites of Nigeria and Niger reveal that the northern Aïr ring-complexes are mid-Palaeozoic in age, and are significantly older than their southern counterparts. It is possible that northward movement of the African plate across one or more hot-spots may be responsible for these age trends (Bowden *et al.*, 1976).

Whatever the ultimate causes of their origin, the Palaeozoic and younger ring-complexes of the Sahara are of considerable economic and biological significance. Well endowed with minerals such as tin, they are often the only sources of water and plant foods in an otherwise barren landscape, and sometimes house galleries of magnificent prehistoric rock engravings and animal paintings.

The Tertiary granites of Cameroon follow a line of weakness in the earth's crust which is also evident in the extrusive or surface manifestations of Tertiary igneous activity. These are the volcanoes which extend from volcanic islands in the Gulf of Guinea north-east across the Sudan (Fig. 3.6). Another line of volcanoes runs from north-west Libya and Tibesti to intersect the Cameroon–Sudan lineament about where Jebel Marra volcano is situated in the western Sudan (Vail, 1972). According to Francis and co-workers about 8000 km^3 of material was erupted during the formation of Jebel Marra, compared with about 3000 km^3 at Tibesti (Francis *et al.*, 1973).

In the Hoggar, Tibesti and Jebel Marra massifs, the Tertiary to Recent vulcanism was accompanied (and probably preceded) by significant uplift of the underlying basement rocks—500 metres at Jebel Marra, 1000 metres in Tibesti and the Hoggar. The immediate consequence of this late Cainozoic uplift was the creation of major upland catchments in an increasingly dry environment, and it is to the rivers that once flowed from these Saharan massifs that we now turn.

3.5. DESICCATION AND CLIMATIC FLUCTUATIONS

The late Cainozoic desiccation of the Sahara was the final outcome of changes in the atmospheric circulation brought about by several independent but roughly synchronous geological events.

The slow northward movement of the African plate during the previous hundred million years meant a migration of northern Africa from wet equatorial into dry tropical latitudes.

Uplift of the Tibetan plateau during the late Tertiary and Quaternary helped to create the easterly jet stream which now brings dry subsiding air to the Ethiopian and Somali deserts.

A further important factor that at first glance seems unrelated to the late Tertiary and Quaternary expansion of the Sahara desert was the progressive build-up of continental ice, initially in Antarctica, later in the Northern Hemisphere. As the temperature gradient between the equator and the poles became steeper, Trade Wind velocities increased, as did their capacity to mobilize the alluvial sands of the Sahara into dunes.

Cooling of the ocean surface, particularly in the intertropical zone where two-thirds of our rain falls today, may also have been a powerful factor in reducing tropical precipitation, dependent as it is upon efficient evaporation from the equatorial oceans.

From about 2.5 million years onwards, the great tropical inland lakes of the Sahara began to dry out (Williams, 1982). The tropical fauna and flora of the well-watered Saharan uplands became increasingly impoverished, and a once efficient network of major rivers became partially obliterated by wind-blown sands. A pattern of climatic oscillations now became established, with each cycle about 100,000 years in duration. Nine-tenths of the cycle involved a gradual build-up towards peak aridity, windiness and cold, followed by a rapid but short-lived return to milder and wetter conditions.

The last arid cycle ended about 12,000 years ago, and is evident in the line of vegetated dunes, shown on Fig. 3.7, which extend well south of the present southern limit of active dunes. For the next 6000 years,

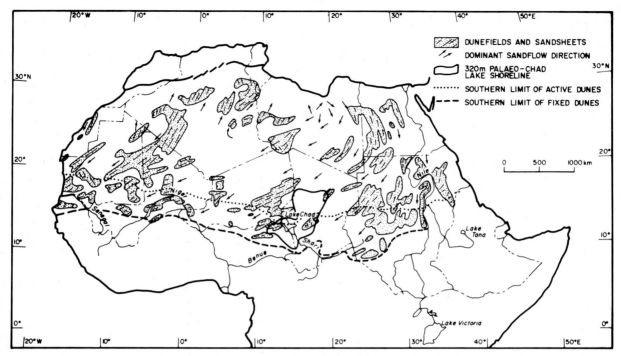

FIG. 3.7. Quaternary dunefields and sandplains of the Sahara. Dunefields and dominant dune trends generalized from Grove (1980, Fig. 1.1) and from Vail (1978, Fig. 8). Southern limits of fixed and active dunes after Mainguet and Callot (1978, Fig. 3), with some amendments. Sandflow directions after Cooke and Warren (1973, Fig. 4.3), Mainguet and Callot (1978, Fig. 11) and Vail (1978, Fig. 8). Extent of 320-m palaeo-Chad lake after Grove (1980, Fig. 1.1).

from about 12,000 to 6000 years ago, the Sahara was once more a land of lakes. The dead rivers flowed anew. This was the time when fish and crocodiles swam in lakes around the Hoggar, the Aïr and other Saharan massifs. Thereafter a return to aridity, accentuated towards 4000 years ago, brought about the great exodus of Neolithic herders from the desert—a forerunner to the recent tragic events in the Sahel.

3.6. CONCLUSIONS

The location of major uplands, stable plateaux and depositional basins within the Sahara reflects the continuing influence of geological events of very great antiquity. Towards the end of the Tertiary, localized uplift and volcanic activity added the finishing touches to the creation of the Saharan landscape. From then on, increasing aridity has converted a savanna land adequately endowed with lakes and rivers into a parched wilderness, one-fifth of which is now covered in wind-blown sand. The long-term climatic prognosis is for a return to the hyper-arid conditions of 18,000–12,000 years ago, when intertropical aridity was world-wide.

REFERENCES

BEUF, S., BIJU-DUVAL, B., DE CHARPAL, O., ROGNON, P., GARIEL, O. and BENNACEF, A. (1971) *Les Grès du Paléozoique inférieur au Sahara.* Pub. Inst. Fr. Pétrole. Technip, Paris.
BLACK, R. and GIROD, M. (1970) Late Palaeozoic to Recent igneous activity in West Africa and its relationship to basement structure. In *African Magmatism and Tectonics*, edited by T. N. CLIFFORD and I. G. GASS, pp. 185–210. Oliver & Boyd, Edinburgh.

BOWDEN, P., VAN BREEMEN, O., HUTCHINSON, J. and TURNER, D. C. (1976). Palaeozoic and Mesozoic age trends for some ring complexes in Niger and Nigeria. *Nature*, **259**, 297–9.

BROCK, A. (1981) Paleomagnetism of Africa and Madagascar, in *Paleoreconstruction of the Continents*, edited by M. W. MCELHINNY and D. A. VALENCIO. Geodynamics Series, Amer. Geophys. Union, Washington; Geol. Soc. Amer., Boulder.

CLARK, J. D., WILLIAMS, M. A. J. and SMITH, A. B. (1973) The geomorphology and archaeology of Adrar Bous, central Sahara: a preliminary report. *Quaternaria*, **17**, 245–97.

CLIFFORD, T. N. (1970) The structural framework of Africa. In *African Magmatism and Tectonics*, edited by T. N. CLIFFORD and I. G. GASS, pp. 1–26. Oliver & Boyd, Edinburgh.

COOKE, R. U. and WARREN, A. (1973) *Geomorphology in Deserts*. B. T. Batsford Ltd., London.

FABRE, J. (1974) Le Sahara: un musée géologique. *La Recherche*, No. 42, vol. 5, 140–52.

FAURE, H. (1962) *Reconnaissance géologique des formations post-paléozoiques du Niger oriental*. Mém. Bur. Rech. Géol. Min. (1966), 47, 630 pp.

FRANCIS, P. W., THORPE, R. S. and AHMED, F. (1973) Setting and significance of Tertiary–Recent volcanism in the Darfur Province of Western Sudan. *Nature Phys. Sci.* **243**, 30–2.

GROVE, A. T. (1980) Geomorphic evolution of the Sahara and the Nile. In *The Sahara and the Nile*, edited by M. A. J. WILLIAMS and H. FAURE, pp. 7–16. Balkema, Rotterdam.

MAINGUET, M. M. and CALLOT, Y. (1978) *L'Erg de Fachi-bilma (Tchad-Niger)*. Mémoires et Documents, Serv. Docum. Cartog. Géogr., vol. 18, CNRS, Paris.

MALEY, J. (1980) Les changements climatiques de la fin du Tertiaire en Afrique: leur conséquence sur l'apparition du Sahara et de sa végétation. In *The Sahara and the Nile*, edited by M. A. J. WILLIAMS and H. FAURE, pp. 63–86. Balkema, Rotterdam.

SERVANT, M. (1973) *Séquences continentales et variations climatiques: Evolution du bassin du Tchad au Cénozoique supérieur*. D.Sc. thesis, Univ. Paris, 348 pp.

VAIL, J. R. (1972) Jebel Marra, a dormant volcano in Darfur Province, Western Sudan. *Bull. Volc.* **36**, 251–65.

VAIL, J. R. (1978) Outline of the geology and mineral deposits of the Democratic Republic of the Sudan and adjacent areas. *Overseas Geol. Miner. Resour. No. 49*, 1–68.

WILLIAMS, M. A. J. (1982) Quaternary environments in Northern Africa. In *A Land between Two Niles*, edited by M. A. J. WILLIAMS and D. A. ADAMSON, pp. 13–22. Balkema, Rotterdam.

CHAPTER 4

Soils

C. W. MITCHELL

Geography Department, University of Reading

CONTENTS

4.1. CONCEPTS AND PAST SURVEYS

Saharan soils resemble those of other regions in being independent natural bodies with properties derived first from the broad climatic zone and secondly from the modifying effects of local factors. They are *zonal* where the reflect the climatic datum, *azonal* where local factors preclude its development, and *intrazonal* where they dominate it.

Datum sites reflect the dominating zonal effect of heat and aridity. Interruptions occur where local topography either exposes inert bare rock or sand, or else concentrates moisture into low sites in sufficient quantities for soil profiles to develop to a greater extent than in surrounding areas. Because of the absence of water and vegetation, pedological processes are feeble and small moisture differences emphasized. This enhances the importance of topographic and lithological factors to an extent which often obscures broad zonal climatic variations and makes geomorphology the overriding key to local soil distributions. This assists interpretation because geomorphological features in deserts are generally recognizable both on the ground and on aerial photographs and other remotely sensed imagery.

All Saharan nations have undertaken soil mapping, but have generally concentrated on their less arid and more populous parts. Small-scale international compilations are rare, but there are some useful sources. These include the map and cross-sectional diagrams of north African landforms by Raisz (1952) of which simplified versions are shown on Figs. 4.1 and 4.2, the terrain analogue studies of north-west and north-east Africa (U.S. Army Waterways Experiment Station, 1958, 1962), Furon's book on the geology

FIG. 4.1. Physiography of the Sahara (after Raisz, 1952).

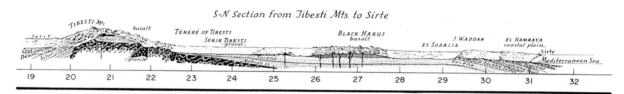

FIG. 4.2. Cross-section of the Sahara from Tibesti to the Gulf of Sirte (after Raisz, 1952).

of Africa (1963) and Petrov's *Deserts of the World* (1976). Other works of value are summarized by Lustig (1970) and Dutil has made a comprehensive survey (1971).

After initial syntheses of soil surveys in the post-war period, Hoore, mainly using French and Belgian terminology, published a soils map of Africa in 1964. In this, Saharan soils consist of rock debris, desert detritus, and weakly developed soils, almost total reliance being placed on geomorphic criteria in the extremely arid parts of the Sahara where water and vegetation have least effect. Dregne reviewed this and the work of soil surveys in Saharan countries in 1970. The FAO/Unesco *Soil Map of the World* at a scale of 1:5,000,000 (1974) presents the most recent comprehensive synthesis of research and national soil surveys, and its nomenclature is used in this chapter. It was, however, completed before the general availability of satellite photography and so its boundaries do not always closely reflect the physiography.

4.2. CLIMATIC ZONATION

The Saharan climate ensures that the soils will be of pedocalic Red Desert type but includes internal variations which influence their distributions. In the extremely arid core area, except for the better watered islands of Hoggar and Tibesti (Meigs, 1952; Unesco, 1979), physical weathering and eolian action are

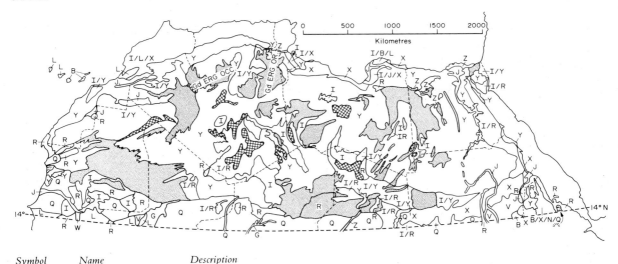

Symbol	Name	Description
SOILS WITH STRONG PROFILE DEVELOPMENT		
B	Cambisols	Soils with lower horizon altered by weathering
G	Gleysols	Gleyed soils
L	Luvisols	Soils with an alluvial clay horizon
V	Vertisols	Dark coloured cracking clays
W	Planosols	Soils with an impermeable lower horizon
Z	Solonchaks	Saline soils
SOILS WITH WEAK PROFILE DEVELOPMENT		
Q	Arenosols	Light coloured soils on coarse unconsolidated materials
X	Xerosols	Desert soils with weak horizonation
Y	Yermosols	Desert soils with very weak horizonation
SOILS WITHOUT PROFILE DEVELOPMENT		
I	Lithosols	Shallow soils over rock
J	Fluvisols	Soils on recent alluvium
R	Regosols	Other soils on unconsolidated materials

Fractional symbols indicate complexes

 bare rock

mobile sands

FIG. 4.3. Soils of the Sahara (simplified from FAO/Unesco, 1974).

overwhelmingly dominant. The scanty rainfall can only effect some surface movement of lime and seldom forms appreciable horizons. Northwards from this core, there is a fairly rapid increase in the rainfall, which because it falls mainly in a colder winter period and thus evaporates less, is quickly reflected in the sub-Mediterranean vegetation and soils, rendering the north African steppe zone relatively narrow. By contrast, the sahel zone in the latitudinal belt south of the core area is wider due to the more gradual increase in rainfall, and is displaced equatorwards because, due to the fact that the rainfall maximum is in summer when evaporation is especially high, more is required to sustain vegetatation and influence soil profiles.

On datum sites, therefore, which neither lose nor gain appreciable water through surface flow, it is possible to discern a general north–south zonal transition (Fig. 4.3) from (a) the clearly horizontal *Luvisols* in the northern Maghreb and red and brown *Cambisols* near to the coast from Tunisia to Egypt, with some organic darkening of the surface soil over redder colours derived from iron movement in the subsoil, through (b) the well-developed clay, lime and gypsum horizons of the semi-arid *Xerosols* to (c) the faintly developed *Yermosol* profiles of the central Sahara. Southwards from this the increasing rainfall gives (d) a wider zone of *Xerosols* and *Arenosols* in the Sahel. Luvisols (e) only reappear in the savanna zone generally south of the 500-mm isohyet and Cambisols in the uplands of Ethiopia.

4.3. PHYSIOGRAPHIC EVOLUTION

4.3.1. Pre–Pleistocene

Within these broad zones, physiography, shown in simplified form on Figs. 4.1 and 4.2, determines the detailed pattern of soil distributions. Most of the Sahara is formed of a relatively level shield of Precambrian rocks which occur at the surface over wide areas. The main interruptions are due to the large-scale tectonic fractures which gave rise, during the Hercynian orogeny, to the Hoggar (3000 m) and Tibesti (3415 m) mountain masses. These have crystalline and metamorphic cores, ringed by dissected plateaux of crystalline rocks (Aïr, 1800 m; Adrar des Ifoghas, 570 m) and Paleozoic sandstones (Tassili N'Ajjer, 1906 m; Mouydir, 1680 m; Ennedi, 1450 m).

The Carboniferous submergence of the western part of the shield deposited limestones, and the Cretaceous submergence of the eastern part under the Tethys sea laid the Nubian sandstones which now cover thousands of square kilometres and provide the source material for a number of the *ergs* (sand seas). Hoggar and Tibesti were subsequently modified by the Tertiary volcanic extrusions which created the Ethiopian Highlands, Jebel Marra (Sudan), Jebel Es Soda and Jebel Harug (Libya) and smaller outcrops elsewhere.

Otherwise, the Sahara has remained essentially a level plateau, between 200 m and 500 m in elevation, with lowest areas near the west coast, round the Gulf of Sirte, and in the Bodélé ('Pays-Bas') area of Chad. The topography is mainly governed by wide areas of eroded and deflated low-dip sediments. Plateau surfaces have been dissected into tablelands with cliffed edges, often fault controlled, which have been frayed into complex arrangements of promontories, re-entrants and inselbergs whose erosion provides the loose materials which are washed into the lowlands to form the colluvial and alluvial soils.

The alignment of some faults can sometimes be traced in the physiography and give rise to a wide lateral extension of the same *soil catenas* (toposequences) over many miles, sometimes visible on satellite photography. Examples are the linked system of south-facing escarpments running from the Qattara depression and Siwa in the west along the southern sides of the Hamadas of El Hamra, Tinghert and Tademait to end in the El Hank escarpment of Mauritania (Fig. 4.1), and the escarpments of the Hamadas of Dra, Daoura, Guir and Bet Tuadjin which face north to the Atlas Mountains across the fault depression of the 'Sillon Sud-Atlasique'. Figure 4.2 shows a representative S–N cross-section.

4.3.2. Pleistocene and Recent

Although the pedological consequences of Pleistocene climatic changes remain little known, investigations of the morphology and genesis of marine and river terraces, wadi systems and lake beds, and of paleosols and archaeological sites have revealed some of the main features. In the northern Sahara, pluvial periods appear to have been marked by accelerated erosion and its associated sedimentation, and the initiation of some karstic phenomena. Some soils were rubefied (reddened by iron movement) and others were impregnated with salts, gypsum and calcium carbonate. The latter often led to case-hardening into gypcrete and calcrete probably towards the termination of pluvial periods. Gypcrete mainly occurs in the neighbourhood of the chotts but calcrete is widespread, though confined to the zone north of 32°–33°N. In the central Sahara, the pluvial periods led to the formation of considerable areas of ferruginous tropical soil which have hardened into shields and extend as far as 26°–27°N. Calcareous crusts were formed in marshes, paleovertisols (buried or preserved vertisols) on plateaux and brown soils showing some profile movement of clay and iron ('brunification') on crystalline rocks in the cooler and moister altitudinal zone above 2000 metres. On the southern side of the Sahara, considerable vertisolization is recognizable up to 19°N although its present northern limit is 13–14°N (Dutil, 1971). The drier interpluvial periods were marked along the coast by beach terraces thought to be associated with the higher sea levels resulting from the reduced polar ice caps. Inland, eolian action increased at the expense of fluvial and led to intensified deflation and erg formation.

The effect of Recent water and wind action has been the continuation of large-scale winnowing and sorting of Saharan surface materials. Boulders and stones, moved only short distances from where they were formed, armour the surfaces. Gravel is moved farther and covers wider areas. Finer materials are washed into natural hollows but are sorted to leave the coarsest particles along the flow channels and to deposit the finest materials in areas of slack water.

This fluvial sorting is increased by wind action but in an opposite sense. Clay particles are usually too strongly bound together to be moved. Silt particles between about 0.002 mm and 0.05 mm are more cohesive than sand, but once dislodged, are carried in suspension often to high altitudes and over long distances until altered winds, rain, hills, or vegetation cause the aerodynamic check which deposits them as loess, often far beyond the margins of the Sahara.

It is, however, the sand fraction which forms the most characteristic and best-known feature of the Saharan landscape. The particles between 0.05 mm and 1 mm in diameter, and especially those between 0.1 mm and 0.4 mm, are large enough neither to cohere nor to remain long suspended in normal winds but are small enough to be readily transported by them. They are mainly formed of quartz, partly coated with iron oxide, but also include heavier minerals reflecting their parent rocks. Near beaches or playas, they may contain calcium carbonate and gypsum, but these, because of their softness, diminish away from the source.

4.4. DISTRIBUTION OF MAIN SOIL TYPES

The sequence of sites described below is generally from high to low in elevation and is accompanied by an increasing availability of soil moisture and a change from coarse to fine in texture. Paradoxically, this does not reflect a comparable sequence of favourability for plant growth. Smith (1949) has shown from a study of tree species in the Sudan that away from the immediate vicinity of running water and where salinity is low, the most critical factor is the availability of moisture in the root zone resulting from the rapidity of infiltration of ephemeral rainfall. In order of favourability the sites were rocky and stony plains and hill slopes, sand hollows, dunes, and plains, clay plains, and increasingly elevated and sloping impermeable sites. In open desert, therefore, coarser textures often increase the favourability of sites for the growth of plants.

4.4.1. Rocks and hamadas

The surfaces formed by the processes described above provide the environment of the soils (Fig. 4.3). In general, the highest sites give rise to bare rock or shallow *Lithosols*, where it has a shallow stony mantle.

Next in elevation are the wide level plateaux covered by difficult bouldery terrain known as hamadas. This name is applied both to the rock hamadas of the structural tablelands and to the boulder hamadas on the colluvial footslopes below them. They are composed of material broken *in situ* or transported only short distances and concentrated at the surface by deflation, water sorting, upward migration of coarse

TABLE 4.1. Analyses of Selected Saharan Topsoils (<2 mm fraction)

Soil no.	Site type	Location	Soil class	Depth
1 (Fig. 4.4)	Nubian sandstone hamada	Murzuk, Libya	Yermosol	0–15
2 (Fig. 4.5)	Sandsheet	Ubari erg, Libya	Regosol	0–15
3	Coastal marsh	Buerat el Hsun, Libya	Solonchak	0–15
4 (Fig. 4.6)	Salt flat	Sebha, Libya	Solonchak	0–20
5 (Fig. 4.7)	Clay plain	Ed Dueim, Gezira, Sudan	Vertisol	0–30
6	Daya floors (32 sites)	⎱ Near Errachidiya,	Fluvisol	0–30
7	Surrounding calcareous hamadas (31 sites)	⎰ Morocco	Yermosols	0–13

Notes: Libyan analyses by Military Engineering Experimental Establishment, Christchurch, Dorset: pH on saturation extract, E.C. c 1:1 paste; Sudan analyses from Greene (1952), mechanical fractions are 2–0.2, 0.2–0.02, 0.02–0.002 and <0.002 mm; Moroccan analyse from Mitchell and Willimott (1974): pH on saturated paste, E.C. on 1:1 paste, mechanical fractions coarse and medium sand 0.2–2 mm fine sand 0.2–0.02 mm, silt and clay <0.2 mm.

particles or a combination of these. The effect is to yield a relatively stable surface because of the progressive strengthening of its stony 'armour'. The character of this surface is determined by the mode of breakdown of the underlying rock. In the Libyan Sahara, the black basalts give rounded boulders up to 1 metre across and Nubian sandstones a surface of dark-toned fragments of varying size and much sand. The underlying soil contains a zone of weathered rock and fine material. This is often shallow, but can, as on the Hamada of Murzuk (Fig. 4.4; soil 1 on Table 4.1), be more than a metre deep. These soils are generally *Haplic Yermosols*, or where substantial clay translocation occurs, *Luvic Yermosols*.

Exposed non-calcareous stones on hamada surfaces are subject to impregnations which harden and protect their surfaces, notably incrustations with iron and manganese oxides (German: *Schutzrinde*). These salts are concentrated near the surface by evaporating water as hydroxide gels which are then transformed

FIG. 4.4. Yermosol on Nubian sandstone, Hamada of Murzuk, Fezzan, Libya (see Table 4.1). Blackboard 20 × 13 cm.

Mechanical analysis								
Coarse sand (2–0.6 mm)	Med. sand (0.6–0.2 mm)	Fine sand + (0.2–0.06 mm)	Silt + (0.06–0.002 mm)	Clay (<0.002 mm)	pH	EC mm Hos/cm at 125°C	% CaCO₃	% SO₄
6	6	60	28		8.1	0.78	4.90	0.025
13	40	47	0		9.7	2.13	0.41	0.003
2	15	31	52		8.2	36.2	22.56	1.121
1	3	20	76		7.1	61.7	2.44	1.212
8		10	13	68	8.6	(0.05% salts)		
13.3		51.3		35.4	8.14	1.08	10.15	0.005
18.6		46.8		34.5	8.12	10.25	14.54	0.299

by heating into a hard crust which grows into the rock surface. The outermost polish is due to a fine coating of silica. The amount of such *case hardening* appears to increase with rainfall and the content of ferromagnesian minerals and is especially common on igneous rocks and sandstones (Fig. 4.4).

Compact limestones yield pavements with loose slabs while softer limestones give smaller fragments of bedrock and flints. The calcareous material shows both the small-scale pitting and channelling on exposed surfaces known as *tafoni*. Enough fine material is usually preserved between these and shallow bedrock for them to develop into *Calcic Yermosols*. Where water can gather in low places, larger karstic hollows known as *dayas* are formed. These can be up to a few metres deep and vary from a few tens of metres to several kilometres in diameter, and can act as basins for temporary fresh-water ponds after rain. They are filled with inwashed, relatively stoneless, and non-saline alluvial soil to a depth of a metre or more, whose surface is lowered by deflation and presents a strong contrast to the surrounding stony hamada surface (soils 6 and 7 on Table 4.1). In the Moroccan and Algerian Sahara the dayas provide virtually the only cultivable land away from the main wadis and provide some barley for the Berber nomads in years of good rainfall (Clark *et al.*, 1974; Mitchell and Willimott, 1974).

The dissection of the hamada surfaces has removed the finer material by colluviation on to footslopes and by alluviation into the network of valleys and terminal basins. Around the margins of uplands there is usually a piedmont bajada of detrital material transported partly by gravity and partly by occasional floods and traversed by incised anastomizing channels which carry ephemeral storm water. This material contrasts with channel deposits in being generally unstratified although it becomes finer in texture outwards, varying from boulders and stones near to the mountain front to gravel and finer materials towards the distal edge.

4.4.2. Reg and serir

The finer materials are moved by the drainage network into the intermontane valleys where they form extensive colluvial aprons, alluvial fans, bajadas, valley terraces, deltas and playas. The upper stony areas take the form of hamadas but lower down on the gently sloping debris aprons between channels the characteristic surface is that known as *reg* in the western and *serir* on the central and eastern Sahara. This is a smooth plain covered with a fine gravel, usually only one or two stones thick, set on or in a matrix of finer material comprising varying mixtures of sand, silt, or clay. It resembles a gravel path extending to the horizon in all directions. The soil profile has less coarse material than the surface, sometimes lacking it altogether in the upper part of the profile. The formation appears to be generally due to the progressive concentration of coarser material at the surface by deflation of the fines, but locally may be assisted by flood deposition of surface gravel and destruction of coarser materials within the profile by weathering.

Pedologically, such soils are usually Yermosols with minimal horizon development, though they may show slight eluviation of clay and accumulation of calcium carbonate or gypsum at shallow depth. Where these features are more marked due to somewhat higher rainfall they become Xerosols.

FIG. 4.5. Sandsheet near Sebka, Libya (see Table 4.1). Blackboard 20 × 13 cm.

4.4.3. Dunes

Eolian sands are selectively removed and deposited in locations generally close to their places of origin because, since these are normally alluvial lowlands, they tend also to be aerodynamic traps. In some areas, however, a strong steady wind may transport them across uplands into neighbouring basins. An example of this is the Issaouen N' Ighargharen (27°N6½°E) which, although a relatively small basin, contains probably the highest dunes in the world (over 300 m). The sand is deposited mainly in ergs made up of coalesced complexes of geometrically arranged sand mountains called *draa* (aligned with the wind and rounded) and *oghrouds* (pointed peaks), but also many smaller areas of dunes and sand sheets (Fig. 4.5; soil 2 on Table 4.1). These mobile sands usually show a degree of internal sorting with the finer particles preferentially moved towards the downwind end and towards the upper parts of dunes and draa.

Moving sands often pose a threat by invading agricultural areas. This occurs on the largest scale north of Al Miniya and in the Kharga and Dakhla oases in Egypt, in the great bend of the Nile near Ed Debba in the Sudan and where the Grand Erg Occidental presses the Wadi Saoura against the eastern escarpment of the Hamada of Guir in Algeria (Gautier, 1935). The inhabitants of the Souf Oasis can only maintain their date orchards by excavating the bottom of circular hollows in high dunes to water table (Bataillon, 1955).

Dune sands cannot be regared as soil in the normal sense except where the rainfall has provided enough vegetation to arrest their movement. Large areas of fixed dunes, such as the Kordofan qozes, along the southern margins of the Sahara are thought to have been formed during a drier period of the Pleistocene. They can be categorized as Regosols and are sometimes cultivated.

4.4.4. Alluvial and saline soils

From a practical point of view, the most important soils of the Sahara are the areas of fine-textured alluvium in valleys and basins. Where non-saline they can generally be categorized as *Fluvisols*. Soil textures reflect the gradient of the valley which transported them and the conditions under which they were deposited. In higher confined reaches these valleys are stony. Across alluvial fans they are gravelly and

FIG. 4.6. Vertisol, Sudan Gezira. Table 4.1 gives analyses for similar soil. Staff is marked in 10-cm steps.

sandy. When enlarged into seasonal watercourses, they show a lateral differentiation between channel, bar and backslope deposits. Perennial rivers such as the Nile and Niger traverse their floodplains depositing fine sand and silt on levees and in channels and clays in backswamps, leaving a complex stratification of alluvial deposits of different textures which contrasts markedly with the unsorted mixtures of different particle sizes characteristic of colluvial deposits.

Alluvial deposits likewise reflect the chemistry and mineralogy of their catchments, those from the Atlas Mountains being calcareous, those from the Nubian sandstone siliceous, and those from igneous rocks rich in both silica and alumina, which can lead to the formation of black cracking clays called Vertisols.

Vertisols occur on plains subject to seasonal inundation. They are widely distributed along the southern margins of the Sahara, notably north of the river Niger in Mali and east of Lake Chad, but their widest distribution is in the central and eastern Sudan where they form a level featureless plain of more than 750,000 square kilometres. They consist mainly of montmorillonitic clays derived from the augite-rich Ethiopian Highlands and Basement Complex outcrops. They tend, however, to increase in percentage of clay southwards towards the mountains. This runs counter to the expected coarsening of alluvial textures in that direction and lends support to Smith's suggestion (1949) that the material has been subjected to chemical change leading to the increasing synthesis of montmorillonite with increasing rainfall.

In that part of the Sudan plains known as the Gezira, Vertisols have been studied because of their value for irrigation. The characteristic soil profile (Fig. 4.6; soil 5 on Table 4.1) changes from dark brown in the top 60 cm to grey to 120 cm and then to yellowish-brown below this. Textures tend to be relatively uniformly 50–70% clay with coarse sand and gravel accounting for less than 10% and large stones being practically absent. The high clay content gives high potential fertility but also severely restricts the infiltration of irrigation water. Salinity is relatively low though tends to rise to a slight maximum between about 50 cm and 1 m.

The alluvial soils of Egypt are derived from a similar source and likewise chemically and mineralogically reflect their origin from Ethiopia and the Lakes Plateau of Central Africa. They are *Calcareous Fluvisols*.

The material is a blackish brown mud generally 8–10 metres deep which is soft, plastic and sticky when wet but hard and relatively impermeable when dry. Coarse sand and gravel are absent and the texture changes from almost equal proportions of fine and, silt and clay at Wadi Halfa (northern Sudan) to a marked dominance of clay at the expense of the fine sand fraction at Cairo (e.g. Ball, 1939).

Because alluvial soils are often fine-textured and impermeable and lie in low sites, they tend to concentrate both surface water and ground water whose evaporation leads to the deposition of soluble salts to form *Solonchaks* (Arabic: *sebkha*). The trend of concentration in mineralized waters with increasing solubility is magnesium → calcium → sodium and carbonate → sulfate → chloride, so that most salty horizons consist predominantly of NaCl although much larger amounts of calcium sulfate are often present deeper in the profiles. The shallow, seasonally inundated basins in which they often form are known as playas and may show a sorting of salts in order of increasing solubility towards their centres.

In the Sahara there are many playas of all sizes and origins. Lake Chad is a large tectonic depression in the Basement Complex shield, the Chott Fedjadj a foundered anticline and the Chott Djerid a syncline. Some of the salt pans in volcanic areas of Hoggar and Tibesti are dried crater lakes, and the coastal marshes of Mauritania (aftouts), of Tauorga, Libya (Table 4.1, soil 3), and the lakes of the Egyptian Delta are depositional basins. They sometimes show traces of past evolution either in the filling of the basin with sediment, or by annular segmentation due to shoreline changes. The oases of the Egyptian desert, partly formed by eolian deflation, are fed by relatively fresh ground-water which supports areas of cultivation but they are salinized around their margins where water pressure is inadequate to maintain downward through-flow in the soil. The Qattara depression has a similar origin but a totally salinized floor.

The surface soil of a playa mainly reflects its current hydrology, especially the balance between surface- and ground-water recharge. Where the former is dominant the surface tends to be smooth with a polygonal mosaic of fine cracks known as *takyr*. The crust is relatively non-saline, but immediately overlies a highly saline friable crumb. An example from Libya is illustrated on Fig. 4.7 with analyses given

FIG. 4.7. Solonchak, Wadi Irawan near Sebks, Libya. Blackboard 20 × 13 cm.

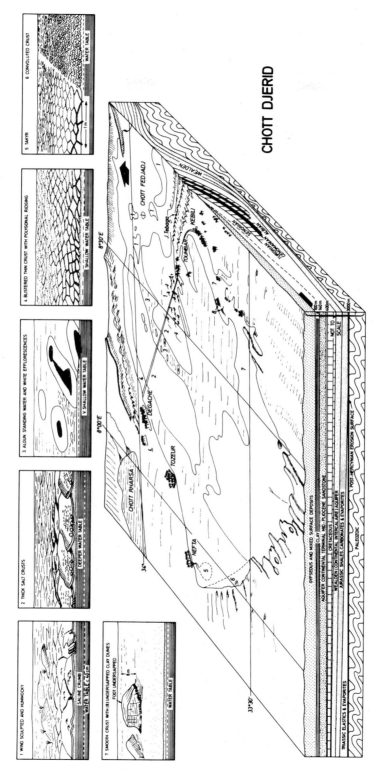

Fig. 4.8. Physiography and surface detail, Chott Djerid, Tunisia.

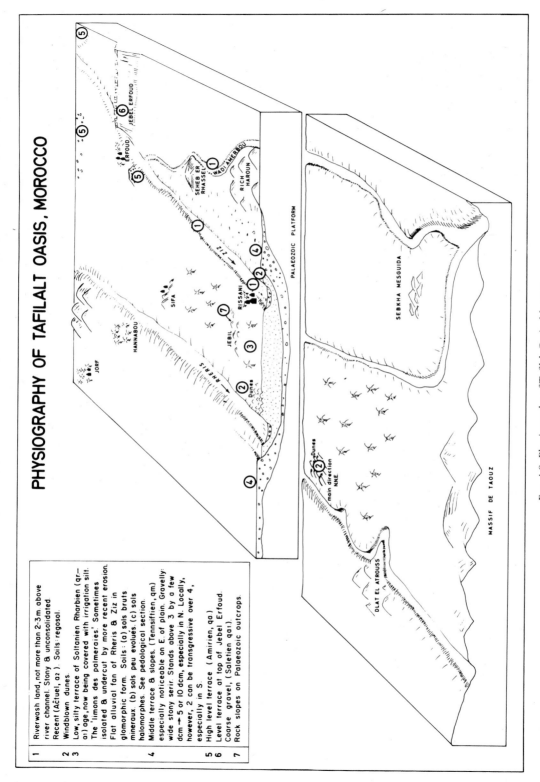

PHYSIOGRAPHY OF TAFILALT OASIS, MOROCCO

1 Riverwash land, not more than 2-3 m above river channel. Stony & unconsolidated. Recent (Actuel, a₂). Soils: regosol.

2 Windblown dunes.

3 Low, silty terrace of Soltanien Rharbien (qr—a) age, now being covered with irrigation silt. The 'limons des palmeraies.' Sometimes isolated & undercut by more recent erosion. Flat alluvial fan of Rheris & Ziz in glomorphic form. Soils: (a) sols bruts mineraux. (b) sols peu evolués (c) sols halomorphes. See pedological section.

4 Middle terrace & slopes. (Tennsiftien, qm) especially noticeable on E. of plain. Gravelly: wide stony serir. Stands above 3 by a few dcm → 5 or 10 dcm, especially in N. Locally, however, 2 can be transgressive over 4, especially in S.

5 High level terrace (Amirien, qa)

6 Level terrace at top of Jebel Erfoud.

7 Coarse gravel, (Saletien qa1).

   Rock slopes on Palaeozoic outcrops.

FIG. 4.9. Physiography of Tafilalt Oasis, Morocco.

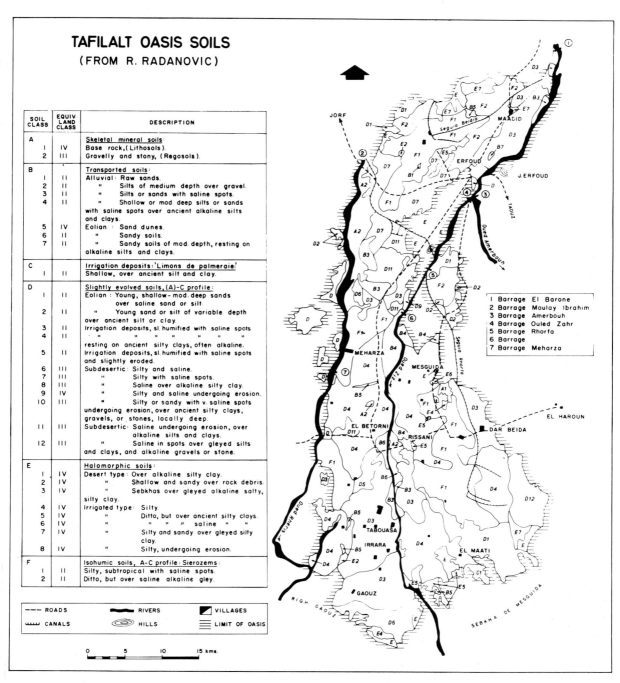

TAFILALT OASIS SOILS
(FROM R. RADANOVIC)

SOIL CLASS	EQUIV. LAND CLASS	DESCRIPTION
A		Skeletal mineral soils:
1	IV	Base rock, (Lithosols).
2	III	Gravelly and stony, (Regosols).
B		Transported soils:
1	II	Alluvial : Raw sands.
2	II	" Silts of medium depth over gravel.
3	II	" Silts or sands with saline spots.
4	II	" Shallow or mod. deep silts or sands with saline spots over ancient alkaline silts and clays.
5	IV	Eolian : Sand dunes.
6	II	" Sandy soils.
7	II	" Sandy soils of mod. depth, resting on alkaline silts and clays.
C		Irrigation deposits: 'Limons de palmeraie.'
1	II	Shallow, over ancient silt and clay.
D		Slightly evolved soils, (A)-C profile:
1	II	Eolian : Young, shallow-mod. deep sands over saline sand or silt.
2	II	" Young sand or silt of variable depth over ancient silt or clay.
3	II	Irrigation deposits, sl. humified with saline spots
4	II	" " " " " " resting on ancient silty clays, often alkaline.
5	II	Irrigation deposits, sl. humified with saline spots and slightly eroded.
6	III	Subdesertic: Silty and saline.
7	III	" Silty with saline spots.
8	III	" Saline over alkaline silty clay.
9	IV	" Silty and saline undergoing erosion.
10	III	" Silty or sandy with v. saline spots undergoing erosion, over ancient silty clays, gravels, or stones, locally deep.
11	III	Subdesertic: Saline undergoing erosion, over alkaline silts and clays.
12	III	Saline in spots over gleyed silts and clays, and alkaline gravels or stone.
E		Halomorphic soils:
1	IV	Desert type: Over alkaline silty clay.
2	IV	" Shallow and sandy over rock debris.
3	IV	" Sebkhas over gleyed alkaline salty, silty clay.
4	IV	Irrigated type: Silty.
5	IV	" Ditto, but over ancient silty clays.
6	IV	" " " saline " "
7	IV	" Silty and sandy over gleyed silty clay.
8	IV	" Silty, undergoing erosion.
F		Isohumic soils, A-C profile: Sierozems:
1	II	Silty, subtropical with saline spots.
2	II	Ditto, but over saline alkaline gley.

1 Barrage El Barone
2 Barrage Moulay Ibrahim
3 Barrage Amerbouh
4 Barrage Ouled Zahr
5 Barrage Rhorfa
6 Barrage
7 Barrage Meharza

---- ROADS   RIVERS   VILLAGES
ⵃⵃⵃ CANALS   HILLS   LIMIT OF OASIS

0   5   10   15 kms.

FIG. 4.10. Soils of Tafilalt Oasis, Morocco.

on Table 4.1, soil 4. Where salt content is especially high the surface may flake and expose the underlying crumb to deflation. A longer annual surface flooding which causes leaching may lead to the formation of a rough channelled and hummocky microrelief known as *gilgai*, as near Sebha, Libya (Perrin and Mitchell, 1969).

Where ground-water is dominant the surface may accumulate a salt crust or a soft puffy surface sometimes called 'self-rising ground' and if it is continuously at shallow depth and occasionally covers the surface a thick saline crust may develop. This often cracks into polygonal slabs which curl up at the edges and develop a microrelief by permitting further efflorescence along the cracks and trapping windborne material along the edges. The most outstanding examples of this type of surface are the large dry lakes known as *chotts* along the southern margins of the Atlas Mountains in Tunisia and eastern Algeria. These are fed mainly by artesian ground-water from further south via a large number of springs (Arabic: *aioun*), which occur in and around the chotts. Some of these features on Chott Djerid are illustrated on Fig. 4.8. Immediately to the east of the chott, artesian pressure creates moundsprings whose water is sufficiently non-saline to irrigate the constellation of oases around Kebili and Tozeur.

4.5. CASE STUDY: TAFILALT OASIS, MOROCCO

Tafilalt Oasis lies in a tectonic basin fed by two rivers: the Ziz and Rheris before their confluence into the Daoura south of Taouz. Irrigation has recently been improved by the completion in 1971 of the Hassan Addakhil Barrage on the Ziz north of Errachidiya. The soils are formed on a series of Pleistocene terraces of decreasing altitude and age, and almost all the irrigation is confined to that of Soltanien–Rharbien age (Fig. 4.9). Its soils were surveyed in detail and assigned to land classes by Radanovic (1966) whose map is reproduced in simplified form on Fig. 4.10.

The soil classification recognizes five main types: skeletal mineral soils (Lithosols and Regosols), transported soils, irrigation deposits, slightly evolved soils with (A)C profiles, halomorphic soils and isohumic soils with AC profiles. Subdivisions are based essentially on the depth, textural stratification, salinity and alkalinity of the profiles. The land classes were (i) very good, (ii) good, but with a slight shallowness or salinity limitation, (iii) usable but with more serious limitation including erosion risk and (iv) unusable.

The main irrigated crops are dates, wheat, maize, lucerne and vegetables, and the main animal production is of sheep and goats. No cultivation is possible outside the irrigated areas.

ACKNOWLEDGEMENTS

I'thank all the following for help in preparing this chapter: Mrs. Pamela Dixon for the typing, Mrs. Sheila Dance and Mr. Brian Rogers for fair-drawing the illustrations and Mrs. Sarah Wheeler for the photographic work.

REFERENCES

BALL, J. (1939) *Contribution to the Geography of Egypt*. Government Press, Bulaq, Cairo.

BATAILLON, C. (1955) *Le Souf: étude de géographie humaine*. Institut de Recherches Sahariennes, Université d'Alger, E. Imbert, Alger.

CLARK, D. M., MITCHELL, C. W. and VARLEY, J. A. (1974) Geomorphic evolution of sediment-filled solution hollows in some arid regions (Northwest Sahara). *Zeitschrift für Geomorphologie*, Suppl. 20, pp. 130–139.

DREGNE, H. E. (1970) Appraisal of research on surface materials of desert environments. In *Deserts of the World*, edited by W. G. McGINNIES, B. J. GOLDMAN and PATRICIA PAYLORE. University of Arizona Press.

DUTIL, P. (1971) *Contribution a'l'étude des sols et des paleosols du Sahara.* Thesis for doctorate, Faculté des Sciences de l'Université de Strasbourg.

FAO/Unesco (Food and Agriculture Organization and United Nations Educational Scientific and Cultural Organizations) (1974) *Soil Map of the World* (1:5,000,000), Vol. 1, Legend and Sheets VI-1 and VI-2 covering the Sahara.

FURON, R. (1963) *Geology of Africa.* Oliver & Boyd, Edinburgh and London.

GAUTIER, E. F. (1935) *Sahara, the Great Desert.* Columbia U.P., New York.

GREENE, H. (1952) The soils of the Sudan. In *Agriculture in the Sudan*, edited by J. D. TOTHILL, pp. 144–175. Oxford University Press.

HOORE, J. L. D' (1964) *Soil Map of Africa* (1:5,000,000). Commission for Technical Cooperation in Africa, Lagos, Nigeria.

LUSTIG, L. (1970) Appraisal of research on geomorphology and surface hydrology of desert environments. In *Deserts of the World*, edited by W. G. MCGINNIES, B. J. GOLDMAN and PATRICIA PAYLORE. University of Arizona Press.

MABBUTT, J. A. (1977) *Desert Landforms*, Volume 2 of *Systematic Geography.* M.I.T. Press, Cambridge, Mass.

MEIGS, P. (1952) *World Distribution of Arid and Semi-arid Homoclimates*, with maps UN392 and UN393. Unesco, Paris.

MITCHELL, C. W. and WILLIMOTT, S. G. (1974) Dayas of the Moroccan Sahara and other arid regions. *Geogr. J.* **140**, 441–453.

MORRISON, A. and CHOWN, M. C. (1965) Photographs of the western Sahara from the Mercury MA4 Satellite. *Photogrammetric Engineering*, **31**, 350–362.

PERRIN, R. M. S. and MITCHELL, C. W. (1969) *An Appraisal of Physiographic Units for Predicting Site Conditions in Arid Areas.* MEXE (Military Engineering Experimental Establishment), Report No. 1111, Christchurch, Dorset.

PETROV, M. P. (1976) *Deserts of the World.* Israel Program for Scientific Translations, Wiley, New York.

RADANOVIC, R. (1967) *Amenagement de la region du Tafilalt.* Rapport Générale Preliminaire, Office National des Irrigations, Rabat, Morocco, by Energoprojekt, Belgrade.

RAISZ, E. (1952) *Landform Map of North Africa.* Cambridge, Mass.

SMITH, J. (1949) *Distribution of Tree Species in the Sudan in Relation to Rainfall and Soil Texture.* Sudan Government, Ministry of Agriculture Bulletin No. 4.

UNESCO (1979) *Map of the World Distribution of Arid Regions*, with explanatory note constituting Man and Biosphere. MAB Technical Notes 7, Paris.

U.S. ARMY ENGINEER WATERWAYS EXPERIMENT STATION (1958) *Analogs of Yuma Terrain in the Northwest African Desert.* Technical Report 3-630, Vicksburg, Missisipi.

U.S. ARMY ENGINEER WATERWAYS EXPERIMENT STATION (1962) *Analogs of Yuma Terrain in the Northeast African Desert.* Technical Report 3-630, Vicksburg, Mississippi.

CHAPTER 5

Microclimates

J. LARMUTH

Wolfson College, Cambridge

CONTENTS

'It is man who creates the desert: The climate only provides the right conditions.'
Le Houérou

5.1. INTRODUCTION

Deserts are considered to be areas where the biological availability of water is low. This limitation is found in polar regions where the water is locked in ice and snow, in high mountain areas (partly for the same reasons but also due to rapid drainage and wind drying) and in climatically dry areas at low altitudes. A division must therefore be made between hot and cold deserts. The Sahara is a hot desert straddling the Tropic of Cancer and occupying nearly 10 million km^2 of Africa north of the Equator, roughly 20% of the world's total arid and semi-arid land surface.

The creation, and maintenance, of hot deserts is due to the combined action of sun, wind, low rainfall and the hand of man. Hot desert physical processes are all, ultimately, sun-driven, the sun dominating life at the desert surface due to high incoming radiation levels during the day, and rapid loss of heat by outgoing radiation at night. The solar regime results in elevated daytime surface temperatures, the recorded maximum being 82.5°C in the Red Sea Hills. Highest measured air temperature was 58°C in Libya, and daily ranges of temperature may also be high, 41 degrees in Death Valley (U.S.A.) but, in the Sahara these do not often exceed 30–35 degrees.

Tables of temperature and rainfall are given by Cloudsley-Thompson (1962) and Kassas and Imam (1957) for the eastern Sahara.

A small, exposed, organism at the desert surface is obviously in a very hostile environment. It is subject to direct incoming radiation from the sun, which includes both short-wave (SW) and long-wave (LW)

57

infrared components, reflected SW from, and LW emitted by, the surface and nearby objects, and it may also generate surplus metabolic heat. An organism is able to lose heat by LW emission from its surface, by conduction to the ground although this is negligible for most animals, convection and evapotranspiration, the latter also being restricted in small organisms because of limitations on 'expendable' water. Radiation loads may be reduced by burrowing, seeking shade, by behavioural adaptation (e.g. orientation) and by altering timing of activity to avoid maximum loads. Size is also important in that smallness leads to more rapid gains and losses of heat, and loss of corporal water, than is the case for larger animals or plants.

As the atmosphere above deserts is usually clear it does not filter out biologically damaging ultraviolet (UV) radiation to the same degree as is the case for most other terrestrial environments. At the top of the earth's atmosphere one square metre receives slightly more than 1360 Watts but, after passing through the atmosphere, this figure is reduced to about 1100 W m^{-2} at the surface in desert areas. Radiation at the desert surface is received mostly in the wavelengths between 0.3–3.0 micrometres (μm), composed of UV (0.3–0.4 μm), visible light (0.4–0.7 μm) and IR (0.7–3.0 μm). Solar radiation 'lost' in the atmosphere is scattered and absorbed by dust, water vapour and cloud in the lower atmosphere, after about 50% of the UV component has been removed by the ozone layer of the upper atmosphere. Atmospheric scattering of solar radiation leads to some of it being reflected back into space. Long-wave IR radiation is absorbed by water vapour in the atmosphere for wavelengths greater than 3.0 μm. Scattering processes give the sky its familiar blue colour, and some 15–25% of radiation reaching the desert surface is 'blue sky', or diffuse, radiation scattered downward. Energy transmitted by radiation is proportional to frequency, so that SW (high frequency) radiations carry a significant fraction of the total energy arriving at the surface. Of this total energy, 50% is in the visible part of the spectrum, both direct and diffuse, and it is this energy which is considered to be biologically useful.

The relative motions of sun and earth lead to daily and seasonal variations in intensity of radiation at the surface of the earth. When the sun is at its azimuth (i.e. due south at local noon) the path taken by rays through the atmosphere is shortest, but at sunrise and sunset the pathlength is longer because rays pass through the atmosphere obliquely. Similarly, declination (change of latitude with season) of the sun results in, smaller, changes in pathlength. These daily and seasonal changes in pathlength result in nearly sinusoidal variations in radiation intensity.

Desert surfaces are more reflective than those which are wet or well vegetated. Albedo is the term used for this reflection and, for deserts, is between 25–40%, varying with solar altitude and the nature of the surface. It is greatest at sunrise and sunset, lowest at midday.

Levels of radiation received at a horizontal surface depend on the cosine of the zenith angle (angle between a line from a point on the surface to the sun and from the same point vertically upwards to the zenith) or the sine of the solar altitude above the real horizon, so that direct solar radiation early, or late, in the day cannot have the same effect as midday sunshine. Combined with higher albedo and greater atmospheric pathlength, the desert surface can be a pleasantly warm environment early or late in the day.

At night the surface loses heat rapidly to a clear sky. Air temperature may fall to 15–20°C and ground temperature may be a few degrees lower still. If a significant amount of dust is suspended in the lower atmosphere it acts as a 'blanket' and tends to prevent temperatures falling to low levels by reducing the rate of loss of heat by radiation.

The word 'microclimate' is generally understood to refer to a 'small climate' but, as smallness is relative, this is not a sufficient definition. A geographer, for example, may consider the climate of a prairie or forest as a microclimate, relative to the large (macro-) climate of an entire country. On the scale of an individual organism this is irrelevant, particularly so when the organism may be concealed within a cavity below a stone, or some similar place. For our purposes 'microclimate' will be understood to imply an envelope of humidity, temperature, radiation, wind and other physical factors in which an organism may find itself at a moment in time. Microclimates are therefore continually changing, and common examples would be a shadow behind a stone or beneath a bush, a space below a stone, a burrow or crevice or the underside of a leaf. Dimensions are therefore mostly 1–10 cm, these would be exceeded by the size of a space occupied by a camel or bush, but are true for the majority of desert organisms.

Few people who have not visited a desert area can appreciate that the 'typical' desert may have 80–90% of its surface consisting of stones, gravel and solid rock, while the remainder may be a mixture of sand, usually as dunes, salt- and clay-pans.

A useful model of the desert biolayer is the concept of a perforated sheet, representing exposed ground with high radiative and thermal characteristics, while the holes in the sheet may be compared with cooler areas under stones, mouths of burrows or shade below plants. The biolayer may be only a few centimetres in depth, but escape from its extreme conditions becomes possible by burrowing into the substrate, by climbing or flying above the hot layer, or by seeking refuge within a 'hole'. Most food resources are found within the layer and, consequently, most activity occurs there. Broadly, microclimates may be separated into three groups occurring above, within or below the biolayer. Each group having certain features in common.

The discussion which follows is mostly concerned with sandy soils which support most plant and animal life in deserts.

5.2. RADIATION

Solar radiation is at its most intense during summer, it may be reduced to 25% of summer levels in winter and to 20% of its level on any day when obscured by cloud. Reduced levels do not impose as severe a stress as do conditions of maximum radiation. Survival of an organism often depends very much on the most extreme conditions encountered, and it is these with which we are most concerned.

Radiation arriving at the desert surface penetrates to a depth which depends on the size, colour and mixture of particles of which the soil is composed. Interaction of exponential extinction of radiation by absorption with depth, usually less than 1 cm, and upward reflection and IR radiation within and from the soil, result in maximum temperatures occurring 0.3–0.5 cm below the soil surface. At sunrise and sunset penetration is effectively zero, but is maximum at midday and, as soil temperatures depend on penetration, and consequent absorption, the rate of change of temperature in the early morning is low. By about 08.00 (local solar time) the sun has risen to 30° above the horizon with penetration of radiation into the soil becoming significant. From 09.00 to 11.00 the rate of rise of temperature just below the soil surface is 5 degrees or so per hour. From 13.00 to 19.00 temperature falls at a rate diminishing from about 4 degrees per hour to 1 degree per hour and then remains at, or below, this value for the rest of the night. The 'desert day' may be typified by surface temperatures of about 20°C or less at sunrise, rising to 60–70°C at midday and, by early evening, cooling to 30–35°C. At 21.00 the temperature is still about 25°C but is falling slowly to its pre-sunrise level. Wind, and infrequent rain, may have dramatic effects on this 'normal' regime, and are discussed in the appropriate sections below.

Dry sandy soils have low thermal conductivity and therefore act as good insulators. During daytime extremes at the surface, the temperature at a depth of 10 cm will only change some 5–10 degrees, superimposed on a daily mean of 25–27°C in summer, and at 50 cm the change is not usually more than about 1 degree.

At night the surface is cooling, but sand at a depth of 10–20 cm is still warm and will transfer heat to the colder layers above and below. A deeper layer will continue to gain heat so that, with special recording instruments, a temperature 'wave' can be seen to travel slowly downwards into the soil during a 24-hour period. Depending on the physical nature of the soil, maximum temperature may be detected between 30–50 cm when the surface is at its coolest, the converse also being true with the soil at this depth being coolest at about the time of midday extremes at the surface.

Orientation of a surface has a marked effect on its thermal characteristics. North- and west-facing exposures always being cooler than those facing east or south, also, the greater the dip of the surface, up to about 35°C, the more the difference. Of importance to organisms is the fact that an exposure facing between north and east will be heated, and dried if moisture is present, early in the day. As air temperature remains high until mid-afternoon the north-east exposure will have little chance to cool significantly, it will effectively experience two thermal maxima during the day, the first due to early direct heating, the second in the afternoon period of high air temperature, with a fall of only a few degrees between the two. An organism on a north-east facing site will therefore suffer an extended period of high temperatures and dry conditions.

The most favourable microclimate occurs where the exposure faces north-west. Here, an organism will be relatively cool during the morning, and evaporation of any moisture is delayed, but will experience a single peak of temperature, both air and surface, during the afternoon. However, as the duration of exposure to high temperature is often of far greater consequence to survival, the single elevated afternoon peak may be preferable to more prolonged conditions at less favourable sites.

Direct solar radiation also results in two important biological effects, the first being the interrelated heating and drying of an exposed organism. The second, cellular damage due to direct UV penetration of tissues, needs to be more thoroughly investigated. Most desert organisms are adapted to the conditions of their environment, which include high levels of UV radiation, and are likely to have evolved protection in cases where the activity pattern of the organism may result in its exposure to UV.

5.3. WIND

The pattern of air movement in the desert is normally fairly consistent from one day to the next. During the morning period 07.00–12.00 the air becomes increasingly warm with very small-scale turbulence close to the surface. By about midday this has increased to the point where light winds occur, persisting for the remainder of the day. Falling temperature during the late afternoon and evening may cause some reduction in air movements before midnight. From the early hours of the morning until after sunrise, continued cooling of the desert surface results in wind engendered by flows of air between widely separated surfaces of different physical and morphological character (e.g. sand, reg, plains and mountains) with a net movement of air on to the desert from adjacent areas. This produces winds of low velocity (0.5–1.0 m s^{-1}) often remaining constant in direction for extended periods of time. Changes of direction cause rapid, but irregular, oscillations of temperature during this period. Humidity fluctuates with these directional changes depending on the source of incoming air. In Fig. 5.1a, recordings made by a thermohygrograph (inside a stone building for protection, and therefore indicating generally lower temperatures and higher humidities than would be found outside) show small-scale thermal fluctuations from midday until after sunrise (upper trace). The 'spikes' between midnight and 06.30 represent changes in wind direction. In the open desert, low-altitude air flows of this nature may be propagated over many tens, or even hundreds, of miles. Alterations of wind direction at night are the result of differential cooling of areas, which may be widely separated, due to variations in temperature due to radiative 'colour' and other physical properties of surface materials.

Occasionally, more violent winds occur. During the afternoon high winds normally result from large-scale turbulence creating sand- or dust-storms. Airborne sand grains collide, and frictional heat generated by innumerable collisions raises air temperature. Following a storm of this type, there is often a period when air temperature falls quickly because the heated air has risen rapidly into the atmosphere. Changes of humidity usually follow changes of air temperature. The typical 'spike and trough' of two small, afternoon, dust-storms can be seen in Fig. 5.1e, together with minor changes in humidity.

Strong winds of extra-desert origin in the evening may cause changes in temperature but, almost invariably, result in a rapid rise in humidity due to the greater moisture content of the incoming air,

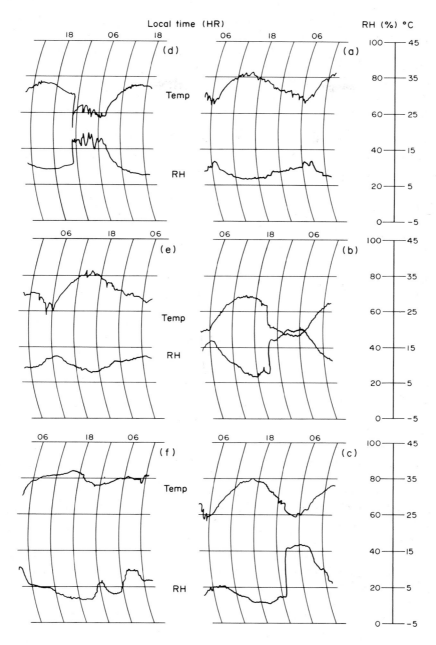

Fig. 5.1. Time-corrected traces of thermohygrograph recordings showing: A, the effect of changing wind direction on temperature (upper trace) and humidity (lower trace); B, the effect of cool evening wind of extra desert origin (High Atlas mountains); C, effects of strong wind of extra desert origin (non-mountain); D, the consequences of a sudden rainstorm; E, the 'spike and trough' event characteristic of small afternoon dust-storms; F, the Sirocco, Harmattan or Khamsin, unpleasantly high temperature and low humidity, sometimes for several days. Further explanation is given in the text.

Note: Scale of figure approx. ×1.5.

Fig. 5.1c. Windspeeds are not as great as those found during sandstorms, consequently less dust or sand becomes airborne. In passing across the desert surface cooler, more humid, air is warmed so that local falls in temperature often do not exceed 2–3 degrees.

Sometimes incoming air may be derived from a very cool and relatively humid source (e.g. a mountain area) and, entering the desert, originate much greater changes than normal, Fig. 5.1b. In this instance incoming air was derived from the High Atlas mountains some 80 miles north of the recording station. The traces of Fig. 5.1b should be compared with those of Fig. 5.1d where the onset of windy conditions was accompanied by a short, sudden, rainstorm. The dramatic deflection of the temperature curve is characteristic of rain.

When strong winds blow from another desert area high air temperatures, with depressed humidity, may continue for one or more days. These conditions are particularly unpleasant as even night-time temperatures may not fall below 35–40°C for several nights in succession, Fig. 5.1f. Many local names are given to these Saharan winds, because of their obvious divergence from normal conditions, Irifi, Chergui, Ghibli, Sirocco, Harmattan, Khamsin. The latter is actually derived from the arabic word for the number fifty, and folklore has it that there is a period of that number of days in the year when these winds may blow, following the spring Equinox.

Above a certain wind velocity, and depending on grain size, sand becomes airborne, forming a moving carpet of stinging sand close to the surface. Airborne grains travel at the same velocity as the wind and kinetic energy, proportional to velocity squared, may be high. Many small organisms depend for their survival upon the integrity of a delicate, waxed, surface layer which is impervious to water. The sandblasting effect of high-energy grains can rapidly destroy the wax layer resulting in severe stress, or death, to the organism.

The temperature of air moving across a hot sand surface at velocities up to 1.0 m s^{-1} affects surface temperature. Moving air at 25°C, crossing a surface at 48°C results in a fall of surface temperature of about 10 degrees. If the moving air is at 35°C virtually no change occurs, but at 45°C surface temperature may actually rise another 5 degrees because heat transfer to the air depends on the difference of temperature between the surface and the medium, and to some extent on the velocity of the medium. The hot air cannot accept much heat from the surface, which is therefore unable to cool normally.

5.4. RAIN

Winter rain can be gentle and, because the surface temperatures of soils are relatively low at this time of year, does not have an effect comparable with that of violent summer rains. The record of a sudden summer rainstorm at 23.00 is displayed in Fig. 5.1d for a downpour that lasted less than 1 minute. Immediate, dramatic changes of temperature and humidity are typical of heavy summer rains. At the time of recording, wind continued throughout the night, gusty and variable in direction, so that temperature and humidity underwent considerable fluctuations, mainly due to moisture being evaporatively cooled during the time it remained in the sand. The effects lasted until mid-morning the following day, indicated by generally lower temperature and elevated humidity. Water only penetrated some 0.5 cm into the soil and, after midnight, had mostly evaporated leaving the spaces between the grains moist. As temperature falls under these conditions further condensation will occur in the interstitial spaces. This moisture is then evaporated by air movement at the surface resulting in further cooling. Moisture is continually brought to the upper layers by surface tension forces, thus allowing the process to continue until the water has gone. The latent heat of vaporization of water is high and most of the energy required in this instance was drawn from heat contained in the upper 5–6 cm of sand. Within minutes of the cessation of rain the temperature at 5 cm was falling, and at 1 cm had already dropped nearly 5 degrees. Air and sand surface temperatures had both become 20.9°C within seconds of the onset of rain.

Daytime rains tend to initiate relatively greater changes because of higher soil and air temperatures, but often the effects are less prolonged. Solar radiation is more rapidly absorbed by moist soil due to the lowered albedo and, consequently, water evaporates more quickly. The surface will remain at more or less the same depressed temperature while water is present, as incoming direct and diffuse energy is taken up as latent heat. Under these conditions almost no water is retained in the soil so that plants derive very little benefit from sudden, heavy, rains.

5.5. SURVIVAL BELOW THE BIOLAYER; CAVES, BURROWS AND SAND

Caves and burrows have features in common, size excepted, in that the matrix material in which they are found has considerable thermal mass and, in both cases, limited rates of water vapour movement. In a cave this is only significant through the mouth and in soil or sand exchange is by slow diffusion to, or from, the spaces between the grains of the surrounding matrix.

There has been only a small number of investigations in these closed habitats, mostly due to inaccessibility and problems of instrumentation.

Studies carried out in caves include the classic work in Wadi Digla, Egypt, by Williams in 1923. As a result of his work, Williams was able to show that, away from the area of the entrance, relative humidity was higher in summer than in winter due to interplay of changing seasonal temperatures outside with almost constant conditions some 10 m into the cave. The vapour pressure difference between these points was such that a net flow of atmospheric moisture into the cave occurred during the day, and reversed direction at night. The seasonal humidity within the cave was also reversed with respect to conditions outside. Inside the cave minimum RH was 42% compared with 16% at the mouth on an August day. At night these values were 64 and 84% respectively.

At a depth of 50–60 cm in soil or sand the daily temperature becomes more or less constant and has been measured in jerboa burrows by several workers. Records for the midday period indicate temperatures ranging from 25° to 31°C [Williams (1923) 31°C at 75 cm depth; Eissa et al. (1975) 25°C 70 cm into a burrow with an external air temperature of 45°C; Kirmiz (1962) 25.7–26.8°C between 12.00 and 16.00 local time in Maryut, Egypt, at a depth of 1.0 m]. Similar results were found by Vorhies (1945), Schmidt-Nielsen (1950) and Pierre (1958).

Humidity is difficult to measure with any accuracy in burrows but some records have been obtained. During the course of his work in Kuwait, Eissa found July RH at 72% in burrows sampled when external air yielded only 12%, and 78–80% was found in the Maryut study by Kirmiz for the same times of day as the temperature recordings by him, quoted above.

Sand-swimming animals, e.g. skinks, beetles, camel-spiders and others, display a lower level of metabolic activity than small mammals, yet they will be surrounded by a small zone of more humid air in the interstitial spaces among the sand or soil grains close by, generated by their own respiratory losses. The zones involved are likely to be highly localized (e.g. round the head of a lizard) but will serve to reduce the net loss of moisture from the animal by enabling it to breathe more humid air.

Obviously, these conditions are virtually impossible to measure, because any attempt to do so would upset them, rendering the record invalid. However, Pierre (1958) managed to obtain a figure of 50% RH for bulk sand in the Sahara, which he explained as moisture due to rain being held by surface tension and other forces, and to the presence of ground-water. Cloudsley-Thompson (1961) compares the 50% found by Pierre with 35% measured by himself in the sand of the Red Sea Hills. Any small, localized, increase in these values would be advantageous to a sand-swimming animal, certainly an improvement over the low RH of daytime conditions above the surface.

Animals retreating below the biolayer take advantage of two physical factors. The ability to enjoy a higher RH has already been discussed above. Temperature at a depth in soil or sand is out of phase with that at the surface, and is normally minimal, dependent on depth, at about the time that extreme conditions exist at the surface. A thorough examination of burrow use by a North American desert lizard was carried out by Porter *et al* in 1973 who showed that early, and late, in the year *Dipsosaurus dorsalis* was active during a single midday period but from April to September activity occurred progressively earlier in the day and the burrow was increasingly used to avoid midday high-summer conditions. The lizard re-emerged in early evening during this period. As autumn progressed emergence times converged, becoming unimodal again in September.

5.6. SURVIVAL ABOVE THE BIOLAYER; SHADE AND AIR

Above open desert surfaces air is at its driest, and radiation at its most intense, for an extended period of a summer day. Few flying creatures are able to withstand the stresses of these conditions because of the high levels of metabolic heat output whilst flying. Birds will usually only make short flights from one bush to another and, of the larger insects, only locusts are adapted to flight in these circumstances.

In 1972 Heinrich and Bartholemew found that a large sphinx moth, *Manduca sexta*, could only fly for 1 minute at an air temperature of 40°C, even though it is well adapted to desert conditions. Speed of flight is equivalent to cooling in a forced draught and has a marked effect on body temperature in insects (Parry, 1951), but even this cooling is not sufficient in desert conditions at elevated temperatures.

Due to greater surface area to volume ratios a small insect is at a distinct advantage in this respect. Research by Digby (1955) indicated that flies had an excess body temperature directly proportional to the radiation load to which they were subject, with the implication that metabolic heat was quickly dissipated. However, for most insects flight must cease during the midday period.

Larger woody plants are more or less self-shading in that the outer leaves protect the 'inner' plant, and also create shade on the ground below. In measurements made on a bush, *Salsola gaetola*, in the Sahara, Larmuth (1979) found that the interior of the bush was 7 degrees cooler than the open shade temperature of 37.5°C. A bush or tree is a suitable refuge for birds, and confers additional advantage with increasing height above the hot surface. This shade is not ideal for insects, however, because the air is still dry, even though cooler, and proves too expensive on limited resources of body fluids for such small animals. The same worker found that spaces between the lower stems in clumps of esparto grass, *Lygeum spartum*, were crammed with beetles (57 in one plant) because these small crevices offered both a lower temperature, and a more closed, and therefore humid, environment. External air movements had little effect within the spaces.

Other insects, Orthoptera in particular, often climb on to a bush thus distancing themselves from the hot air nearer the surface, and by orienting head-on to the sun minimize the body area exposed to radiation load. Parry, in 1951, showed that orientation behaviour could reduce radiation loading by as much as 50% and, also, that orientation could advantageously be made laterally with respect to wind direction. At temperatures approaching 40°C grasshoppers from the North American deserts were shown by Pepper and Hastings (1952) to control their temperature by evaporative cooling although, as with many desert animals, this must be a last resort.

Large mammals, gazelle, camel, donkey and man also use the shade of trees to avoid excessive heat loads and, below trees and bushes, the entrances to scorpion and lizard burrows are often found for similar reasons.

5.7. SURVIVAL IN THE BIOLAYER; SUN AND STONE

The desert is a place of night, of hunter and hunted, life and death, whose record in written on the stillness of the morning sands. Few species venture on to the daytime surface but, for those doing so special behavioural modifications enable the avoidance of lethal temperatures, but any reduced risk of potential predation must be offset against the greater environmental hazards.

In 1971 Edney observed the activities of tenebrionid beetles and found that if they alternated between 30 seconds shade and 10 seconds exposed to the sun their mean body temperature remained constant within less than ±1 degree at 33°C. If the regime were to be reversed (10 s shade, 30 s sun) mean body temperature rose to 38°C. Constancy of temperature could be achieved because of thermal time-lag which, for the largest beetles, was about 7 minutes. A beetle head-on to the sun could achieve a reduction of about 4 degrees compared with a beetle oriented laterally.

Rates of water loss from beetles increase dramatically at temperatures greater than 30–35°C (Ahearn and Hadley, 1969). Hamilton (1975) indicated that tenebrionid beetles burrow a few centimetres into sand, on appropriately oriented slopes, to avoid excessive temperatures and, it is possible, that this behaviour could be used as an alternative to shade in more open terrain. Body temperatures of 43°C were tolerated voluntarily by these beetles, 5 degrees below their lethal temperature at 48°C. Periods of activity were observed to be 100% before 10.00 and after 17.00, but were reduced to 20% or less near midday. In cooler conditions of spring and autumn activity may occur 100% of the time.

An excellent study in the Algerian Sahara by Délye (1968) indicated just how critical extreme surface conditions were for ants enduring very short exposures.

Similar observations have been made on desert lizards. Early in the day, maximum exposure to the sun is chosen by basking on east-facing slopes, or by climbing on to bushes or rocks, until a body temperature between 30–40°C has been attained. As temperature increases further the lizard begins to make intermittent use of the shade cast by bushes. The hotter the desert becomes the faster the crossings between patches of shade until conditions are no longer tolerable and the animal retreats to its burrow. Regulation within ±2.5 degrees of the mean body temperature of about 37°C is maintained for about 80% of the time by means of this form of behavioural thermoregulation (Schmidt-Nielsen, 1964). Whilst crossing open spaces lizards may run with the body held high to reduce the rate at which heat is gained from the surface but, in patches of shade the animal will often flop on to the surface in order to lose heat by conduction to the cooler ground.

Not all organisms are sufficiently motile to take advantage of the temperature–time domain. Desert snails aestivate in summer, during which time the opening of the shell is closed by a tough impermeable epiphragm, and the snail withdraws to the upper whorls of the shell. The lower whorl contains only air and acts as an insulator between the body of the snail and the ground beneath. Under these conditions the snail's body has been shown not to exceed a temperature of 50°C, which it can survive almost indefinitely and, in such circumstances, may remain in the dormant state for many years. The chalky-white shell has a reflectance of about 95% which helps ensure that the body of the snail inside does not rise towards a lethal temperature near 55°C (Schmidt-Nielsen et al., 1971).

Stones on the surface provide protection against both environment and predators for many animals. The maximum temperature on the underside of a stone may not exceed about 40°C and is dependent on the colour of the stone (Larmuth, 1978). A white stone may be as much as 13 degrees cooler on its under surface when compared with a grey stone of the same thickness (2 cm), and an increase in thickness by a factor of 5 only halves this difference. Positions below stones, near the edges, are cooler at the side away from the sun than elsewhere, and the presence of a cavity below a stone reduces the temperature, experienced by the occupant, a further 5 degrees or so.

For plants the most critical period is the summer following germination, as the seedling has had little time to form appreciable thicknesses of bark, or other protection. Neither have its roots grown sufficiently to penetrate to a depth adequate to ensure continuity of water supply. In a study on seedlings in the north-west Sahara, Larmuth and Harvey (1978) found that more than 90% of all seedlings which had survived their first summer were to be found within 0.5 cm of the edge of a stone and, with one exception out of 450 sampled, all were within 1.0 cm of the stone. Nearly 70% of all survivors were on the western side of stones where they were sheltered from direct sun until well into the day, and where any moisture resulting from rain or dew would not be evaporated by the morning sun. These seedlings were in positions where they were able to benefit from even marginal advantages.

REFERENCES

AHEARN, G. A. and HADLEY, N. F. (1969) The effects of temperature and humidity on water loss in two desert tenebrionid beetles, *Eleodes armata* and *Cryptoglossa verrucosa*. *Comp. Biochem. Physiol.* **30**, 739–749.

CLOUDSLEY-THOMPSON, J. L. (1961) Microclimates and the distribution of terrestrial arthropods. *Ann. Rev. Entom.* **7**, 199–222.

CLOUDSLEY-THOMPSON, J. L. (1962) Bioclimatic observations in the Red Sea Hills and coastal plain, a major habitat of the desert locust. *Proc. R. Ent. Soc. Lond.* (A), **37**, 27–34.

DÉLYE, G. (1968) Recherches sur l'écologie, la physiologie et l'éthologie des Fourmis du Sahara. Doctoral Thesis. L'Université d'Aix-Marseille.

DIGBY, P. S. B. (1955) Factors affecting the temperature excess of insects in sunshine. *J. Exp. Biol.* **32**, 279–298.

EDNEY, E. B. (1971) The body temperature of tenebrionid beetles in the Namib Desert of Southern Africa. *J. Exp. Biol.* **55**, 253–272.

EISSA, S. M., EL-ZIYADI, S. M. and IBRAHIM, M. M. (1975) Autecology of the jerboa, *Jaculus jaculus*, inhabiting Al-Jalia desert area, Kuwait. *J. Univ. Kuwait (Sci.)*, **2**, 111–122.

HAMILTON, W. J. (1975) Coloration and its thermal consequences for diurnal desert insects. In *Environmental Physiology of Desert Organisms*, edited by N. F. HADLEY, pp. 67–89. Dowden, Hutchinson & Ross Inc., Stroudsberg, Pennsylvania.

HEINRICH, B. and BARTHOLEMEW, G. A. (1972) Temperature control in flying moths. *Sci. Amer.* **226** (June), 71–77.

KASSAS, M. and IMAM, M. (1957) Climate and microclimate in the Cairo desert. *Bull. Sté. Géog. d'Égypte*, **xxx**, 25–52.

KIRMIZ, J. P. (1962) *Adaptation to Desert Environment: A Study on the Jerboa, Rat and Man.* Butterworths, London.

LARMUTH, J. (1978) Temperatures beneath stones used as daytime retreats by desert animals. *J. Arid Env.* **1**, 35–40.

LARMUTH, J. (1979) Aspects of plant habit as a thermal refuge for desert insects. *J. Arid Env.* **2**, 323–328.

LARMUTH, J. and HARVEY, H. J. (1978) Aspects of the occurrence of desert plants. *J. Arid Env.* **1**, 129–134.

PARRY, D. A. (1951) Factors determining the temperature of terrestrial arthropods in sunlight. *J. Exp. Biol.* **28**, 445–462.

PEPPER, J. H. and HASTINGS, E. (1952) The effects of solar radiation on grasshopper temperatures and activities. *Ecol.* **33**, 96–103.

PIERRE, F. (1958) *Ecologie et Peuplement Entomologique des Sables vifs du Sahara Nord-Occidental.* Centre National de la Recherche Scientifique, Paris.

PORTER, W. P., MITCHELL, J. W., BECKMAN, W. A. and DEWITT, C. B. (1973) Behavioural implications of mechanistic ecology: Thermal and behavioural modeling of desert ectotherms and their microenvironment. *Oecologia (Berl.)*, **13**, 1–54.

SCHMIDT-NIELSEN, B. (1950) Evaporative water loss in desert rodents in their natural habitat. *Ecol.* **31**, 75–85.

SCHMIDT-NIELSEN, K. (1964) *Desert Animals: Physiological Problems of Heat and Water.* Oxford University Press.

SCHMIDT-NIELSEN, K., TAYLOR, C. R. and SHKOLNIK, A. (1971) Desert snails: Problems of heat, water and food. *J. Exp. Biol.* **55**, 385–398.

VORHIES, C. T. (1945) Water requirements of desert animals in the southwest. *Univ. Ariz. Agric. Exp. Sta., Tech. Bull. No. 107*, 487–525.

WILLIAMS, C. B. (1923) Some bioclimatic observations in the Egyptian desert. *Min. Ag. Egypt, Bull.* **29**, and in I.O.B. Symp. *Biology of Deserts* (1954), edited by J. L. CLOUDSLEY-THOMPSON. Institute of Biology, London.

FURTHER READING

BROWN, G. W. (1968, 1974) Ed. *Desert Biology*, Vols. I and II. Academic Press, London.

CLOUDSLEY-THOMPSON, J. L. (1977) *Man and the Biology of Arid Zones.* Edward Arnold, London.

CLOUDSLEY-THOMPSON, J. L. and CHADWICK, M. J. (1964) *Life in Deserts.* Foulis, London.

CRAWFORD, C. S. (1981) *Biology of Desert Invertebrates.* Springer, New York.

GEIGER, R. (1950) *The Climate near the Ground*, 1965 edition. Harvard University Press., Cambridge, Massachusetts.

GOODALL, D. W. and PERRY, R. A. (1979, 1981) Eds. *Arid Land Ecosystems*, Vols. I and II. International Biological Programme Nos. 16, 17. Cambridge University Press, Cambridge.

GOUDIE, A. and WILKINSON, J. (1977) *The Warm Desert Environment.* Cambridge University Press, Cambridge.

LOUW, G. N. and SEELY, M. K. (1982) *Ecology of Desert Organisms.* Longmans, London.

MONTEITH, J. L. (1973) *Principles of Environmental Physics.* Edward Arnold, London.

ROSENBERG, N. J. (1974) *Microclimate: The Biological Environment.* Wiley, New York.

CHAPTER 6

Flora

G. E. WICKENS

Royal Botanic Gardens, Kew, Richmond, Surrey

CONTENTS

6.1. INTRODUCTION

The Sahara is our largest desert. Its western, northern and eastern flanks are limited by the natural barriers of the Atlantic, Mediterranean and Red Sea, to the south it merges with the savannas of tropical Africa. The Atlas ranges and the Ethiopian massif guard the north-western and south-eastern extremities, and between them at about 1000-km intervals are the Ahaggar, Tibesti and Marra massifs. During more humid periods of the Quaternary, these Saharan massifs were the hydrographic centres from which rivers radiated to the interior depressions and to the Nile and Niger systems. Today the Sahara is traversed by only one river, the Nile (which is also our longest river). By far the greater part of the Sahara lies between 200 and 500 m asl, and is free of any surface water as well as being very sparsely vegetated.

To the north, between the limits of the desert and the Mediterranean Sea is an attenuated and often very depauperate Mediterranean vegetation. Ecologically, the southern limits of this Mediterranean vegetation are approximately that of the 100-mm isohyet, and this limit appears to be generally accepted by ecologists.

The southern transition, from desert to semi-desert scrub and thorn savanna, is more debatable. The difficulty lies in the interpretation of the term 'sahel', which is often given to the transitional vegetation. The term is derived from the Arabic 'sahil', meaning coast or plain. In North Africa 'sahel' is used in reference to the coastal plains of Morocco, Tunisia and Algeria (the term is also used for a river bank, such as the high ground bordering the Nile).

Its first application to the south of the Sahara is uncertain; it may possibly have been introduced by French colonists during the late 19th century. The first published application of the term appears to have been by Chevalier (1900), who used it primarily as a phytogeographical concept but, because of certain ecological implications contained within its original usage, there have risen considerable misunderstandings as to its meaning. In its phytogeographical context it was used to describe the belt of vegetation separating the Sahara proper from the deciduous savanna woodlands to the south. It was later interpreted in various ecological contexts, and was given an even more restricted political meaning by the mass media when reporting on the so-called 'Sahel Drought' of the late 1960s and early 1970s.

Grove (1978) interpreted the term ecologically and adopted a mean value of 200 mm rainfall as the northern limits of the Sahel. However, I believe that between 75–100 mm is a better demarcation; the much higher rainfall figures that in the past have been given for West Africa reflect the higher population density and the degradation of the vegetation by man and his livestock. The 100-mm isohyet is also the limit set by the editor of this volume.

It should be pointed out that the distinction between phytogeographical and ecological concepts are not clear cut; there is a grey area where the two overlap. Whereas phytogeography is mainly concerned with the flora, or common pool of species at regional macroclimatic level, and the floristic changes resulting from historical factors in geological time, ecology is concerned with the segregation of the species within this common pool into communities as a result of local environmental factors in historical time.

Thus, in order to understand the present-day flora of the Sahara it is first necessary to understand how the past environmental changes of the Sahara have influenced its evolution.

6.2. PAST HISTORY

The past climatic vicissitudes of the Sahara and the associated changes in the vegetation have led to considerable speculation as to whether the present desert was once a forest. Although the available information on palaeoclimates permits a certain degree of reasoned speculation, the fossil evidence is unfortunately restricted to relatively few localities and cannot be taken as being representative of the Sahara as a whole.

During the late Cretaceous to Paleocene (75–55 m.y.a.) the continent of Africa lay 15–18° to the south of its present position so that the equator then passed through what is now the central Sahara. The available fossil records imply widespread lowland rainforest and high-rainfall savanna over much of the Sahara with presumably montane rainforest on the Precambrian massifs of present-day Hoggar, Aïr and Tibesti. It is possible that there may have been local areas of aridity present due to either high-pressure cells or edaphically arid sites.

By the close of the Oligocene (22.5 m.y.a.) the African plate had moved to what is virtually its present position. This movement, accompanied by uplift resulted in somewhat drier climatic conditions over the Sahara, as a consequence of which savanna began to replace the rainforest. Glaciation in Antarctica and the progressive development of the cold Benguela current during the Pliocene (5 m.y.a.) became an additional contributing factor.

As aridity increased, a seasonally dry climate became increasingly more firmly established. This in turn helped to impoverish the flora and increase the selection pressure for drought-resistant taxa as well as giving rise to increasing areas of desertification. It is probable that a desert climate extended over a major part of the Sahara in the Pliocene. During this period the evidence of pollen spectra from a number of scattered localities in northern Africa suggested that many elements of the present Saharan flora were in fact already present (Malley, 1980).

The present-day distribution of the flora largely reflects the recolonization of much of the desert following the late Pleistocene (20000–15000 B.P.) when the climate was cool and dry and the plant life extremely restricted. The warm, wet climatic conditions of the early Holocene (*c*. 11000–7000 B.P.) would have permitted an expansion in the flora and opportunities for an exchange of species between the Mediterranean and tropical Africa. It is not necessary to presume such an exchange took place on a broad front. It would not have required a very great increase in present-day precipitation over the Saharan massifs for the streams arising there to achieve at least a seasonal flow and thereby establish a pathway for north–south migration. The Mediterranean elements certainly penetrated southwards across the Sahara, sufficiently so for *Quercus ilex* L., *Alnus*, *Tilia* and *Juglans* species to grow on the massifs with *Pinus* at lower altitudes.

As the precipitation gradually decreased and the temperature rose, a more xerophytic flora, which included *Olea*, *Cupressus*, *Celtis*, etc., reached the massifs, and *Cedrus*, *Pistacia* and *Erica* invaded the higher slopes. Because of the alleged affinities of the Saharan *Olea europaea* L. subsp. *lapperrinei* (Bat. & Trab.) Cifferi with subsp. *africana* (Mill.) P. S. Green, Malley (1980) suggests that the *Olea* arrived by way of north-east Africa. However, the distinctions between subsp. *europaea* of the Mediterranean and subsp. *africana* are very ill-defined and it would appear more likely that the *Olea* would have reached the massifs with other Mediterranean taxa with which it is most commonly associated rather than arrive independently from eastern Africa.

By about 2800 B.P. much of the Mediterranean flora had almost entirely disappeared and was replaced by an invasion of *Acacia* species, etc., from the south. The rapid increase in desertification *c.* 500 B.P. brought about the isolation of the vegetation on the Saharan massifs. This isolation in turn led to a certain degree of speciation, especially amongst the Mediterranean element, e.g. *Olea europaea* subsp. *lapperrinei*, *Cupressus dupreziana* A. Camus, etc., resulting in the flora that we know today.

6.3. EXPLORATION

Floristic studies are dependent upon sound taxonomy and representative sampling. In the Sahara, with its sparse flora, harsh climate and often hostile inhabitants, plant collections reflect the ability of their collectors to survive and travel freely. The conditions of colonial rule, whatever people may consider to be their limitations in other respects, did eventually ensure safe travel and was consequently responsible for our present-day knowledge of the Saharan flora.

The following very brief summary of the history of the European colonization of the Sahara is given in order to illustrate the impact on floristic exploration.

European interest in crossing the Sahara dates from the close of the eighteenth century. The first recorded crossing was by the English traveller William George Browne, who, in 1793, followed the ancient caravan route from Egypt to Darfur (Sudan), the 'Darb al-Arbab-in', returning by the same route in 1796. This crossing of the Libyan Desert by Browne along a well known albeit ill-defined route perhaps received less credit than it deserved, possibly due to the inaccuracies found in the narrative of his travels (Browne, 1800).

The organized exploration of the Sahara may be considered to have begun in London in 1788 with the founding of the Association for Promoting Discovery of the Interior Parts of Africa by Sir Joseph Banks and his associates. The sources of the Nile and Niger were the main objectives but the reputed wealth and flourishing trade associated with Timbuktu must have been an added incentive. This is not the place to recount the trials and tribulations of these early explorers; for a popular account see Wellard (1964). Nevertheless, a brief mention of some of the persons taking part helps to put things into perspective.

In 1798 the Association financed the German Frederick Hornemann, who, travelling as a Moslem, set out from Cairo and was last heard of at Marzuq. It was subsequently learnt that he reached Nigeria and was within a day's journey of the Niger when he died. He is credited as being the first European to cross the Sahara since the Romans.

Neither of the above explorers contributed materially to our knowledge of the flora. The first to do so was the French botanist and physician, Alire Raffeneau Delile, who, in 1798, accompanied Napoleon Bonaparte's expedition to Egypt. His collections are now in the Institute de Botanique, Montpellier, with duplicates in Paris and other herbaria; his account of the flora was published in 1813.

In 1822 an Englishman, Major Dixon Denham, and two Scots, Lieut. Hugh Clapperton R.N. and Dr. Walter Oudney, crossed the Sahara from Tripoli to Lake Chad. They attempted to reach the Niger but were forced to retrace their steps to Tripoli. Unfortunately the botanist on the expedition, Oudney, died near Katagun, Nigeria, in January 1923. His plants were brought back and are now in the herbarium of the British Museum (Natural History) and Kew (Keay, 1961; Hamilton, 1980).

It is true that the dangers from the Toureg, disease and thirst made travelling extremely difficult; between 1790 and 1890 of about 200 who set out 165 never returned alive (Wellard, 1964). Still, there were some who survived and brought back plants for the edification of botanists.

One of the best equipped expeditions was that led by James Richardson, which in 1850 left Tripoli en route for Chad. On the death of its leader the German Dr. Gustav Barth took over and eventually reached Timbuktu. Returning to Chad he was met by a fellow countryman, Edward Vogel, who had accompanied another expedition sent out by the British government to find Barth. Vogel made a collection of plants but was unfortunately murdered in 1856. His plants were brought back and are now in the British Museum (Natural History) (Keay, 1961; Hamilton, 1980).

Perhaps the most important and frustrating travels in the Sahara were those of Dr. Gustav Nachtigal, another German, who ranks with Heinrich Barth and Georg Schweinfurth amongst the foremost of the African explorers. Between 1869 and 1874 he wandered between Tripoli and Lake Chad and then eastwards to the Nile. He had the great misfortune of being forced to abandon his collection of plants and geological specimens from Tibesti when fleeing from that inhospitable region. It was 46 years later, in 1916, before another collection was made by Lieutenant-Colonel Tilho (Wickens, 1977).

The French meanwhile had not been idle. Algeria had become a colony of France in 1830, Tunisia a protectorate in 1881. In 1839 official encouragement was given for the scientific exploration of Algeria, culminating in the first flora of that country by Battandier (1888–90) and Battandier and Trabut (1893). In 1882 Dr. E. Cosson and his colleague were encouraged to undertake a series of botanical explorations of Tunisia, resulting in the publication of a flora by Bonnet and Barratte (1896).

These early botanical explorations were extremely hazardous and were often undertaken by military personnel due to the dangers from the Toureg (Quézel, 1965). Their subjection was not finally accomplished until the beginning of this century. The Saharan massifs of Hoggar, Tassili des Ajjer, Adrar des Ifoghas, Aïr, Tibesti, Ennedi and Gebel U'weinat were amongst the last areas to be explored. We now have splendid composite accounts of their floras compiled by Quézel (1954), Leredde (1957), Maire and Volkonsky (1945), Bruneau de Miré and Gillet (1956), Quézel (1958), Gillet (1968) and Osborn and Krombein (1969).

South of the Sahara the French had expanded beyond Senegal to form the French Sudan, an area that was finally pacified in 1900. Our knowledge of this area is based on the extensive collections by the Abbé Auguste Chevalier, now in the Paris Herbarium. Naegele (1958) has described the flora for Mauritania and plant lists have been produced for Chad and Niger by Lebrun et al. (1972) and Peyre de Fabregueq and Lebrun (1976).

Libya had been occupied by Italy in 1911 following the collapse of the Ottoman Empire. An extensive guide to the floristic literature of Libya has been provided by Boulos (1972) as a precursor to a flora now being prepared under the editorship of Jafri and El Gadi (1976–79).

Egypt, following the initial work by Delile (1813), was well served by a series of competent botanists and floras culminating in the over-ambitious production by Täckholm, Täckholm and Drar (1941) and Täckholm and Drar (1950–69), which now, after a period of inactivity, is being completed under the editorship of Professor Hadidi (1981). To the south the standard work on the flora of the Sudan is by Andrews (1950–56).

Readers requiring more information on the botanical exploration of the Sahara are recommended to read the introductions to the various publications cited above; their journeys make very interesting reading.

Hopefully some day, before it is too late, somebody will write an account of the numerous collectors to whom we are so indebted; irrespective of whether their collections were large or small, they deserve recognition.

6.4. THE PRESENT-DAY FLORA

The characteristic climatic features of the Saharan region are high temperatures, often with violent and extreme diurnal fluctuations, and low and usually erratic rainfall. In areas of true desert, which also occur within the region, the precipitation if any, is insufficient to support life. The combined effect of the high temperatures and low precipitation is to produce an environment of low humidity and high evapo-transpiration, factors which can also lead to an increase in surface salinity in certain areas. The result of this harsh environment is a scanty and monotonous vegetation in which xerophytic and ephemeral plants are especially favoured; there is also a preponderence of halophytes.

The flora of the Sahara is estimated to be in the region of 1200 species (Ozenda, 1977). There are about 104 families of angiosperms and 10 of cryptogams (Quézel, 1978). The relative richness of the six largest families is shown in Table 6.1.

TABLE 6.1. Species Richness of the Major Families (data from Quézel, 1978)

Family	Genera	Species	Endemic species
Compositae	80	164	13
Graminae	74	´203	19
Cruciferae	44	73	12
Leguminosae	30	154	21
Chenopodiaceae	23	64	2
Caryophyllaceae	22	73	13

The flora is remarkable for the absence of any truly endemic families. Ozenda (1977) suggests that the Zygophyllaceae is the sole example of a typically Saharo–Sindian family, with some minor extensions north into Europe and south into tropical Africa. Another unusual family is the Neuradaceae, formerly included in the Rosaceae, with two genera (*Grielum* and *Neuradopsis*) in southern Africa and *Neurada* in the Saharo–Sindian Region. The problem of distinction between southern Africa and the Sahara is fairly well documented, e.g. Verdcourt (1969).

Another striking feature is the occurrence of a number of very isolated monotypic genera of both wide and narrow distribution (Davis and Hedge, 1971). These include *Neurada procumbens* L. (Neuradaceae), *Fredolia aretioides* Coss. & Dur. (Chenopodiaceae), *Ammodaucus leucotrichus* Coss. & Dur. (Umbelliferae), *Tourneuxia variifolia* Coss. and *Warionia saharae* Coss. (Compositae), etc. The presence of so many monotypic genera is regarded as suggestive of a distant Tertiary origin with the probable extinction of linking forms.

Among the widespread Saharo–Sindian species whose distributions have been mapped by Lebrun (1980) are *Notoceras bicorne* (Ait.) Caruel, *Anastatica hierochuntica* L., *Savignya parviflora* (Del.) Webb, *Schouwia thebaica* Webb (Cruciferae), *Cornulaca monacantha* Del. (Chenopodiaceae), *Monsonia nivea* (Decne.) Decne. (Geraniaceae), *Silene villosa* Forssk. (Caryophyllaceae) and *Asteriscus pygmaeus* (DC.) Coss. (Compositae). However, as Davis and Hedge (1971) have pointed out, it would be unwise to over-emphasize the wide distribution of the Saharo–Sindian species because many of the genera, e.g. *Fagonia*, *Tamarix*, *Farsetia*, *Zilla*, *Launaea*, *Calligonum*, *Aristida*, are equally well represented by a succession of vicarious taxa across much of the region.

FIG. 1. Chorological divisions of northern Africa (after White (1976) and Lebrun (1980)).

1. Mediterranean Regional Centre of Endemism.
2. Sub-Mediterranean Regional Transition Zone.
3. Saharo–Sindian Regional Centre of Endemism. (a) Oceanic Saharo–Sindian Centre of Endemism. (b) Saharo-montane Archipelago-like Centre of Endemism.
4. Sahel Regional Transition Zone.
5. Sudanian Regional Centre of Endemism.
6. Afro-montane Archipelago-like Centre of Endemism.

6.5. PHYTOGEOGRAPHY

Our phytogeographical interpretation of an area is clearly based on our knowledge of the flora, both taxonomically and for distribution; the latter also reflects the degree of botanical exploration. Thus, for the African continent Schouw (1823) based his phytogeographical map on the known vegetation of the coastal fringe, much of the interior being largely unexplored. Half a century later exploration had improved our knowledge to such an extent that Grisebach (1872) was prepared to mark boundaries across the continent. The following hundred years saw the development of the now traditional hierarchical system of the German and French schools of phytogeography. Their progress has been described by Wickens (1976) and Lebrun (1980).

Phytogeographically the flora of the Sahara belongs to the Saharo–Sindian Region. To the north it is bounded by the Mediterranean Region and to the south by the Sahel Domain of the Sudano–Zambezian Region. As defined by Eig (1931), the Saharo–Sindian Region extends from the Atlantic coast of Morocco and Mauritania eastwards across the Sahara, Sinai, extra-tropical Arabia, southern Iraq, Iran and Baluchistan to the deserts of Sind, Thar and the Punjab. It is a concept that is accepted by the majority of present-day phytogeographers.

According to Eig (1931), Zohary (1962), Davis and Hedge (1971) and others, the Saharo–Sindian flora has been derived from Mediterranean, Sudanian and to a lesser extent, Irano–Turanian stock. This has led to some difficulty in its hierarchical placement; Eig (1931) and Guest (1966) considered it to belong to the Holarctic Kingdom, Engler and Diels (1936) placed it in the Palaeotropical Kingdom. Monod (1939, 1957) and Quézel (1965, 1978) consider the region to be partially in both regions, a judgement with which I am in agreement.

TABLE 6.2. Chorological Divisions of the Sahara (adapted from Lebrun, 1980)

Division	Representative species
Saharo-Sindian regional centre of endemism	*Silene villosa* Forssk. (Caryophyllaceae)
	Cornulaca monocantha Del. (Chenopodiaceae)
	Asteriscus pygmaeus (DC.) Coss. & Dur. (Compositae)
	* *Anastatica hierochuntica* L. (Cruciferae)
	Notoceras bicorne (Ait.) Caruel (Cruciferae)
	Savignya parviflora (Del.) Webb (Cruciferae)
	Schouwia thebaica Webb (Cruciferae)
Oceanic Saharo-Sindian centre of endemism	*Enchinochilon chazaliei* (Boissieu) I. M. Johnst. (Boraginaceae)
	Inula lozanoi Cab. (Compositae)
	Andrachne gruveli Daveau (Euphorbiaceae)
	Teucrium chardonianum Maire & Wilcz. (Labiatae)
	Hedysarum argentatum Maire (Leguminosae)
	Limoniastrum weygardiorum Maire & Wilcz. (Plumbaginaceae)
Western Saharo-Sindian region of endemism	* *Nucularia perrini* Batt. (Chenopodiaceae)
	Stipagrostis brachyathera (Coss. & Bal.) de Winter (Gramineae)
	Crotalaria saharae Coss. (Leguminosae)
Middle Saharo-Sindian region of endemism	*Chiliadenus montanus* (Vahl) Brullo (Copositae)
	Reseda urnigera Webb (Resedaceae)
	Tamarix arborea (Sieb. ex Ehrenb.) Bunge (Tamaricaceae)
	Tamarix mannifera (Ehrenb.) Bunge (Tamaricaceae)
	Fagonia schimperi Presl (Zygophyllaceae)
Saharo-montane archipelago-like centres of endemism	*Solenostemma oleifolium* (Nect.) Bullock (Asclepiadaceae)
	Bidens minuta Brunneau de Miré & Gillet (Compositae)
	Chiliadenus sericieus (Batt. & Trab.) Brullo (Compositae)
	Senecio hoggarensis Bat. & Trab. (Compositae)
	Cupressus dupreziana A. Camus (Cupressaceae)
	Centaurium flexuosum (Maire) Lebrun & Marais (Gentianaceae)
	Erodium oreophilum Quézel (Geraniaceae)
	Salvia chudaei Bat. & Trab. (Labiatae)
	Myrtus nivellei Batt. & Trab. (Myrtaceae)
	Fagonia tenuifolia Hochst. (Zygophyllaceae)

* Monotypic.

Eig (1931), Zohary (1962), Guest (1966), etc., recognize three sub-regions within the Saharo-Sindian Region. (1) A Western sub-region stretching from the Atlantic coast eastwards to the Egyptian frontier in the Libyan Desert. (2) The Middle Saharo-Sindian sub-region which includes Egypt and northern Sudan, Sinai, parts of lower Israel, much of central and southern Jordan, extra-tropical Arabia and lower Iraq. The Libyan Desert appears to have been an effective barrier between these two sub-regions for there are at least 200 species found in the Middle Saharo-Sindian sub-region not encountered further to the west. (3) The remainder of the region is known as the East Saharo-Sindian Region.

However, in 1976 Frank White of Oxford cut across the traditional approach and produced a chorological map based on centres of endemism, whose boundaries were based on those of the yet unpublished revised vegetation map of Africa (White, in press); his ideas stem from our now considerable knowledge of the flora and ecology of Africa.

Our interest lies in the Saharo-Sindian regional centre of endemism, which is bounded to the north and south by the Sub-Mediterranean and Sahel regional transition zones. In many respects the boundaries differ little from the more traditional presentations; it lacks the rigid hierarchical system and places the emphasis on areas of known genetic variability. It has, at present, the unfortunate disadvantage of isolating Africa from a system currently accepted in the rest of the plant world, but that is hardly insuperable.

Our knowledge of the flora of the Saharo-Sindian region as a whole is patchy. The Sahara is relatively well known and sufficiently well documented as a region for the general patterns to be plotted. The flora of Arabia has yet to be studied both regionally and in relation to neighbouring areas. The flora of Pakistan and northern India is again relatively well known. Thus, our knowledge is at present only sufficient for certain generalities to be pointed out.

While Quézel (1978) noted the concentration of endemic genera in the north-west of the Sahara, the relationship of the Sahara with Arabia is virtually unknown. The situation is not helped by the probably unconscious partisan approach by the botanist. Thus the African-orientated botanist would tend to regard Arabia as an unnecessary adjunct to Africa, and vice versa, instead of regarding the two areas as a single unit for study purposes. Until this is done we will not achieve a balanced view of the problem.

Lebrun (1980) from the evidence on his distribution maps finds general agreement with the chorological divisions of White but suggests the desirability of recognizing an Oceanic Saharo-Sindian centre of endemism and a Saharo-montane archipelago-like centre of endemism (the liberal translation from the French is my own adaptation to the terminology favoured by White). Lebrun has also found it necessary to adopt such terms as Western and Middle Saharo-Sindian in order to describe the different distribution patterns, a mode in keeping with White's original concept of a flexible, non-hierarchical system (Table 6.2).

REFERENCES

ANDREWS, F. W. (1950–56) *The Flowering Plants of the [Anglo-Egyptian] Sudan*. Buncle, Arbroath.

BATTANDIER, J. A. (1888–90) *Flore de l'Algérie et Catalogue des Plantes du Maroc. Dicotylédones*. Adolphe Jourdan, Alger.

BATTANDIER, J. A. and TRABUT, L. C. (1893) *Flore de l'Algérie et Catalogue des Plantes du Maroc. Monocotylédones*. Adolphe Jourdan, Alger.

BONNET, E. and BARRATTE, G. (1896) *Catalogue raisonné des plantes vasculaires de la Tunisie*. Imprimerie national, Paris.

BOULOS, L. (1972) Our present knowledge of the flora and vegetation of Libya: bibliography. *Webbia*, **26**, 365–400.

BROWNE, W. G. (1800) *Travels in Africa, Egypt & Syria for the years 1792 to 1798, 1799*. Cadell, Davies, Longmans & Co., London.

BRUNEAU DE MIRÉ and GILLET, H. (1956) Contribution à l'étude de la flore du massif de l'Aïr. *J. Agr. trop. et Bot. appl.* **11**, 221–247, 422–438, 701–760.

CHEVALIER, A. (1900) Les zones et les provinces botaniques de l'Afrique occidentale française. *C.R. Acad. Sci.* **130**, 1205–1208.

CHEVALIER, A. (1938) *Flore vivante de l'Afrique occidentale française*. Mus. Nat. Hist. Nat., Paris.

DAVIS, P. H. and HEDGE, I. C. (1971) Floristic links between NW Africa and SW Asia. *Ann Naturhistor. Mus Wien* **75**, 43–57.

DELILE, A. R. (1813) *Description de l'Egypte. Histoire naturelle*. Imprimerie impériale, Paris.

EIG, A. (1931) Les éléments et les groups phytogéographiques auxiliaires dans la flore palestiniene. *Fedde Rep.* **63**, 470–496.

ENGLER, A. and DIELS, L. (1936) *Syllabus der Pflanzenfamilien.*, 2nd ed. G. Borntraeger, Berlin.

GILLET, H. (1968) Le peuplement végétale du massif de l'Ennedi (Tchad). *Mém. Mus. Nat. Hist. Nat.* N.S. sér. B, **17**, fasc. unique.

GROVE, A. T. (1978) Geographical introduction to the Sahel. *Geogr. J.* **144**, 404.

GUEST, E. R. (1966) *Flora of Iraq. 1. Introduction*. Min. Agric., Baghdad.

GRISEBACH, A. H. R. (1872) *Die Vegetation der Erde nach Klimatischen Anordung.*, 2 vols. W Engelmann, Leipzig.

HADIDI, M. N. El. (1981) Flora of Egypt. *Taeckholmia Add. Ser.* No. 1.

HAMILTON, P. (1981) *Seas of Sand*. Readers Digest Assoc. Ltd., London.

JAFRI, S. M. H. and EL-GADI, A. (Editors) (1976–79) *Flora of Libya*, parts 1–68. Al Faateh Univ., Tripoli.

KEAY, R. W. J. (1961) Botanical collection in West Africa prior to 1860, in *C.R.ive Réunion Pléniere de l'Association pour l'Étude Taxonomique de la Flore d'Afrique Tropicale*, edited by A. FERNANDES, pp. 55–68. Junta de Investigações do Ultramar, Lisbon.

LEBRUN, J.-P., AUDRU, J., GASTON, A. and MOSNIER, M. (1972) *Catalogue des Plantes Vasculaires du Tchad Méridional*. Inst. Élevage Med. Vet. Pays Trop. Étude Bot. No. 1, Maisons-Alfort.

LEBRUN, J.-P. (1980) *Les Bases floristiques des grandes divisions chorologiques de l'Afrique seche*. Thesis Dr. Ingenier, Univ. Pierre et Marie Curie, Paris.

LÉONARD, J. (1980) *Noms de plantes et de groupements végéteaux cités dans Pierre Quézel: La Végétation du Sahara, Du Tchad à la Mauritanie*. Jardin Bot Nat. Belgique, Meise.

LEREDDE, C. (1957) Étude écologique et phytogéographique du Tassili des Ajjer. *Trav. Inst. Rech. sahar. Alger*. Sér. du Tassili 2.

MAIRE, R. and VOLKONSKY, M. (1945) Le passage du Sahara central au Sahara méridional (zone Sahélo-saharienne) entre l'Adrar des Ifoghas et l'Aïr. *Trav. Inst. Rech. sahar. Alger*. no. 3.

MALEY, J. (1980) Les changements climatiques de la fin du Tertiaire en Afrique: leur conséquence sur l'apparition du Sahara et de sa végétation. In *Sahara and the Nile*, edited by M. A. J. WILLIAMS and H. FAURE, pp. 63–86. A. A. Balkema, Rotterdam.

MONOD, T. (1939) *Phanerogames. Contributions à l'étude du Sahara occidental*, Fasc. 11. Libraire Larose, Paris.

MONOD, T. (1957) *Les Grandes divisions chorologiques de l'Afrique*. CSA–CCTA Publ. No. 24, London.

NAEGELE, A. (1958) Contribution à l'étude de la flore et des groupements végéteaux de la Mauritanie. *Bull. Inst. Fr. Afr. Noire*, Dakar 20.

OBORN, D. J. and KROMBEIN, K. V. (1969) Habitats, flora, mammals and wasps of Gebel 'Uweinat, Libyan Desert. *Smithsonian Contrib. Zoo*. No. 11.

OZENDA, P. (1977) *Flore du Sahara*. Ed. 2, CNRS, Paris.

PEYRE DE FABREGUES, B. and LEBRUN, J.-P. (1976) *Catalogue des Plantes Vasculaires du Niger*. Inst. Élev. Med. Vet. Pays Trop. Étude Bot. No. 3, Maisons-Alfort.

QUÉZEL, P. (1954) Contribution à l'étude de la flora et de la végétation du Hoggar. *Inst. Rech. sahar. Univ. Alger Monogr. Rég.* No. 2.

QUÉZEL, P. (1958) Mission Botanique au Tibesti. *Inst. Rech. sahar. Alger Mém.* 4.

QUÉZEL, P. (1965) *La Végétation du Sahara, du Tchad à la Mauritanie*. G. Fischer, Stuttgart (for index see Léonard (1980)).

QUÉZEL, P. (1978) Analysis of the flora of the Mediterranean and Saharan Africa. *Ann. Miss. Bot. Gard.* **65**, 479–534.

SCHOUW, J. F. (1823) *Pflanzengeographischer Atlas zur Erläuterung von Schouws Grundzugen einer Allgemeinen Pflanzengeographie*. G. Reimer, Berlin.

TÄCKHOLM, V., TÄCKHOLM, G. and DRAR, M. (1941) *Flora of Egypt*, Vol. 1. *Fouad I Univ. Bull. Fac. Sci.* No. 17.

TÄCKHOLM, V. and DRAR, M. (1950–69) *Flora of Egypt*, Vol. 2. *Fouad I Univ. Bull. Fac. Sci.* No. 28 (1950), Vol. 3. *Cairo Univ. Bull. Fac. Sci.* No. 30 (1954), Vol. 4. *Cairo Univ. Bull. Fac. Sci.* No. 36 (1969).

VERDCOURT, B. (1969) The arid corridor between the north-east and south-west areas of Africa. In *Palaeoecology of Africa covering the years 1966–1968*, Vol. IV, edited by E. M. VAN ZINDEREN BAKKER Sr., pp. 139–144. A. A. Balkema, Cape Town.

WELLARD, J. (1964) *The Great Sahara*. Hutchinson, London.

WHITE, F. (1976) The vegetation map of Africa—the history of a completed project. *Boissiera*, **24**, 659–666.

WHITE, F. (in press) *UNESCO/AETFAT Vegetation Map of Africa*. UNESCO, Paris.

WICKENS, G. E. (1976) *The Flora of Jebel Marra (Sudan Republic) and its Geographical Affinities*. Kew Bull. Add. Ser. 5, H.M.S.O., London.

WICKENS, G. E. (1977) Gustav Nachtigal, his botanical collections and observations and their consequences. *Veröff. Übersee-Museum Bremen* Reihe C, **1**, 212–216.

ZOHARY, M. (1962) *Plant life of Palaestine*. Chronica Botanica No. 33. Ronald Press, New York.

CHAPTER 7

Plant Ecology

M. KASSAS and K. H. BATANOUNY*

Department of Botany, Faculty of Science, University of Cairo, Giza, Egypt

CONTENTS

7.1. INTRODUCTION

The Sahara is a geographical entity that is often mapped as occupying the northern sector of Africa. From an ecological and botanical point of view the Sahara adjoins in its northern reaches the Mediterranean coastal land which comprises the Meso-Mediterranean region of the western part of North Africa (in Morocco, Algeria and Tunisia). In its southern reaches the Sahara adjoins the semi-arid region of the Sudano–Sahelian steppe. Definition of the precise limits of the Sahara on botanical grounds is admittedly difficult. This is due to two sets of attributes. The first is the notable modification of the desert environment caused by local conditions of topography, e.g. the presence of wadis and other types of run-off collecting areas and of mountain masses. The second is the consequences of overexploitation of the natural plant growth. This is especially evident in fringing regions where desertification is most prominent and thus the Sahara is extended to beyond its natural boundaries.

The Sahara, with its core of extreme aridity, represents the best and most uniform model of north–south zonation in a warm desert. The zonation results from the variation in the amount and seasonality of the rainfall. The vegetation is strongly influenced by winter rainfall in the north, irregular and very scanty

---

* Present address: The University of Qatar, Doha, Qatar.

rainfall in the centre, and predominantly tropical summer rainfall in the south. Hence, three main zones are easily distinguished: the North Sahara desert, the Central extreme desert, and the South Sahara desert.

The North Sahara desert zone starts at 150 mm of winter rainfall. Plant geographically, it is a part of the Saharo–Arabian desert (Kassas, 1967; Zohary, 1973); floristically it belongs to the holarctic region. The vegetation of this zone is characterized by the lack of trees; only dwarf, widely spaced diffusely scattered shrubs are present. In wadis and depressions, the vegetation becomes dense (Barry and Celles, 1972/73).

The Central extreme desert occupies vast areas with rainfall below 10 mm. Such a rainfall cannot support perennial plant cover; only annuals may grow for a short period after the rare event of sufficient rainfall. In addition to the scarcity of rainfall, it is very irregular and can be a centenary event. The floristic composition in this zone indicates that it belongs to the South Sahara. Ozenda (1958) finds that 47% of the species in central Sahara are Saharo-Sindian. Stocker (1976) states that a transitional range, where the species and physiognomic forms of the holarctic region intermingle with those of the palaeotropics, and from where some species penetrate far into the other regions, is situated along a strip with a rainfall of 30–50 mm.

The South Sahara, which belongs plant geographically to the Sudanian region, has elements of the savanna flora in its vegetation. The plant growth is mainly confined to the wadis and their flood plains and is dominated by trees and big shrubs.

7.2. WATER RESOURCES AND VEGETATION TYPES

Apart from their scantiness in the Sahara, water resources are variable. Since the variability increases with aridity, it is most marked in the central Sahara. Edaphic aridity in the Sahara is due to limited water resources. Desert surface soil is almost dry all the year round, except for a few days in the wet season. However, there is often a permanently wet layer in the deep soil strata and in the crevices between rocks that is in the reach of deeply penetrating roots of perennial plants.

The type of vegetation coincides with the pattern of available water resources. Kassas (1970) recognized three types of vegetation: (a) accidental (Kassas, 1966) in the rainless parts where rainfall is not an annually recurring incident; (b) restricted type (Walter, 1963), mode contracté (Monod, 1954) or run-off desert (Zohary, 1962) in arid areas where rainfall, though low and variable, is an annually recurring phenomenon, and perennial vegetation is confined to restricted areas (depressions, runnels and wadis) with relatively adequate water supply; (c) diffuse type or rainfall desert (Zohary, 1962), in less arid areas, where perennial vegetation is more or less widely distributed.

7.3. LANDFORM AND VEGETATIONAL PATTERN

Plant life in the desert shows particularly intimate relations between landforms and plant growth. The significance of landforms is primarily due to their controlling influence on the moisture regime. Physiographic features control the run-off and the processes involved in surface drainage collecting and redistributing the available water. Low parts (collecting areas) receive run-off water, water-borne material and wind-blown sediments. On the other hand, elevated parts have limited water resources and shallow soil, which is usually more coarse-textured than in low parts. A bird's-eye view of any desert stretch will show that the perennial plant growth maps the surface drainage systems.

Topography, geological structure, lithological features of parent rock, geomorphological patterns and features, characteristics of surface deposits, etc., are factors affecting not only the plant and animal life, but also the whole complex of ecological relationships. Through its direct and/or indirect effects on the water revenue of the habitat, the landform affects the desert vegetation. A slight fall in the ground surface, a rock fissure, or a stone block may create a new and special environmental set-up and produce habitats for a characteristic type of vegetation.

Elevation, exposure and angle of slope may increase or reduce the effectiveness of the limited water resources. Landform patterns are associated with the occurrence of various microhabitats which are different as regards their water resources, soil properties, microclimatic conditions and consequently their plant cover (Batanouny, 1973).

7.4. SURFACE MATERIAL AND PLANT LIFE

The features of the ground surface have their perceptible effects on plant life. A compact rock surface affords little possibility of plant growth except for certain lithophytes including lichens. A fissured rock surface will permit the growth of chaemophytes, with their roots extending into the crevices. A rocky surface veneered with residual rock fragments (hamada) provides a habitat with peculiar features different from those of the exposed rock surface. The rock fragments afford no room for seedling growth, but there may be small pockets of soft materials amongst the fragments where seedlings may grow. A rocky surface covered with a duricrust (rock detritus cemented by capillary rising salts) is usually very sterile. An erosion pavement is a rocky surface overlain by a layer of residual mixed rock waste with a surface mantle of lag scree. The plant cover is here confined to the shallow drainage runnels.

A surface mantle of soft material will afford a favourable substratum for plant growth. The material may be alluvium or aeolian sand sheets and dunes, or again a sedentary soil developed on rock.

The properties of the soil are of primary importance in determining the character of vegetation. The importance of the physical attributes of the soil is related to their influence on the water resources. The thickness of the soil layer has a great effect on the moisture regime. Shallow surface deposits will be moistened during the short spells of the rainy season but are eventually dried, while deep soils allow for the storage of some water in the deep layers and thus provide a continuous supply of moisture to the deeply seated roots of perennials. Plants inhabiting shallow soils are usually shallow-rooted therophytes, while deep soils provide room for the growth of perennial vegetation. It has been shown that plant growth is denser in areas with deep soils than in areas with shallow soils (Batanouny and Sheikh, 1972; Batanouny and Hilli, 1973).

The main source of water supply to desert perennials is the moisture retained in the deep layers of soil to which their roots extend. Water loss from deep layers is mainly due to root absorption and transpiration; direct evaporation from the soil is minimal due to the presence of the upper dry layer acting as a protective layer.

Soil penetrability plays an important role in the distribution of the different plant growth. Batanouny and Zaki (1974) concluded that, with the exception of salt marshes, variation in penetrability and soil depth affect the distribution of plant communities as well as the growth of plants and their root penetration and habit. The same authors (1969) observed that root systems of plants of the same species growing in different habitats show different numbers of laterals and vary in horizontal extension and vertical penetration.

7.4.1. Formation of phytogenic hillocks

The dry open desert regions and sand beaches with scanty vegetation are usually swept by violent winds. Obstacles that check the velocity of the wind cause some of its load to be deposited and mounds or ridges to form. The gradual building of these mounds is usually a physical process and only partly due to plant growth. Wind-blown sand and dust and water-borne sediments may accumulate around the plant growth. The plant acts as a trap for sediments.

Numerous desert and marsh plants have the ability to build mounds around their stems. Batanouny and Batanouny (1968 and 1969) give the characteristics of plants forming hillocks as follows:

1. They have the ability to produce adventitious roots from vegetative organs that become buried under the collecting sediments. These roots may be long, thread-like travelling horizontally at a shallow depth over long distances, as in *Arthrocnemum glaucum*. They may be in the form of tufts of very fine roots, as in *Anabasis articulata*.

2. They are capable of producing new shoots replacing those buried by onblowing sand. In this way the plant growth copes with the sand accumulation: plant growth keeps afloat on the sand inundation.

3. They usually have intricately branching shoot system, thus effective in intercepting wind-borne and water-borne materials causing its deposition around the plant body.

4. They have perennating organs buried in the soil or shoots and buds close to or near the soil surface. In other words, they have definite life forms. These include:

(a)   Geophytes with rhizomes, e.g. *Ammophila arenaria, Sporobolus virginicus, Panicum turgidum, Pennisetum divisum, Sprobolus spicatus* and *Juncus acutus*.

(b)   Hemicryptophytes, specially the hemicryptophyta caespitosa or the tussock forming species, e.g. *Stipagrostis plumosa, S. zitteli, S. ciliata, Lasiurus hirsutus* and *Frankenia revoluta*.

(c)   Chamaephytes:
   (i)   Chamaephyta pulvinata (cushion plants), e.g. *Halocnemum strobilaceum, Arthrocnemum glaucum, Suaeda vermiculata, S. fruticosa, S. pruinosa, Salsola tetrandra, Haloxylon articulatum, Anabasis articulata, A. setifera, Cornulaca monacantha, Zygophyllum album* and *Limoniastrum monopetalum*.
   (ii)   Chamaephyta graminida (hard grasses), e.g. *Lygeum spartum*.
   (iii)   Chamaephyta suffrutiscentia (semishrubs), e.g. *Ephdera alata* and *Calligonum comosum*.

(d)   Phanerphytes: Only some nanophanerophytes have the ability to form mounds. Usually these plants have the lower branches growing prostrate, hence able to trap the moving sand and other materials close to the earth's surface. As examples: *Lycium shawii, L. europaeum, Nitraria retusa* and *Tamarix* spp. The most common example in the Sahara is *Ziziphus lotus*, which forms what is locally known as *Nebkas*. These are roundish dune forms which can reach 10 m in diameter and 3 m in height. The Arabic name *sidra* is used both for the plant itself and the sand mass in and on which the thorn bush grows. Mensching and Ibrahim (1977) consider the formation of *nebkas* in greater number on cornfields as an important indicator of an advanced state of desertification.

The formation of mounds has many consequences (cf. Batanouny and Batanouny, 1968 and 1969). These include modification of the rooting habit of the plants, affect the soil properties and the plant–soil–water relationships, the microclimatic conditions and the vegetation.

7.5. WATER ECONOMY OF DESERT PLANTS

The survival of desert plants depends on their capacity to maintain a favourable balance between water uptake and water loss under conditions of severe edaphic and atmospheric drought. If the water loss exceeds the absorption, the balance becomes negative and the plant is put under moisture stress. Such a stress disturbs the normal functioning of all the metabolic processes in the plant. As any economic system, the balance sheet has a debit and a credit side. The credit side is represented by the water sources available to the plant and its ability to absorb this water, while the debit side is the water lost by the plant. Desert plants need to intensify the water absorption and to reduce the water loss. Achievement of these requirements of survival is fulfilled by a complex of mechanisms and adaptations. The mechanisms of drought resistance are not similar in all, even for species growing under the same environmental conditions. Every species has its own potentialities in enduring or resisting drought. Migahid *et al.* (1972) state that even for the same species, the efficiency of the mechanisms of drought resistance varies with the variation of the environmental conditions. Microvariation in the climatic or edaphic conditions leads to suppression or enhancement of the effectiveness of one or more mechanisms.

All the plants growing in arid regions have been called 'xerophytes'. This term needs qualification, because, in any arid region there are habitats, such as the oases, where plants are well supplied with water. We shall discuss the main types of xerophytes and the mechanisms of drought resistance or avoidance.

(i) *Plants capable of withstanding desiccation.* This group of plants is represented by the desert algae and lichens. The desert algae occupy very special ecological niches in the desert where other xerophytes cannot live. These pioneer plants live in crevices and fissures of rocks and below semi-transparent gravels and are adapted to extreme variation in water and temperature. They lose the greatest part of their body water and desiccate to air dryness whenever water is not available, then they become dormant and their metabolic processes are reduced to a minimum. In such a state their reactivity to the environment is much reduced. In the words of Evenari *et al.* (1971) they are physiologically and ecologically quasi-isolated from their surroundings and therefore more resistant to drought and high temperatures.

The strategy which the lichens have evolved is not to conserve water and to use the soil as a reservoir, but to be entirely dependent on the more ephemeral water supplies such as rainfall and dew (Harris, 1976). To do this, they have achieved some remarkable physiological adaptations, not the least of which is an ability to remain dormant whilst dry and to become metabolically active again when water is available. This strategy ensures survival in extremely hostile environments (cf. Lange *et al.*, 1970 a and b). Because of the activation–inactivation cycles, the algae and lichens in the desert have limited periods during which production of dry matter takes place. Therefore the rate of growth is exceedingly slow.

(ii) *Plants escaping drought.* These drought evading plants are annuals whose life cycle is restricted to the rainy season. They represent high percentages of the plant species growing in the Sahara. In more extreme parts of the Central Sahara where complete drought may prevail for a number of consecutive years, the percentage is higher and these plants may be the only ones to appear after the rare rainfalls. Comparing ephemerals from the Moroccan Sahara with ephemerals from Middle Europe, Lemée (1953a) found that there is no difference in their morphological feature or transpiration or their resistance to dehydration. He assumed that their great capacity of water absorption is the main adaptation to the desert climates. However, their adaptational mechanisms are different from those of true xerophytes. The main adaptational trait is their disappearance by the advent of the dry season. Their roots are mostly shallow and they cannot tolerate considerable water deficits. They possess the remarkable ability to regulate their body size according to water conditions of the habitat. Under ample supply of water, annuals flourish and attain considerable sizes. While under dry conditions, when rainfall is succeeded by a long rainless period, they become dwarfed and have a few small leaves, produce a limited number of flowers and fruits in a shorter period than those growing under wetter conditions. They are termed 'ephemerals' on account of the brevity of their life-cycles. Another character of annuals is their mechanism for germination and seed dispersal. Annuals produce tremendous numbers of seeds equipped with dispersal units which aid their distribution. In this way, the offspring have increased chances of reaching sites that are suitable for germination. We may distinguish between winter ephemerals that abound within the northern parts of the Sahara (Mediterranean and Saharo-Arabian affinities) and summer ephemerals that abound within the southern parts of the Sahara (Sudano-Deccanian tropical affinities). Ephemeral growth appears during the rainy season and may be profuse in particularly wet years; it may not appear in particularly dry years. Associations of therophytes have been distinguished by some authors. Near Beni Unif, Lemée (1953b) distinguished associations of *Althea ludwigii–Trigonella anguina, Lotononis dichotoma* and *Asphodelus pendulinus.* Quézel (1965) distinguished associations of *Leysera leyseroides–Trigonella anguina* and *Lotus glinoides–Matthiola livida* in the Hoggar, and *Astragalus eremophilus* and *Bidens pilosa* in Tibesti.

(iii) *Perennial xerophytes*. These plants form the permanent framework of desert vegetation. They can withstand the adverse conditions of the desert environment: shortage of water, scorching temperatures, and likelihood of soil salinity. Though they exhibit a wide range of structure and form, desert perennials may be classified into succulent and non-succulents. The famous succulent cacti of the American deserts are not native in the Sahara, but the succulent species of *Euphorbia* and *Caralluma* are comparable to cacti. Plants with succulent leaves or stems or both growing in the different parts of the Sahara as *Zygophyllum*, *Nitraria*, *Haloxylon*, *Anabasis*, *Salsola*, and others do not belong to this group. The cell sap of the true succulents (cacti) has a very low concentration which does not rise even during the long dry periods when large amounts of water have been lost. On the other hand, the succulence of the latter-mentioned species is attributed to the storage of salts, mainly chlorides. Their cell sap is very concentrated and its concentration is increased by the intensification of the evaporative power of the atmosphere.

7.5.1. Mechanisms of drought resistance and/or avoidance

Adaptations of desert plants to the prevailing aridity may be morphological, physiological or behavioural in nature. These adaptations serve to maintain water balance: reduce loss and/or increase uptake.

7.5.1.1. *Mechanisms of water-loss reduction*

1. *Reduction of the transpiring surface*. Desert plants, active during the dry period, may reduce their water loss by reducing the area of their transpiring surface. This is achieved by the following means:
   (a) Shedding a part of the green cortex, which is the transpiring and photosynthesizing organ, e.g. *Haloxylon salicornicum* and *Anabasis articulata* (Orshan, 1963).
   (b) Shedding, partly or wholly, of branches, e.g. in many members of the family Chenopodiaceae, which are also cortex shedders.
   (c) Leaf shedding is an effective device developed by numerous desert plants. Many desert plants become leafless, particularly during the dry season, e.g. *Zilla spinosa*, *Retama raetam* and *Leptadenia pyrotechnica*.
   Leaf shedding may be acompanied by branch shedding in some plants. In such cases, the body weight of the plant is greatly reduced. The rate of leaflet shedding in *Zygophyllum* spp. increases by the intensification of drought.
   (d) Shedding of winter leaves and retaining small summer leaves is a device of reducing the transpiring surface in the dry season, e.g. *Artemisia herba-alba* and *A. monosperma*.

Shortage of water, a common phenomenon in the desert, results in a negative water balance. This does not mean that the plant will die suddenly, but a part of the transpiring organ will become desiccated. The desiccation of shoots goes hand in hand with the continuation of water deficiency. It may take years for a desert plant to become desiccated to the point at which only a single small branch remains alive. When it rains, the plant regenerates from dormant renewal buds. The production and growth of new branches are proportional to the amount of water supply. Partial death in plants is thus a means of survival.

There are other means of reducing water loss by minimizing the exposed area of the transpiring organs. These include:
   (a) Presence of scaly leaves or leaf sheaths protecting the buds from excessive water loss and further wilting, e.g. grasses.
   (b) Persistence of dead leaves around the stems protecting them from exposure to drastic atmospheric conditions, e.g. grasses.

(c) Folding of leaflets and rolling of leaves reduce the exposed transpiring surface, e.g. *Stipagrostis* spp. (rolling of leaves), *Astragalus* spp. and *Cassia* spp. (folding of leaflets). A rolled leaf has a minimized rate of transpiration to 30–85% (cf. Lemée, 1954).

Many desert plants endowed with xeromorphic characters reduce the transpiring surface by spine formation. Shoots (*Lycium* spp., *Nitraria retusa*, *Zilla spinosa* and *Maytenus senegalensis*), leaves and stipules (*Acacia* spp.) may be metamorphosed to spines.

2. *Reduction of the transpiring rate.* The daily march of the transpiration of desert plants has been investigated by numerous authors (Lemée, 1954; Stocker, 1970, 1971, 1972, 1974; Migahid *et al.*, 1972). The most striking feature of the transpiration as measured by these authors and others is its lack of parallelism with the atmospheric factors. This discrepancy is more pronounced in the dry season. The shape and trend of the transpiration curve are influenced by the prevailing environmental conditions and the water conditions in the transpiring organs. Usually the turning point in the transpiration curve in the dry season is shifted almost 3 hours earlier than that of the evaporating factors of the atmosphere. Despite the continuous rise of the evaporating power of the atmosphere till the afternoon hours, the transpiration rate is reduced before noon. The reduction in the transpiration rate may be attributed to a water deficit caused by excessive water loss in the forenoon hours.

In the dry season, the great evaporating power of the atmosphere favours excessive water loss from the plants. However, the increase of the transpiration rate is not equivalent to that of the evaporation. The magnitude of monthly variation of the mean transpiration rate is not as wide as that of the evaporating power of the atmosphere. Migahid *et al.* (1972) report that the highest mean transpiration rate of *Leptadenia pyrotechnica* in summer is nearly twice the lowest in spring, while the highest mean evaporation rate is about 4 times the lowest.

The efficacy of the combined reduction of transpiring surface, decrease of transpiration rate and morphological–anatomical change of the transpiring organs become obvious if the water loss for a whole plant is calculated for different months of the year. Evenari *et al.* (1971) report that the maximum water output in *Retama raetam* reaches 380 kg/plant in March, while it decreases to 7 kg/plant in October by the end of the dry season.

7.5.1.2. *Intensification of water absorption*

1. *Root system*

(a) *Root extension.* Increased efficiency of moisture absorption is primarily due to the development of an extensive root system which is a common feature of desert plants (Kausch, 1959; Cloudsley-Thompson and Chadwick, 1964; Batanouny and Abdel Wahab, 1973). Deep penetration of roots makes it possible for plants to absorb water from the deep, permanently wet zone. Root extension to pockets in the soil between stones helps in providing the plant with available water (cf. Kausch, 1959). Even old, suberized roots absorb considerable amounts of water (Kramer, 1956).

(b) *Root/shoot ratio.* The preponderance of the root system over the above-ground shoot system facilitates the adjustment of the water balance of the desert plants. The transpiration rate of desert plants may be very high when calculated on the basis of the fresh weight or area of the transpiring shoot, but low when calculated on the basis of the fresh weight or area of the absorbing roots. Each unit of transpiring surface is supplied with water by more roots than is usual in mesophytes. The ratio of shoot height to root axial length (length of the root excluding lateral branches) is 1:7 in *Leptadenia pyrotechnica*.

(c) *Absence of root competition.* Desert vegetation is characterized by openness, resulting in the reduction of competition between the perennials. The exhaustion and withdrawal of available moisture from the soil depend on the density, cover and openness of the vegetation (Abdel Rahman and Batanouny, 1965 a and b). The sparsity of roots per unit volume of soil reduces root competition, which, in turn, minimizes the amount of water withdrawn by the roots per unit volume of soil. These characteristics ensure lesser withdrawal and better conservation of soil moisture. The dry season lasts, in some localities, up to 9 months. During this time the uppermost soil layers are dry and form a protective layer, which eliminates water loss from the soil through evaporation. Water loss in the dry season takes place only through absorption by root and transpiration.

Ephemerals exploit the excess water which the perennials are unable to utilize in a year of good rain. If, as in years of drought, there is no such surplus, the ephemerals simply do not develop.

The root system of a small bush of *Leptadenia pyrotechnica*, 160 cm high, has been found to penetrate to a depth of 11.5 m and has a lateral extension of 10 m (Batanouny and Abdel Wahab, 1973). This root system exploits about 850 m^3 of soil with total available moisture of 23,000 kg. The annual water output by this bush reaches 5700 kg. This means that the available water in the soil occupied by the root system is sufficient to supply the plant for a period of 4 years without replenishment by rainfall.

Reference may be made in this respect to the so-called 'rain roots' which are fine rootlets produced by the woody parts of the root just below the ground surface in response to light showers. They may also be produced during periods of dew formation. These roots are ephemeral and eventually wither, reappearing shortly after a new supply of water has become available. Field observations on many perennials of the Sahara confirm the conclusion of Káusch (1965) who experiments with cacti and shows that 'When roots of cacti which have been kept dry for many months are remoistened, absorbing roots are formed after a few hours. The transpiration rate increases rapidly within a day after remoistening. This is an indication that as soon as the roots appear water absorption begins immediately, and rapid improvement of the water relationships in the plant occurs.'

Reference may also be made to the so-called 'sheath roots' described by Volkens (1887), Price (1938), Killian and Lemée (1956), etc. These are found in desert plants especially grasses, roots seem to exude a mucous material that cements sand particles and forms a sleeve-like structure ensheathing the root. These sheaths seem to protect the small roots against desiccation. *Stipagrostis* spp. and *Panicum turgidum* are among the many grasses having these sheath roots.

2. *Water absorption from relatively dry soil.* Xerophytes have higher osmotic pressure values than mesophytes. The high osmotic pressure of the root and shoot tissues may be considered as a special physiological asset that adds to the efficiency of absorption (Killian and Faurel, 1936). Walter (1972) states that osmotic values may serve as an exact indicator of the water balance and, therefore, of the total water economy of the plant. The osmotic pressure values are closely related to the water supply and decreases with increase in water supply (Batanouny, 1980).

Experimental ecological research in the north-western part of the Sahara was first carried out by Fitting (1911) on plants from Biskra. He was the first to show the importance of the osmotic pressure of the desert plants. He studied some plants from different groups and determined their osmotic values using the plasmolytic method with potassium nitrate and sodium chloride solutions. Some of these plants were further investigated by Killian and Faurel (1933 and 1936) at Beni Unif. The latter used the cryoscopic method. Walter (1964) compared the results obtained by Fitting and by Killian *et al.* The data obtained by Killian *et al.* show that the osmotic values of the plants growing in saline habitats are high, e.g. *Atriplex halimus* (35.5–70 atm), *Limonium delicatulum* (20.7–48.6 atm), *Limoniastrum guyonianum* (49.6–53.4 atm) and *Frankenia pulverulenta* (40.6 atm). Perennial xerophytes have osmotic values lower than halophytes, e.g. *Peganum harmala* (14.9–34.4 atm), *Capparis spinosa* (18.4–21.3 atm) and *Fagonia glutinosa* (9.7–17.3 atm).

7.5.2 Photosynthesis/transpiration

Stomatal transpiration is the most effective process by which plants lose large amounts of water, up to 95% or more of the water absorbed by the roots. Closure of the stomata results in the reduction of water loss, but it means that no carbon dioxide, needed for photosynthesis, will pass through the stomata. On the other hand, opening of the stomata leads to excessive transpiration. The desert plants are between the Scylla of death by lack of water and the Charybdis of death by lack of organic matter (Evenari *et al.*, 1971). In the words of Stocker (1976): 'desert plants have to navigate between two cliffs: the danger of desiccation and the danger of starvation; their problem is, how to reduce water consumption and how to maintain photosynthesis at the same time. Between these two strategies a physiological balance, that can be called the "water-photosynthesis syndrome", must be achieved.'

Stocker (1970, 1972 and 1974) carried out studies on the water relations and photosyntheis of the desert plants in the Mauritanian and Algerian Sahara. In 1976 Stocker gives the diurnal variations of transpiration and photosynthesis of two shrubs that have the same type of malacomorphous, deciduous leaves, but belong to different zones of the Sahara. These are *Ziziphus lotus* (Rhamnaceae) from the Northern and *Cassia aschrek* (Caesalpinaceae) from the Southern Sahara. In spite of the severe atmospheric drought, the transpiration of *Ziziphus* does not follow the potential evaporation in the forenoon hours, and remains on an almost constant level for the rest of the day. In such a manner the plant adjusts its water consumption to an existence minimum, which is needed to prevent the danger of a heat death (Lange, 1962). However, this restriction is not sufficient to keep water relations balanced. Increasing water deficits lead to a deterioration of the water potential, which results in a rapid decrease of photosynthesis; this is obvious between 11 and 12 a.m. Net photosynthesis reaches the compensation point or drops below it and recovers only at decreasing evaporation and transpiration in the late afternoon.

In the South Sahara, this process is quite different with *Cassia*. The water supply is sufficient to keep transpiration in step with evaporation rate until the early afternoon and to keep it on a high level for the rest of the day. The water potential remains favourable, and the afternoon reduction of photosynthesis is of only a short duration. Related to surface area, for *Cassia* the daily rate of transpiration is 3.8 times that of *Ziziphus* and the photosynthetic production is 6.6 times greater.

For a plant, the genetically fixed and limited constitution type is decisive. In the Sahara two types, the holarctic and the palaeotropic, are in competition. Drought resistance is a property of both, although in the palaeotropic type to a much higher degree than in the holarctic one, whereas freezing tolerance is developed only in the holarctic type, and is lacking in the palaeotropic one. The higher aridity in the South Sahara, and the winter cold in the North Sahara, determine the distribution areas of the two constitution types in a 'balance of power' (Stocker, 1976).

7.5.3. Photosynthetic pathways

Recent studies do not appear to support the validity of the concept of a single type of photosynthetic carbon assimilation occurring in all green plants. At least three pathways, namely C_3, C_4 and Crassulacean acid metabolism (CAM), have been characterized (Laetsch, 1974). Plants with C_4 photosynthesis are characterized by a complex of anatomical and physiological properties. These properties include:

(a) Leaf anatomy: Principally, in a transverse section, the leaf of a C_4 plant shows a chlorenchymatous sheath of large, thick-walled cells around the vascular bundles, which again may be surrounded by one or more layers of loosely fitted mesophyll cells. This structural specialization of leaf has been termed 'Kranz-type', and is found in several plants belonging to both monocotyledons and dicotyledons.

(b) The synthesis of C_4 dicarboxylic acids as first photosynthetic products.

(c) Fractionation of the stable carbon isotopes (^{12}C and ^{13}C) due to preferential utilization of ^{12}C and partial exclusion of ^{13}C by plants during carbon assimilation (Smith, 1972). C_3 plants show a greater discrimination against the heavy isotopes of carbon than do C_4 plants. Thus, ^{13}C-values of -9 to -16% indicate the Kranz-syndrome, while ^{13}C-values of -23 to -32% are suggestive of C_3 plants (Bender, 1971).

(d) In C_4 plants, net photosynthesis is generally favoured by high temperatures and high light intensities. Water use efficiency is usually higher in C_4 than C_3 species, i.e. C_4 plants require fewer units of water to fix one unit of CO_2 and produce one unit of dry matter.

It has, therefore, been suggested that the C_4 pathway is an adaptation to tropical and arid zones. Thus, along a geographical gradient of increasing aridity, the proportion of C_4 types in plant communities may increase (Winter *et al.*, 1976). However, Ziegler *et al.* (1981) in their studies on the photosynthetic pathway types of desert plants find that most of the plant species have solved the special problems of the desert habitat not by aquisition of a special photosynthetic CO_2 fixation, but by other adaptations.

Winter *et al.* (1976) investigated the ^{13}C values for forty-five graminaceous species from the Northern Sahara, at the foot of the Saharan Atlas. Their results indicate a clear relationship between carbon isotope discrimination and phytogeographical distribution of the grasses. From the twenty-eight grass species identified as C_3 plants, twenty-four are mainly Mediterranean types and these occur seldom in the Saharan regions. By contrast, nearly all grasses with C_4 pathway are Saharo-Arabian, Sudanian and partly tropical. Main exceptions are the widespread *Imperata cylindrica* and *Cynodon dactylon*. The *Aristida* (*Stipagrostis*) species analysed are dominant members of the Saharan grass flora (Kassas, 1967). Among the C_4 grasses may be mentioned: *Andropogon distachyus*, *Cymbopogon schoenanthus*, *Aristida* (*Stipagrostis*) *acutiflora*, *A. adscensionis* var. *pumila*, *A. caerulescens*, *A.* (*Stipagrostis*) *ciliata*, *A. obtusa*, *A. plumosa*, *A. pungens*, *Panicum turgidum*, *Pennisetum divisum*, *P. setaceum*, *Setaria verticillata* and *Enneapogon scaber*. Among the plants recorded in the Sahara and found in other desert areas, Ziegler *et al.* (1981) identified many C_4 plants including *Anabasis articulata*, *A. setifera*, *Calligonum comosum* and *Zygophyllum simplex*.

7.5.4. Proline accumulation

Perennial desert plants are under the impact of water stress throughout their lives except for short spells in the rainy season. Water stress induces certain biochemical processes which may be considered of ecological importance under water-deficit conditions. Among the characteristic responses to water stress is the accumulation of proline (cf. Stewart *et al.*, 1980 and Batanouny *et al.*, 1981). Greater amounts of proline in wilted leaves compared to non-wilted leaves have been reported in many plant species suggesting that the phenomenon is a general one. The study of the diurnal changes in proline content of desert plants by Batanouny and Ebeid (1981) shows that different species have different values of accumulated proline under the same environmental conditions and there is a remarkable coincidence of the proline increase with the water-content decrease. The same authors have noted that the curves showing the diurnal march of proline content in *Zygophyllum quatarense* and *Francoeuria crispa* are mirror images of those showing the march of the water content. Watering, through its effect in eliminating water stress, results in reducing the accumulation of proline.

Many suggestions on ways in which the capacity for proline accumulation could confer adaptation to drought have been argued. Possible adaptive roles for proline accumulation in water-stressed leaves have been proposed by Stewart and Hanson (1980). Proline accumulation might be involved in osmoregulation, or it acts as a desiccation protectant, or may provide a convenient source of energy and nitrogen in immediate post-stress metabolism, or it is a regulation of the cellular water structure.

7.6. SALINES AND HALOPHYTES

Salines do not occupy vast areas in the Sahara, as compared to the area occupied by the Sahara itself. Salines are mainly present in the Northern Sahara. These abound in depressions south of Constantine and Tunisia and extend south through the Wadi Rhir, in south Oranis through the Wadi Saoura and its affluents. In the oases and their vicinities, there is salt-affected land where halophytes grow. Salines are of limited occurrence in the Central and Southern Sahara. Quézel (1971) gives a map showing the distribution of vegetation zones in the Sahara. In 1965 Quézel gave a detailed account of the halophilous vegetation in the Sahara. This account can be summarized as follows:

I. The halophilous and halo-gypsophilous vegetation of the Northern Sahara (Wadi Saoura). The following associations were recognized:
    (a) Hyperhalophytic associations in the sebkhas which are dry in summer: association of *Halocnemum strobilaceum*.
    (b) Hyperhalophytic associations on soils wet all the year around: association of *Arthrocnemum indicum* and association of *Salicornia arabica* and *Phragmites communis* spp. *pungens*.
    (c) Halogypsophilous associations: association of *Salsola sieberi* var. *zygophylla* and *Zygophyllum cornutum*, and association of *Zygophyllum geslini* and *Traganum nudatum*.
    (d) Association of *Suaeda vermiculata* and *Salsola foetida*.
    (e) Association of *Suaeda fruticosa* var. *longifolia* and *Limonium delicatulum* in the old palm plantations.
II. The halophilous vegetation of the Occidental Sahara (excluding Wadi Saoura) and the Oceanic Sahara. This includes:
    (a) The group of *Limoniastrum ifniense* and *Nitraria retusa* in the wadis of the oceanic and Occidental Sahara.
    (b) Vegetation of *Zygophyllum waterlotti*, *Z. fontanesii*, *Salsola foetida* and other halophytes in the littoral oceanic Sahara.
    (c) Association of *Zygophyllum gaetulum* or *Z. waterlotii* in the sebkhas of the Occidental Sahara.
III. The halophilous and halo-gypsophilous vegetation of the Central and Southern Sahara.
    (a) Numerous *Tamarix* spp. (*T. balabsae*, *T. boveana*, *T. trabutii* and *T. gallica*) and *Juncus maritimus*, *Cyperus conglomeratus* and *Phragmites communis* in the Central Sahara.
    (b) Association of *Sporobolus acutus* and *Hyphaene thebaica* in the Southern Sahara.

The problem of salinity is aggravated in the Sahara by the inadequacy of rainfall. The plants growing in salines are termed 'halophytes'. These plants exhibit behavioural uniformity in the different climatic regions and the same genus or even the same species may occur in different climatic regions. Plants growing in these salines are subjected to severe climatic and edaphic aridity, in addition to the physiological drought (Schimper, 1903), which is a common phenomenon in salines of different climatic regions. Several attempts have been made to classify the halophytes (Stocker, 1928; Steiner, 1935; Chapman, 1942; and Waisel, 1972). Walter (1973) defined true halophytes as plants that store large quantities of salt in their organs without thereby undergoing any damage. They may even benefit from the salt if its concentration is not too high. The salts involved are usually sodium chloride, but can occasionally also be sodium sulphate or organic sodium salts. The osmotic effect of the salt concentration in the soil has to be balanced by an equally high salt concentration in the cell sap. The salts in the cells have an effect on the protoplasm and are toxic to salt-sensitive species, which for this reason cannot survive on saline soils. Even the euhalophytes have an upper limit, varying from species to species, for the concentration of the salt tolerated in the cell sap. If this rises too far the plants wilt. In the case of Chenopodiaceae, this rise is accompanied by their turning red due to the formation of nitrogen-containing anthocyanin.

Halophytes exhibit different responses to salinity and are endowed by various adaptive mechanisms. Combination of more than one mechanism may be observed in many halophytes. It seems proper to classify halophytes according to their response to salinity. These responses include:

1. *Salt secretion*. Salt-secreting halophytes are usually non-succulent plants with salt glands. These glands are special structures which are able to secrete excess salts from the plant body (cf. Batanouny and Abu Sitta, 1979). They occur on the photosynthetic organs. The well-known tamarisk (*Tamarix* spp.), of which there are many species in the Sahara, has salt glands. Salt dust rains down if the branches of these trees are shaken. In the early morning, drops of salt solution fall down from the canopy on to the soil surface. This occurs specially after nights with dew or with high air humidity. Salt secretion is a mechanism by which the plants get rid of the excessive salts in their tissues, hence regulating the mineral content of their body. Among the salt-secreting plants in the Sahara may be mentioned: *Limoniastrum* spp., *Limonium* spp., *Tamarix* spp., *Aeluropus* spp. and *Frankenia* spp.

2. *Removal of salts by salt-accumulating bladders*. Salt-accumulating bladders are well known in some *Atriplex* spp., e.g. *Atriplex mollis* (Berger-Landefeldt, 1959). Salts accumulate in these bladders; their concentration in the bladder is higher than in the mesophyll cells. Though such bladders usually function for only a short period, it was reported by various investigators that their role seems to be important (Waisel, 1972).

3. *Removal of salt-saturated organs*. Shedding of salt-saturated organs is another mechanism by which some halophytes can regulate their salt content. In species as *Juncus maritimus*, leaves are shed after being loaded with undesirable salts.

It is notable that some succulent halophytes and xero-halophytes discard portions of their fleshy cortex, thus releasing large quantities of salts from the plants and enabling their survival (Chapman, 1968). Among these plants may be mentioned the succulent chenopodiaceous plants such as *Halocnemum* and *Arthrocnemum*.

4. *Succulence*. Some halophytes can tolerate high concentrations of salts in their cell sap due to an increase in succulence (Steiner, 1935). Succulence is developed from high uptake of chlorides. This causes a swelling of the protein and, therefore, leads to a special ionic hydration of the protoplasm. This results in cell hypertrophy due to water uptake or in other words, a succulence of the organs. Numerous succulent halophytes are recorded in the salines in the Sahara, including *Salsola* spp., *Arthrocnemum* sp., *Halocnemum* sp. and *Salicornia* sp.

REFERENCES

ABDEL RAHMAN, A. A. and BATANOUNY, K. H. (1965a) The water output of the desert vegetation in the different microhabitats in wadi Hoff. *J. Ecol.* **53**, 139–145

ABDEL RAHMAN, A. A. and BATANOUNY, K. H. (1965b) Transpiration of desert plants under different environmental conditions. *J. Ecol.* **53**, 267–272.

BARRY, J.-P. and CELLES, J.-C. (1972/73) Le problème des divisions bioclimatiques et floristique au Sahara Algérien. *Naturalia monspeliensia*, sér. Botan. **23–24**, 5–48.

BATANOUNY, K. H. (1973) Soil properties as affected by topography in desert wadis. *Acta Bot. Acad. Sci., Hung.* **19**, 13–21.

BATANOUNY, K. H. (1980) Water economy of desert plants. In *Pollution and Water Resources*, Columbia Univ. Seminar Series, edited by G. J. HALSI-KUN, pp. 167–177. Pergamon Press, Oxford, New York.

BATANOUNY, K. H. and ABDEL WAHAB, A. M. (1973) Eco-physiological studies on desert plants. VIII. Root penetration of *Leptadenia pyrotechnica* (Forssk.) Decne in relation to its water balance. *Oecologia* (Berl.) **11**, 151–161.

BATANOUNY, K. H. and ABO SITTA, Y. (1979) Eco-physiological studies on halophytes in arid and semi-arid zones. I. Autecology of the salt-secreting halophyte *Limoniastrum monopetalum* (L.) Boiss. *Acta Bot. Acad. Sci., Hung.* **23**, 13–31.

BATANOUNY, K. H. and BATANOUNY, M. H. (1968) Formation of phytogenic hillocks. I. Plants forming phytogenic hillocks. *Acta Bot. Acad. Sci., Hung.* **14**, 243–252.

BATANOUNY, K. H. and BATANOUNY, M. H. (1969) Formation of phytogenic hillocks. II. Rooting habit of plants forming phytogenic hillocks. *Acta Bot. Acad. Sci., Hung.* **15**, 1–18.

BATANOUNY, K. H. and EBEID, M. M. (1981) Diurnal changes in proline content of desert plants. *Oecologia* (Berl.), **51**, 250–252.

BATANOUNY, K. H. and HILLI, M. R. (1973) Phytosociological study of Ghurfa desert, central Iraq. *Phytocoenologia*, **1**, 223–249.

BATANOUNY, K. H. and SHEIKH, M. Y. (1972) Ecological observations along Baghdad–Huseiba road, western desert of Iraq. *Feddes Repertorium*, **83**, 245–263.

BATANOUNY, K. H. and ZAKI, M. A. F. (1969) Root development of two common species in different habitats in the Mediterranean subregion in Egypt. *Acta Bot. Acad. Sci., Hung.* **15**, 217–226.

BATANOUNY, K. H. and ZAKI, M. A. F. (1974) Edaphic factors and the distribution of plant associations in a sector in the Mediterranean zone in Egypt. *Phyton* (Austria), **15**, 193–202.

BENDER, M. M. (1971) Variations in the $^{13}C/^{12}C$ ratios of plants in relation to the pathway of photosynthetic carbon dioxide fixation. *Phytochemistry*, **10**, 1239–1244.

BERGER-LANDEFELDT, U. (1959) Beiträge zur Ökologie der Pflanzen nordafrikanischen Salzpfannen. *Vegetatio*, **9**, 1–47.

CHAPMAN, V. J. (1942) The new perspective in the halophytes. *Quart. Rev. Biol.* **17**, 291–311.

CHAPMAN, V. J. (1968) Vegetation under saline conditions, in *Saline Irrigation for Agriculture and Forestry*, edited by H. BOYKO, pp. 201–216. Dr. W. Junk, The Hague.

CLOUDSLEY-THOMPSON, J. L. and CHADWICK, M. J. (1964) *Desert Life*. Foulis, London.

EVENARI, M., SHANAN, L. and TADMOR, N. (1971) *The Negev, the challenge of a desert.* Harvard University Press, Cambridge, Mass.

FITTING, H. (1911) Die Wasserversorgung und die osmotischen Druckverhältnisse der Wüstenpflanzen. *Zschr. Bot.* **3**, 209–275.

HARRIS, G. P. (1976) Water content and productivity of lichens. In *Water and Plant Life*, edited by O. L. LANGE, L. KAPPEN and E.-D. SCHULZE, pp. 452–468. Springer, Berlin, Heidelberg, New York.

KASSAS, M. (1966) Plant life in deserts. In *Arid Lands*, edited by E. S. HILLS, pp. 145–180. Methuen, London, Unesco, Paris.

KASSAS, M. (1967). Die Pflanzen der Sahara. In *Sahara*, edited by C. KRÜGER, pp. 162–181. Schroll, Wien, München.

KASSAS, M. (1970) Desertification versus potential for recovery in circum-Saharan territories. In *Arid Lands in Transition*, edited by H. E. DREGNE, pp. 123–142. Amer. Assoc. Advanc. of Science, Publication 90.

KAUSCH, W. (1959) *Der Einfluß von edaphischen und klimatischen Faktoren auf die Ausbildung des Wurzelwerkes der Pflanzen unter besonderer Berücksichtigung einiger algerischer Wüstenpflanzen.* Habilitationsschrift, Techn. Hochschule, Darmstadt.

KAUSCH, W. (1965) Relationship between root growth, transpiration and CO_2-exchange in several cacti. *Planta* (Berl.), **66**, 229–238.

KILLIAN, CH. and FAUREL, L. (1933) Observations sur la pression osmotique des végétaux désertiques et subdésertiques de l'Algérie. *Bull. Soc. Bot. Franc.* **80**, 775.

KILLIAN, CH. and FAUREL, L. (1936) La pression osmotique des végétaux du Sud Algérien. *Ann. Physiol.* **12**, 5.

KILLIAN, CH. and LEMÉE, G. (1956) Les xérophytes. Leur économie d'eau. In *Encyclopedia of Plant Physiology*, edited by W. RUHLAND, pp. 787–824. Springer, Berlin, Göttingen, Heidelberg.

KRAMER, P. J. (1956) Roots as absorbing organs. In *Encyclopedia of Plant Physiology*, edited by W. RUHLAND, pp. 188–214. Springer, Berlin, Göttingen, Heidelberg.

LAETSCH, W. M. (1974) The C_4 syndrome: a structural analysis. *Ann. Rev. Plant Physiol.* **25**, 27–52.

LANGE, O. L. (1962) Über die Beziehungen zwischen Wasser- und Wärmehaushalt von Wüstenpflanzen. *Veröff. Geobot. Inst., Rübel* (Zurich), **37**, 155–168.

LANGE, O. L., SCHULZE, E.-D. and KOCH, W. (1970a) Experimentell-ökologische Untersuchungen an Flechten der Negev-Wüste. II. CO_2-Gaswechsel und Wasserhaushalt von *Ramalina maciformis* (Del.) Bory. am natürlichen Standort während der sommerlichen Trockenperiode. *Flora* (Jena), **159**, 38–62.

LANGE, O. L., SCHULZE, E.-D. and KOCH, W. (1970b) Experimentell-ökologische Untersuchungen an Flechten der Negev-Wüste. III. CO_2-Gaswechsel und Wasserhaushalt von Krusten- und Blattflechten am natürlichen Standort während der Trocken-periode. *Flora* (Jena), **159**, 525–538.

LEMÉE, G. (1953a) Contributions à l'étude écologique de la végétation des confins Saharo-marocains. *Desert Research*, Proceedings Int. Symp. Jerusalem, pp. 302–306.

LEMÉE, G. (1953) Les associations à thérophytes des dépressions sableuses et limoneuses non salées et des rocailles aux environs de Béni Ounif. *Vegetatio*, **4**, 137–154.

LEMÉE, G. (1954) L'économie de l'eau chez quelques graminées vivaces du Sahara septentrional. *Vegetatio*, **5/6**, 534–541.

MENSCHING, H. and IBRAHIM, F. (1977) The problem of desertification in and around arid lands. *Applied Sci. and Development*, **10**, 7–43.

MIGAHID, A. M., ABDEL WAHAB, A. M. and BATANOUNY, K. H. (1972) Eco-physiological studies on desert plants. VII. Water relations of *Leptadenia pyrotechnica* (Forssk.) Decne growing in Egypt. *Oecologia* (Berl.), **10**, 77–91.

MONOD, TH. (1954) Modes 'contracté' et 'diffus' de la végétation Saharienne. In *Biology of Deserts*, edited by J. L. CLOUDSLEY-THOMPSON, pp. 35–44. Inst. Biol., London.

ORSHAN, G. (1963) Seasonal dimorphism of desert and Mediterranean chamaephytes and its significance as a factor in their water economy. In *Water Relations of Plants*, Symp. Br. Ecol. Soc., edited by A. J. RUTTER and E. H. WHITEHEAD, pp. 206–222. Blackwell, London.

OZENDA, P. (1958) *Flore du Sahara Septentrional.* C.N.R.S., Paris.

PRICE, J. R. (1938) The roots of some north African desert grasses. *New Phytol.* **10**, 328–340.

QUÉZEL, P. (1965) *La Végétation du Sahara*. G. Fischer, Stuttgart.

QUÉZEL, P. (1971) Flora und vegetation der Sahara. In *Die Sahara und ihre Randgebiete*, Bd. 1, Physiographie, edited by H. SCHIFFERS, pp. 429–476. Weltforum Verlag, München.

SCHIMPER, A. F. W. (1903) *Plant Geography upon a Physiological Basis*. Clarendon Press, Oxford.

SMITH, B. N. (1972) Natural abundance of the stable isotopes of carbon in biological systems. *Bio. Sci.* **22**, 226–231.

STEINER, M. (1935) Zur Ökologie der Salzmarschen der nordöstlichen Vereinigten Staaten von Nordamerika. *Jahrb. Wiss. Bot.* **81**, 94–202.

STEWART, C. R. and HANSON, A. D. (1980) Proline accumulation as a metabolic response to water stress. In *Adaptation of Plants to Water and High Temperature Stress*, edited by N. C. TURNER and P. J. KRAMER, pp. 173–189. John Wiley & Sons, New York.

STOCKER, O. (1928) Das Halophytenproblem. *Ergeb. Biol.* **3**, 265–353.

STOCKER, O. (1970–1972) Der Wasser- und Photosynthesehaushalt von Wüstenpflanzen der mauretanischen Sahara. I–III. *Flora*, **159**, 539–572; **160**, 445–494; **161**, 46–110.

STOCKER, O. (1974) Der Wasser- und Photosynthesehaushalt von Wüstenpflanzen der südalgerischen Sahara. I–III. *Flora*, **163**, 46–88, 89–142, 480–529.

STOCKER, O. (1976) The water-photosynthesis syndrome and the geographical plant distribution in the Saharan deserts. In *Water and Plant Life*, edited by O. L. LANGE, L. KAPPEN and E.-D. SCHULZE, pp. 506–521. Springer, Berlin, Heidelberg, New York.

VOLKENS, G. (1887) *Die Flora der ägyptisch-arabischen Wüste. auf Grundlage anatomisch-physiologischer Forschungen*. Borntraeger, Berlin.

WAISEL, Y. (1972) *Biology of Halophytes*. Academic Press, New York, London.

WALTER, H. (1963) Water supply of desert plants. In *Water Relations of Plants*, Symp. Br. Ecol. Soc., edited by A. J. RUTTER and E. H. WHITHEAD, pp. 199–205. Blackwell, London.

WALTER, H. (1964) *Die Vegetation der Erde in Öko-physiologischer Betrachtung*, Bd. 1. G. Fischer, Jena.

WALTER, H. (1973) *Vegetation of the Earth, in relation to Climate and the Eco-physiological Conditions*. Transl. by Joy Wieser. Heidelberg Science Library, Vol. 15. The English Univ. Press Ltd, London and Springer, New York, Heidelberg, Berlin.

WINTER, K., TROUGHTON, J. H. and CARD, K. A. (1976) ^{13}C values of grass species collected in the Northern Sahara desert. *Oecologia* (Berl.), **25**, 115–123.

ZIEGLER, H., BATANOUNY, K. H., SANKHLA, N., VYAS, O. P. and STICHLER, W. (1981) The photosynthetic pathway types of some desert plants from India, Saudi Arabia, Egypt and Iraq. *Oecologia* (Berl.), **48**, 93–99.

ZOHARY, M. (1962) *Plant Life of Palestine*. Ronald Press, New York.

ZOHARY, M. (1973) *Geobotanical Foundations of the Middle East*, 2 vols. G. Fischer, Stuttgart.

CHAPTER 8

Soil Fauna

A. H. EL-KIFL and SAMIR I. GHABBOUR *

Department of Natural Resources, Institute of African Research and Studies,
Cairo University, Giza (Cairo), Egypt

CONTENTS

8.1. INTRODUCTION

Soil fauna studies have only been carried out during the last few decades, but it is now known that the inhabitants of the soil may play a vital rôle in its properties. Many animal phyla are represented in temperate region soils. In desert soils, however, members of only four phyla, including Annelida, are at all common. These can exist in varying soil structures and under varying edaphic circumstances. The following account of Protozoa, Nematoda and Arthropoda is based on a study of these phyla in sandy and semi-sandy soils of the Egyptian desert and in newly reclaimed areas.

Although some soil animals exhibit obvious morphological adaptations to their environment, others do not show any apparent modifications. The majority of the soil microfauna are no different from corresponding species living elsewhere. Nevertheless, in general, soil animals are adjusted to their mode of life as follows: (a) They can exist successfully in complete or partial darkness; consequently, the majority are blind, for eyes are useless in such a habitat. (b) Most soil animals are either small enough to

---

* Section 8.1 is contributed by El-Kifl, with references at the end of the section, and Section 8.2 is contributed by Ghabbour, with references at the end of that section. Before his retirement, Prof. El-Kifl was Dean of the Faculty of Agriculture, Al-Azhar University, Nasr City, Cairo, Egypt.

pass through existing soil passages or strong enough to burrow new ones. (c) Temperature fluctuations in the soil are less intense than on its surface, reducing thermal stress on its inhabitants. (d) Subterranean animals are less exposed to be preyed upon by surface-dwelling carnivores. (e) Soil-inhabitant animals can tolerate certain special problems created by their environment such as desiccation, excess of carbon dioxide, loss of oxygen content, etc.

Soil animals acquire certain morphological or physiological modifications for adjusting themselves to soil life.

1. Modification for burrowing does not depend on the period of time passed by the animal in the soil or its ability to penetrate it. Subterranean ants and termites do not possess modified legs. In general, morphological modifications in soil animals can be summarized as follows: (a) Reduction or total loss of sight. (b) Absence or suppression of accessory processes and protruding hairs or bristles. Reduction or absence of legs occurs in burrowing reptiles and amphibians. (c) Loss of pigmentation is common. (d) Anchoring devices are possessed by certain soil-dwellers, as ant-lions. (e) Flat or depressed bodies enable the animal to enter through narrow passages. Soft-bodied animals are able to squeeze themselves through cracks much narrower than their bodies.

(2) The physiological adaptations enable the soil animals to overcome numerous difficulties they encounter in their habitat. Examples of such adaptations are: (a) Adjustment of the reproductive cycles of subterranean animals to the habitat. Several devices for protecting both eggs and young are well known. (b) Uniform temperature in lower soil layers and other constant conditions permit continuous breeding. (c) The typhlosole of earthworm guts increases the digestive area to cope with the large quantities of material ingested. (d) Blind soil animals possess very sensitive olfactory and tactile receptors. (e) The water content is adjusted in the bodies of soil animals by several physiological means. Desiccation as well as flooding may be fatal. (f) Soil microfauna can tolerate drought by encysting, or by entering into a resting stage (suspended animation = anabiosis). (g) Soil animals are adapted to tolerate reduced oxygen and an excess of carbon dioxide.

Soil animals have been classified according to their feeding habits into:

1. Primary feeders which include: (a) phytophagous: which feed on green plants; (b) zylophagous: or wood-feeding animals; (c) truly saprophagous which feed on dead vegetable matter only; (d) humivorous or soil-swallowing species.

2. The secondary feeders which include: (a) Predators which attack other animals. The forms which in turn attack primary predators are called 'tertiaries' which can be attacked by 'quartaries', and so on. (b) Parasitic species which parasitize soil animals. (c) Fungivorous or fungus-eaters. (d) Low secondaries which feed on animal waste (necrophagous) or animal remains (carrion-feeders). Coprophagous animals include vertebrate faeces-eaters.

3. Aphagous forms which do not feed in the soil but are periodic or temporary soil-inhabiting animals (e.g. pupae and hibernating forms).

8.1.1. Protozoa

In a culture solution, Protozoa normally die or encyst in a temperature range of 44–54°C. Encysted forms die in a range of 70–72°C. The following experiment was carried out in Nasr City, near Cairo. Two areas were examined for the existence of Protozoa, one was pure desert sand while the other was newly-reclaimed sandy soil, irrigated for about one year with fresh water containing mud from the Nile water. The two areas were in a district where the temperature may reach 68°C on the sand surface at 14.00 h in summar. Culture media were made up and inoculated with samples from the two areas. The results were as follows:

1. In pure sand (down to 10 cm) only ciliates were demonstrated. This indicates that ciliates were more

tolerant to pure sand than other Protozoa. The following species were identified: Colpodidae: *Tillina magna* Grubber and *T. canalifera* Grubber; Metopidae: *Bryometopus sphagni* (Penard); Oxytrichidae: *Amphisiella oblonga* Schewiakoff; Vorticellidae: *Vorticella campaniola* Ehrenberg.

2. In newly-reclaimed sandy soil, the number of species was comparatively very much larger. Ciliates still dominated, followed by rhizopods and then flagellates. The existence of such populations proved that the high soil temperatures at Nasr City did not cause sterilization but only induced encystment. The species encountered were:

    (a) Ciliata Holophryidae: *Prorodon teres, Lacrymaria* sp. and *Enchelys simplex*; Amphileptidae: *Amphileptus clapardeni*; Tracheliidae: *Branchioecetes gammari*; Loxodidae: *Loxodes magnus*; Paramecidae: *Paramecium caudatum*; Colpodidae: *Tillania magna, T. canalifera, Colpoda cuculus, Colpidium colpoda, Glaucoma scintillans, Uronemopsis kenti, Philasterides armata* and *Cohnilembus pusillus*; Pleuronematidae: *Pleuronema coronatum*; Metopidae: *Bryometopus sphagni*; Oxytrichidae: *Uroleptus piscis, Urostyla grandis, Amphisiella oblonga*; Euplatidae: *Euplotes patella, Aspidisca castata*; Lagenophryidae: *Lagenophrys* sp.; Vorticellidae: *Vorticella campanula, Intranstylum invaginatum*.

    (b) Rhizopoda Amoebidae: *Amoeba* sp.; Thecamoebidae: *Rugipes bilzi*; Cochliopodiidae: *Cochliopodium pilimbosum*; Centropyxidae: *Centrophyxis constricta*; Difflugiidae: *Difflugia lebes*; Nebelidae: *Nebela collaris*; Gromiidae: *Gromia fluviatilis, Chalamydophrys minor*.

    (c) Flagellata Bodonidae: *Bodo caudatus*; Monadidae: *Monas vivipara, Tetramytus pyriformis*.

It appears from this experiment that Protozoa can exist in Nasr City desert, both in pure sand and in newly-reclaimed areas, where the sub-surface soil temperature does not rise to the sterilization level in summer (70°C). In desert conditions, Protozoa cannot survive in completely dry sand, but they can live even in water membranes covering sand granules. Daily dew, as well as rain in winter, may provide desert sand with sufficient humidity for them to survive in favourable localities.

8.1.2. Nematoda

In newly-reclaimed areas (4–6 years) of Nasr City the following nematode genera were demonstrated (El-Kifl et al., 1973): Mononchidae: *Mylonchulus, Prionchulus, Anatonchus*; Dorylaimidae: *Dorylaimus, Eudorylaimus*; Haplolaimidae: *Rotylenchoides*; Cephalobidae: *Cephalobus, Eucephalobus*; Rhabditidae: *Mesorhabditis, Rhabditis*; Monohysteridae: *Monohystera*; Axenolaimidae: *Cylindrolaimus*.

Pure sand in the desert of Nasr City contained no nematodes. Newly-reclaimed areas were contaminated by the transportation of earth and organic manure from distant sources. Irrigation water, as well as transplanted seedlings were other sources of contamination. Estimation of annual and monthly population means showed that nematode species became more numerous in proportion to the length of the reclamation period. Peaks occurred during March and April in spring, and during October in autumn. The lowest nematode means occurred during summer and winter. Newly-reclaimed areas, though having smaller nematode populations when compared with the well-established cultivated areas, developed quickly, doubling in a short period. Free-living nematode species were more common than the free-living stages of parasites.

In Tahreer (Freedom) Province, the desert sand of which has been reclaimed for more than 20 years, the following nematode genera were found in association with the roots of various plants: *Aphelenchoides, Criconemoides, Ditylenchus, Helicotylenchus, Hemicriconemoides, Heterodera, Haplalaimus, Paratylenchus, Tylenchulus, Xiphinema, Aphelenchus, Potylenchulus, Tylenchorhynchus, Melaidogyne* and *Paratylenchus*. The number of species and population sizes of the last three genera were the greatest. Sandy soils were evidently more suitable to the root-knot nematodes than other soil types.

The Nile Valley is overcrowded with its human population, and there is a demand to reclaim desert areas bordering the Valley. In a recent investigation into the sources of nematode infestations in newly-reclaimed irrigated lands (Farahat, 1976), the following results were obtained.

1. The nematode species of the following genera were found to be in association with the perennial propagation stock plants: *Aphelenchoides, Criconemoides, Ditylenchus, Helicotylenchus, Hemicriconemoides, Hemicycliophora, Haplolaimus, Meloidogyne, Paratylenchus, Pratylenchus, Rotylenchulus, Tylenchorhynchus, Tylenchulus* and *Xiphinema*.

2. Windbreaks of Australian pine trees (*Casuarina* sp.) harboured the nematode *Meloidogyne* sp. which spread from their roots to neighbouring plants.

3. Weeds of common occurrence in newly-reclaimed areas were found to be hosts of nematode species belonging to the following genera: *Heliocotylenchus, Hoplolaimus, Meloidogyne, Rotylenchulus* and *Tylenchorhynchus*.

4. Irrigation as well as drainage water acted as an important factor in nematode dispersal. The larvae of species representing most of the above-mentioned genera were able to survive in such water for long periods.

In an area in Tahreer Province reclaimed for 7 years, Oteifa and Shaarawi (1962) found that various citrus plants were susceptible to infestation with *Tylenchulus semipenetrans* Cobb, to an extent of about 40% of the trees examined. The neighbouring non-citrus trees were free of infestation.

8.1.3. Arthropoda

The arthropod soil fauna of the cultivated Nile Valley has been subject to numerous investigations, both qualitative and quantitative. The following remarks refer to the species likely to be encountered in sandy and semi-sandy areas—that is in the desert land bordering both sides of the valley.

1. Land isopods are confined to humid areas, and do not exist in sand or sandy districts.

2. Predaceous, myriapods exist in humid desert areas bordering cultivated localities and oases where they can obtain their prey. Species of Lithobiidae and Scolopendridae are examples.

3. Certain arachnids are more tolerant of arid environments than other arthropods: (a) Araneae (spiders) are predators which often find their prey in cultivated areas or meadows, but seldom in sandy areas. (b) Scorpion species prefer arid and desert environments. They can withstand wide ranges of temperature and humidity which other arthropods cannot tolerate. Species of *Buthus, Scorpio* and *Androctonus* are typical examples. In Egypt, most scorpion species live in desert or semi-desert areas bordering the Nile Valley or surrounding oases. These species represent mainly the following genera: *Buthus, Butheotus, Parabuthus, Heterometrus, Nebo* and *Turus*. Their classification depends on the morphology of the two last abdominal segments, together with the poisonous gland and its sting (see Chapter 13). (c) Pseudoscorpions are mainly found in cultivated areas where they encounter their prey among decayed organic matter. They may exist in ant nests in more arid areas. (d) Other arachnids, especially Acarina, are mainly confined to cultivated areas. In reclaimed areas, additional soil arachnids are found which may have been transported from other sources.

4. Collembola. The collembolan fauna of the Sahara is not well known, especially that of Tunisia, Libya and Egypt. In all, about 170 species are known from the countries of North Africa from Morocco to Libya (Morocco 88, Algeria 103, Tunisia 10, Libya 10). Of these, 46, or 27%, are of wide distribution, 15, or 8.8%, are Holarctic, 52 are European, 29 Mediterranean, and a number are endemic. Endemic species number 7 in Morocco, 17 in Algeria and 4 in Tunisia. About 48% of species are therefore European or southern European. Very few species of North Africa originate from the eastern Mediterranean, and only one is of Ethiopian origin, showing the barrier effect of the Sahara. In Egypt, about 15 species have been described, but recent collections at least double this number. One-third is cosmopolitan, another third European or Mediterranean: 8% exclusively Mediterranean, 8% Mediterranean/African and 1%

endemic (Thibaud, in press). A collection from Khartoum revealed about 22 species, of which 7 genera are also represented in Egypt (Greenslade, 1981). Collembola can resist drought on the soil surface, and in waterlogged soil. This helps in their dispersal. Most species are polyphagous, but some are geophagous. They are good indicators of soil conditions. Soil-dwelling Collembola can resist desiccation to some extent, but not cave-dwelling species (Thibaud, in press). Greenslade (1981) distinguishes six types of adaptation of Collembola to arid environments: (a) morphological and physiological adaptations that lead to tolerance of high temperatures and low humidity, (b) behavioural adaptations that minimize the desiccation risk, (c) resistant stages or reduction of activity for long drought periods, (d) hatching of eggs under moist conditions, and (e) very short life history.

5. Insects. Soil-dwelling insects are rare in sandy and semi-sandy areas. They include ants, beetles and ant–lion larvae. For instance, in the Omayed Biosphere Reserve soil animals are concentrated under shrubs of *Thymelaca* and *Anabasis* with higher densities on their leeward sides (from 6 to 21/m²). Lowest densities are found in summer (10/m²) and the highest in spring (22/m²). Sand cockroaches (*Heterogamia syriaca*) are the most abundant species, followed by ants (45 and 23% respectively). Predators comprise 10% of the population. Sand cockroaches are adapted by internal physiology to resist desiccation, and ants by social organization (external nest physiology). They represent two diametrically opposed survival strategies: the sand cockroach is individualistic and opportunistic, taking chances as they come and undergoing great risks in summer when they die in large numbers. Ants are careful and harvest and share seeds as they appear through the year. They feed their young till adulthood, show division of labour and a caste system, preserve nest galleries by excavating them after rain, etc. When LaFontaine gave us the fable of the ant and the grasshopper, he did not stop to ask, 'why does the grasshopper survive in spite of its frivolity?' The sand cockroach of the Omayed site represents the grasshopper. Adaptations of this species to the desert environment are seen in the accumulation of fat during the hot, dry summer months (Ghabbour and Mikhail, 1977). It is capable of maintaining a very low level of water loss by evaporation—between 20 and 40°C in dry air. Sand cockroaches are also capable of recharging their reserves of body water from ambient air with a relative humidity of 100% to as low as 75%. In nature, this possibility is available when the animals hide in the seclusion of the soil at a depth of 50 cm, or at night, when they emerge on the surface to feed on litter. Because of its shield–like dorsum and its fringe of hairs, the animal can maintain an immediate microclimate under its shield that is more favourable than the surrounding general desert atmosphere (Vannier and Ghabbour, 1983).

The basis of life of sand cockroaches is litter, but they can scarcely obtain their water requirements from dry litter. At least the deeper soil layers (30–60 cm) remain in a state of 100% relative humidity for most of the year, even in summer if winter rain is adequate. Because the soil mesofauna (a few mm in length) are highly mobile, this suitable condition of the deep layers enables them to hide there and withstand the heat of the day. Conditions at the surface become tolerable at sunset, when dew begins to form, and when temperature inversion between the upper and deeper soil layers occurs. This temperature inversion may be a factor inducing soil mesofauna to move upwards. Dew formation is sufficient to permit feeding on litter by some groups of soil mesofauna during the night hours. Such short periods of activity occur even in summer, and especially in July and August when relative humidity exceeds 90% during the night. It is so effective that sand cockroaches can have an egg-laying (reproductive) season in August, which is the hottest month of the year. The combined mechanical and chemical breakdown of litter, and the enzymes they add to the breakdown products, help to cause a burst of activity of bacterial decomposition when the rains come in autumn and the soil is sufficiently moist for continued microbial activity (El-Ayouty, Ghabbour and El-Sayed, 1978).

Although it is difficult to discern the role of soil mesofauna in litter breakdown and the decomposition of soil organic matter, as distinct from that of soil microbes (bacteria and fungi), it is possible by the use of certain statistical methods to show that, when soil moisture is depleted at a depth of 50 cm due to low rainfall, soil mesofauna populations decline, and organic matter accumulates. This indicates that soil microbes may not be as important as is widely held, in decomposing organic matter on the surface. During the same period, soil moisture at the surface remained constant at the usual values prevalent in the desert, implying that there was no change especially adverse to surface microbes (Ghabbour and Mikhail, in preparation).

8.1.4. References

EL-AYOUTY, E. Y., GHABBOUR, S. I. and EL-SAYYED, N. A. M. (1978) Role of litter and the excreta of soil fauna in the nitrogen status of desert soils. *J. Arid Environments*, **1**, 145–155.
GHABBOUR, S. I. and MIKHAIL, W. Z. A. (1977) Variations in chemical composition of *Heterogamia syriaca* Sauss. (Polyphagidae, Dictyoptera), a major component of the soil fauna of the Mediterranean coastal desert of Egypt. *Rev. Biol. Ecol. méditerr.* **4**, 89–104.
GREENSLADE, P. (1981) Survival of Collembola in arid environments: observations in South Australia and the Sudan. *J. Arid Environments*, **4**, 219–228.
THIBAUD, J.-M. (1983) Variations sur les insectes collemboles, particulièrement ceux d'Afrique du Nord et d'Egypte. *Cairo Univ. Afr. Stud. Rev.* (in press).
VANNIER, G. and GHABBOUR, S. I. (1982) Effect of rising ambient temperature on transpiration in the cockroach *Heterogamia syriaca* Sauss. from the Mediterranean coastal desert of Egypt. In *New Trends in Soil Biology*, 8th Int. Colloquium of Soil Zoology, 30 August–2 September 1982, Louvain-la-Neuve, Belgium, 441–453.

8.2. EARTHWORMS

8.2.1. 'Rain' worms in the rainless desert

It may seem strange to talk about earthworms in a book about the Sahara desert, where one might expect to find accounts of animals adapted to the harsh, hot and dry environmental conditions of the largest of all deserts. Yet, there *are* earthworms in the Sahara, either on its fringes, or in especially favoured habitats such as oases and the highlands. Even these limited areas present inhospitable conditions to all forms of life, which all have to become adapted in order to survive. The existence of permanent populations of earthworms is remarkable, the more so when we recall the name given to them in German 'Regen-würmer', or 'rainworms'.

Not only do the oligochaete faunas of these favoured localities, where there is some water, contain elements from different zoogeographic regions, but the various oases support different genera and species. The distribution of species is therefore sporadic and not uniform. Earthworms do not have a drought-resistant stage in their life-cycles and therefore require moisture throughout their lives; a few species can, however, undergo periods of quiescence or diapause when the soil temporarily dries up; but if this continues for long the worms eventually die. They may migrate from one place to another, however, especially when forced out of their burrows by excessive rain or when emerging to copulate. Their movements are very slow, but, given enough time on a geological scale, they may disperse over great distances. If the route they take is flanked by impassable terrain, then their course is linear and along a 'corridor'. They may be transported in some cases by natural agents, such as logs, floating plants, or the like, or inadvertently by human activities. It is possible in some instances to judge whether an earthworm is naturally present at a certain locality or has been transported there passively, by natural agents or by human action.

8.2.2. The origins and characteristics of earthworms

Earthworms have most probably evolved from marine ancestors whose descendants are still present today on continental shores, especially in tidal flats. These are the Polychaeta or 'bristle worms'. These worms have a flattened body divided into many similar segments. Each segment has two lateral appendages which are not themselves segmented and carry tufts of bristles. This distinguishes the Polychaetes from animals like centipedes or millipedes which also have a flattened segmented body but whose appendages are segmented (phylum Arthropoda). Polychaeta and Oligochaeta (the class to which earthworms belong) are both classes of the phylum Annelida. It is believed that, some time in the remote geological past, in coastal wetlands where fresh water merged with seawater over large flat expanses, there were ample opportunities for polychaetes to adapt to wetland conditions and later on to move to wet soil habitats so that earthworms, as we know them, now finally emerged. The sequence passes through a series of aquatic oligochaetes such as we find today in shallow or moderately deep fresh-water habitats. But earthworms, in turn, have returned in some cases to fresh water and even to continental coastal waters. In almost all continents there exist widespread but locally rare marine 'earthworms'; aquatic oligochaetes which have to burrow into tidal flats in a manner similar to that of their fresh-water relatives rather than of their polychaete ancestors. Polychaetes construct a U-shaped tube with the anterior end near one opening and the posterior end near the other opening. Because they are predaceous, they protrude their anterior ends (which are supplied with eyes, jaws and sensory organs) to catch passing prey. Aquatic oligochaetes, on the other hand, burrow into mud to eat it. They swallow the mud and digest whatever organic matter it may contain—living (bacteria, small eggs and small animals) and non-living (decaying organic matter)—as terrestrial earthworms do. They differ from the latter, however, in that they do not form extensive underground networks of burrows, but only burrow with the anterior ends of their bodies into the mud below the water surface, leaving the posterior parts of the body undulating in the water to absorb dissolved oxygen for respiration through the thin body surface. This permits diffusion and the exchange of respiratory gases between the water and the worm's blood. In some species, special extensions of the skin with very thin walls facilitate gaseous exchange. These gills are vulnerable parts of the body and are quickly withdrawn into the mud at the least sign of danger such as touch or vibration.

Earthworms have lost most of their bristles as an adaptation to life in the soil. They have also lost their eyes, jaws and sense organs. They have only retained some specialized sensitive cells on the skin. They have also lost their lateral segmental appendages and become cylindrical, tube-like, organisms tapering at both ends. They thus resemble strings or thin ropes and, in the western Sudan are called 'Hable El-Wata' meaning ropes of the soil.

Earthworms obtain the oxygen necessary for respiration from the oxygen dissolved in the water of the soil. They possess an integument that allows diffusion and the exchange of respiratory gases, as if they were actually aquatic organisms. This makes evaporation and the loss of water from the body through the body wall very easy in dry soil or air. In moist air, worms may remain alive for somewhat longer, so that they can survive long enough on humid nights to accomplish significant journeys to different localities.

Since they are hermaphrodites, earthworms reproduce by mutual exchange of spermatic fluid. Each individual secretes a tough sac (called the cocoon) which protects the eggs. This sac is deposited in the soil. After some time the young worms emerge and begin a way of life similar to that of their parents.

8.2.3. The earthworms of the Sahara

The presence of a certain group of species of earthworms in the Sahara may be due as much to their special ability to invade this region of the earth and to maintain a permanent population there, as to the opportunities they have been given in the history of the Sahara and circum-Saharan territories, to invade the desert from neighbouring regions. The presence of these species can therefore be explained by events of geological as well as of human history, and can also help to throw further light on these events. Different regions of the Sahara contain different groups of species, indicating histories that experienced different events.

In the region of the Atlas Mountains (northern Morocco, northern Algeria and northern Tunisia), as well as in western Morocco and the Western Sahara (former Spanish Sahara), one can distinguish a distinct region with species undoubtedly originating from the Iberian Peninsula and to a lesser extent from the region of Sardinia and Sicily.

According to Omodeo (1954), the Western Sahara contains only two species: the ubiquitous *Allolobophora caliginosa* and the aquatic *Eiseniella tetraedra*, both of the family Lumbricidae, which is the dominant family of Oligochaeta in European soils. Morocco and northwest Algeria have, in addition, *Eisenia rosea* (which extends to Tunisia, Libya and Egypt) and *Octolasium complanatum*, which is found in Tunisia but not in Libya or Egypt. Both belong, again, to the Lumbricidae and are found in southern Europe. Moreover, another pair of species, *Eisenia foetida* and *Octolasium lacteum*, are found in Morocco and north-west Algeria, and also occur in Europe except for Sardinia, Sicily, Calabria, Greece and Turkey. This indicates an unequivocal migration from the Iberian Peninsula across the Straits of Gibraltar, with a two-way movement along the Atlantic coast to the Western Sahara, and along the Mediterranean coast to north-west Algeria. All six of the aforementioned species occur in the Canary Islands.

Another migration route from the central Mediterranean to Tunisia is attested by the presence of *Helodrilus festai* in northern Tunisia as well as in southern Sardinia, Sicily and Calabria. Another species, *Hormogaster redii* (family Hormogastridae), occurs in northern Tunisia and north-east Algeria, as well as in Sardinia, Corsica, Sicily and western Italy. *Eophila möbii* and *Dendrobaena byblica* (occurring in Portugal) do not occur in Morocco (or rather have not been found), but occur in northern Algeria. Species of the genus *Lumbricus* have never been found in Africa, although they are present in Spain. In the eastern Mediterranean, *Lumbricus* spp. do not occur in Turkey, the Island of Rhodes, Lebanon, nor in Palestine. It may be assumed, therefore, that they are restricted to temperature conditions.

The species *Eisenia rosea* is of several different forms. It occurs in Spain and Morocco (indicating a migration across the Straits of Gibraltar) and in Sardinia, Tunisia, Tripolitania, as well as in Siwa Oasis. However, the form occurring in Siwa resembles the Sardinian form (El-Duweini and Ghabbour, 1968a), suggesting that this series of appearances may result from migration via Sicily.

Thus, the two major routes of entry into the Sahara of European earthworms have been across the Straits of Gibraltar and the Sicily Straits. Most species came from the first and went to the Atlantic coast and the Atlas; a few came from the second and travelled a small distance westwards and a much longer distance eastwards to the western edge of the Nile Delta. Examples of the latter are *Eisenia rosea* in Siwa, and *Eiseniella tetraedra* at Burg El-Arab, 53 km west of Alexandria (new record). Other species occurring in north-west Africa, and coming by the same migration route, may include *Octolasium cyaneum*.

Eisenia rosea, *Eiseniella tetraedra* and many species belonging to the genera *Bimastos*, *Dendrobaena* and *Allolobophora* occur in Palestine, and *Octolasium* occurs in Lebanon. Yet, only two species of this large array have managed to cross the Sinai Desert and establish themselves in the Nile Delta. These are *Eisenia rosea* f. *bimastoides* and *Allolobophora jassyensis* var. *orientalis*. They are not, however, found in Upper Egypt. They represent a minor penetration of the European Lumbricidae into Africa from its north-eastern corner. Thus all European earthworms which managed to enter into the Sahara remained confined to its northern rim, except one species which we have already mentioned, *Allolobophora caliginosa*. The southernmost limit of these species, apart from the last, is the Western Sahara on the Atlantic border as well as the Canary Islands. (A single record of *Eisenia foetida* in Luxor, Upper Egypt, does not significantly change this picture.) To the east they are found as far as the Oasis of Siwa and the Mariut to the west. It is remarkable that no earthworms have been recorded from a region from which a reasonably varied earthworm fauna might be expected, that is, the Gebel Akhdar of Cyrenaica.

At any rate, the line connecting the Canary Islands, the Western Sahara, the southern edge of the Atlas chains, Tripoli, the Oasis of Siwa, Burg El-Arab, and the southern tip of the Nile Delta, represents the southern limit of Lumbricidae, a typically Palaearctic family. This line may also, perhaps, represent the southern limit of the Palaearctic realm itself. Lumbricidae have proved that they could cross a sea barrier, but the desert has been too formidable. Five of the lumbricid species occurring in north-west Africa are entirely Holarctic, none is Euro-American or typically continental European, three are Palaeo-endemic-European, and two are circum-Mediterranean. The three Palaeo-endemic-European species are confined to north Tunisia and north-east Algeria. It appears that the North Saharan species, therefore, are extensions of south European species rather than of holo-European species, which explains why they can tolerate the warmer North African climate (Omodeo, 1961).

The other major source of oligochaetes inhabiting Saharan territories is from the south, from the Ethiopian realm. Here we can distinguish four routes. One is of minor importance from the Ethiopian Highlands and along the Blue Nile, but it stops at Wad Medani. The second is from Kenya, and is represented by a single species which occurs as far north as Siwa Oasis. This is *Allonais paraguayensis* (var. ?) related to *A. paraguayensis* var. *aequatorialis* of Kenya (Cernosvitov, 1938). The third is from the Lakes plateau and is represented by the genus *Gordiodrilus* which occurs in southern Sudan (several species), in Selima Oasis in northern Sudan (*G. zanzibaricus*), in Siwa Oasis in north-west Egypt (*G. siwaensis*) and in Tarhuna, near Tripoli (*G. pampaninii*). Another very closely related species, *Nannodrilus staudei*, occurs in Egypt in the Baris Oasis of the Kharga group of oases, in the vicinity of Cairo, and in the western fringe of the Nile Delta. The three genera *Pygmaeodrilus, Gordiodrilus* and *Nannodrilus* belong to the family Ocnerodrilidae. An unknown species of the genus *Pygmaeodrilus* occurs in Siwa Oasis. This belongs to a heterogeneous group of species which inhabits the regions of Lake Albert, L. Tanganyika, L. Mweru, Mt. Meru, L. Stefanic, Mt. Kilimanjaro, Ethiopia and the Zambezi and Orange Rivers (Jamieson, 1957). The northernmost locality for a species of this group (apart from *Pygmaeodrilus* sp. in Siwa Oasis) is *P. aequatorialis* of Lake Albert. The other species-group of *Pygmaeodrilus* is considered homogeneous and occupies the more restricted area of the Albert–Victoria drainage system. Thus the former group, the heterogeneous one, is more widespread and overlaps with the Ethiopian Plateau area of species, from which the species of the Blue Nile have descended, and overlaps also with the Kenya area from which was derived the *Allonais* inhabiting Siwa Oasis.

The fourth source of Saharan earthworms is West Africa. These are represented by *Chuniodrilus ghabbouri* occurring in the Dahshur marshes in the vicinity of the Dahshur Pyramid, south-west of Cairo. This species is related to a fairly large group of species of *Chuniodrilus* occurring in western Liberia, the Ivory Coast and Gambia (Jamieson, 1969). No intermediate localities for the genus are yet known. On the other hand, two closely related genera, *Nannodrilus* and *Gordiodrilus*, with a probable origin in southern Sudan (El-Duweini and Ghabbour, 1968b), are represented by a number of species found in West Africa, but of limited distribution. Those belonging to the genus *Gordiodrilus* occur in Cameroon, and those belonging to the genus *Nannodrilus* in Liberia.

The aquatic genus *Alma* is represented in Egypt by two species. The larger and more abundant, *Alma nilotica*, extends south to the region of Khartoum and, strangely enough, also occurs in Bahriya Oasis (Ghabbour, 1976). The other, *A. stuhlmanni*, has an extremely restricted distribution. In fact, it is probably represented in Egypt only by a single isolated population in the vicinity of Cairo. Yet this species is widely distributed in Gambia, Cameroon, Zaire, Uganda and Tanzania (Jamieson and Ghabbour, 1969). In the Bahr El-Ghazal and Bahr El-Jabal areas the genus is represented by *Alma emini* (Ghabbour, 1976). This species is adapted to life in anaerobic swamps by special respiratory structures at the hind end of the body. *Alma nilotica* is provided with gills on the surface of its posterior end. It is probable that *A. nilotica* is a more ancient species of Tertiary Palaearctic origin (Brinkhurst and Jamieson, 1971, pp. 156–159), which could have given rise, in the reaches of the central Sudan, to *A. emini* (Jamieson, *in* Brinkhurst and Jamieson, 1971, p. 152 and p. 789). The more widespread distribution of *A. stuhlmanni* may be due to its eurythermal (wide tolerance of temperature) character. Finally, West African species seem to have populated the Gebel Marra area and the reaches of the White Nile—as appears to be the case from a cursory collection of oligochaetes from this area.

To complete the picture, there remains the distribution pattern of *Allolobophora caliginosa* which is the only earthworm species to have penetrated deep into the Sahara, most probably by human help. It occurs in the oases of Augila, Kufra, Siwa, Bahriya, Dakhla, Kharga and Selima. It also occurs on the north-western coast of Egypt in newly irrigated lands (Ghabbour and Shakir, 1982), in Suez and in the gardens of the Red Sea Monasteries (St Anthony). Kollmannsperger (quoted by Niethammer, 1971) found it in large numbers, but of dwarf size, in the mountains of the Hoggar, Aïr, Tibesti and Ennedi. Only in the northern Sahara is this species accompanied by *Eisenia rosea* with which it has some overlap in ecological requirements.

Pheretima is a south-east Asian genus that probably came into Egypt from the 19th century onwards with the introduction of exotic ornamental plants. It requires a high moisture level and is found in gardens, where it is represented by five or six species, all in the Nile Valley and the Cairo region. It is represented in Sudan by one species, *Pheretima elongata*, which occupies the Nile Basin from arid Dongola to humid Malakal, and has penetrated to Gebel Marra (Ghabbour, 1976).

Thus, to sum up, the earthworms of north-west Africa are mostly European Lumbricidae, with a small pocket of the ancient European Hormogastridae, and do not exhibit any endemism. These European Lumbricidae have extended into the Western Sahara and the Canary Islands, and to Tripoli and Siwa Oasis in the east. They came across the Straits of Gibraltar and the Straits of Sicily. Another minor route brought two species to the Delta from the Balkans to Asia Minor and the Levant, across the Isthmic Desert of the Sinai Peninsula. Only *Allolobophora caliginosa*, the species with the highest tolerance to desert conditions, has been able to penetrate deep into the oases of the eastern Sahara and the mountain massifs of the central Sahara. Omodeo (1952), however, believes that *A. jassyensis* var. *orientalis* originated in the Caucasus and migrated thence to Romania across the Ukraine to the Island of Rhodes, across Anatolia and then to the Nile Delta from the Levant.

A rather high proportion of endemism is shown by species of Ethiopian origin, by contrast with European species. We have *Gordiodrilus pampaninii* endemic in Tripoli and *G. siwaensis* endemic in Siwa. *Nannodrilus standei* is endemic in Egypt, and *Alma nilotica* in the main Nile of Egypt and the Sudan. Endemism is also exhibited at the subspecies level of *Allonais paraguayensis* (var. ?) in Siwa. *Pygmaeodrilus* sp., also in Siwa, is likewise probably endemic. The four routes of entry of Ethiopian oligochaetes into the Sahara were all down the Nile; those from the Ethiopian Plateau did not go very far, those from Kenya reached Siwa, those from the Lakes Plateau reached both Siwa and Tripoli, and finally those from West Africa reached southern Sudan, Gebel Marra, the White Nile and the vicinity of Cairo. The genus *Alma*, probably of a very ancient (Tertiary) European origin, ascended the Nile to southern Sudan, spread to equatorial Africa, and then redescended the Nile back to the vicinity of Cairo. *Pheretima*, of Asian origin, occupies the whole Nile Valley from Cairo to Malakal and has penetrated westwards as far as Gebel Marra. These two sources reflect the zoogeographic regions of the Saharan inland waters enunciated by Timm (1980): the Holarctic (Euro-Siberian) in the north and the Ethiopian in the south and the Nile Valley.

8.2.4. The periods of earthworm migrations in the Sahara

There is no doubt that attempts to explain how earthworms arrived at the places they now occupy in the Sahara must take account of the Pluvial periods that affected parts of the Sahara at certain times in its geological history, unless human interference can suffice for an explanation in every case as it does with the ubiquitous *Allolobophora caliginosa*. This species has been described as following in the wake of the farmer, and *Pheretima* can be described as following in the wake of the horticulturist.

Criticism has sometimes been expressed at any explanation of earthworm distributions on the basis that the worms, or their cocoons, can be carried in soil adhering to plants, cattle hooves, birds' feet, etc. Although such criticism may be justified in some cases, accidental migration cannot explain the entire distribution pattern of the earthworms of the Sahara, on account of the vast distances that would have to be crossed in dry, hot air before a suitable habitat was reached. If such accidental migration had occurred in the Sahara, one would expect to find a homogeneous and unified fauna in the oases, as these are more or less similar to each other in ecological conditions, and differ only in distance from potential sources of introduction and in geological history. A species such as *Alma nilotica* is about 25 cm long and 5 mm thick; its cocoons are 6 cm long and 5 mm thick—there would be no way for it or its cocoons to be carried accidentally and remain alive over a distance of about 400 km between the Nile and Bahariya Oasis. The only possible explanation for its presence there is to postulate a connection between the Nile Valley and the oasis. The absence of *Alma* spp. from Selima Oasis, which is also relatively close to the Nile, coupled with the presence there of the East African *Gordiodrilus zanzibaricus*, shows that there must have been a connection with the Nile Valley also, but not at the same time or under the same set of conditions. Again, the closely related *Gordiodrilus siwaensis* exists in Siwa but not elsewhere in the Egyptian oases, nor does the genus exist in the Egyptian Nile although it is represented in Tripoli. Furthermore, the closely allied *Nannodrilus staudei* exists in Baris Oasis as well as in the western Delta and the region of Cairo. These are all subtropical (*Alma*) or tropical (*Gordiodrilus* and *Nannodrilus*) species, so it is interesting to note that of the twenty-two genera of mosses found in Egypt, only one is of tropical origin—the genus *Philonotis* (Imam and Ghabbour, 1972). This occurs in Bahariya Oasis, Fayum, Giza (west of Cairo), and the western Delta. But the remarkable thing about this genus is not only that it is the only tropical genus of the Egyptian moss flora, but that it is the only one that harbours earthworms among its root-like structures, in all the localities in which it occurs. One of these earthworms is *Nannodrilus*. Thus, the association between earthworms and tropical plants is old and must have been effective in transporting earthworms downstream in matted vegetation whenever conditions allowed.

Natural movement of earthworms over the Sahara (with the exception of the anthropochorous species of *Allolobophora caliginosa* and *Pheretima*) must have occurred gradually through wet soil during Pluvial periods.

In regard to transportation by birds, it has been suggested that cocoons may cling to birds' feet or feathers. Löve (1963) remarks that 'wild birds usually preen themselves meticulously before taking off on a flight, and it seems especially so before a migratory flight (Hochbaum, pers. comm.)'. Omodeo (1963) adds that earthworms and their cocoons are found underground, and have little chance of getting stuck to the feet of most birds. Migratory birds use the oases of the Sahara as resting stations during their travels, yet these oases have distinct earthworm faunas.

Migration of earthworms in the Sahara, even in Pluvial periods, was limited to the Mediterranean coast, the Nile Valley, and the oases adjacent to the coast or to the Nile. Such migrations seem unlikely during the post-Pleistocene moist spells which were of limited ecological significance, but rather took place in the earlier Pleistocene Pluvial phases. These phases were earlier than the dry mid-Würm period which must have left *Eisenia rosea* isolated in Siwa, together with the Ethiopian *Gordiodrilus*. The absence of the latter from the Delta (which must have been the source of the Siwan individuals) may be due to the temporary submergence of the Delta during a part of the early or mid-Holocene. When the Delta re-emerged, the western coast was desiccated, so that reunion of the earthworm faunas of the Delta and Siwa was not possible. The Delta was repopulated by fresh introductions from the south. In historical times, the basin irrigation system of Upper Egypt and Nubia obliterated the indigenous fauna, which was partially preserved in the marshes of the Delta and the vicinity of Cairo.

To return to Siwa, Jamieson (1957) postulated that *Pygmaeodrilus* evolved in the Lakes Plateau and then spread from its original home after the formation of the Albertine Rift in the Middle Pleistocene (*ca.* 200,000 years B.P.; Said, 1981). Thus the *Pygmaeodrilus* must have reached there during a Pluvial subsequent to that Rift, the Abbassia Pluvial (probably corresponding to the Riss/Würm interval) or the later Mousterian/Aterian (Said, 1981). The Siwan worms of tropical African origin must have entered during one or both of these Pluvials. For these taxa to advance along the coast as far as Tarhuna (*Gordiodrilus pampaninii*), rainfall must have been at least 500 mm. Pickford (1937) and Lee (1959) assume this is the limit for indigenous earthworms, depending upon temperature. In southern Darfur, earthworms exist in areas of temporary rainpools at the 600–700-mm isohyets, where the mean annual temperature is 28–29°C. The present mean annual temperature on the Mediterranean coast is 25°C; there is winter rainfall, so that its effectiveness is enhanced. Butzer (1958) assumes that mean annual temperature on the coast was 4–5°C lower in the Würm period, so that conditions approached the 500 mm/21°C found to limit earthworms in South Africa (Pickford, 1937).

It can be assumed that rainfall over the coast in the Abbassia and/or the Mousterian/Aterian Pluvials was 500 mm; that is 400 mm above present average, and hence 400 mm (or about this amount) further inland (a matter of 20–30 km). Calcium carbonate concretions at Bahig (23 km south of the coast and 42 km west of Alexandria), at a depth of 6 metres, and a thickness of 1–2 metres, indicate this amount of rainfall (G. Aubert, pers. comm., 1981). The site of Tripoli might have had 800 mm (it has 400 mm now). Dakhla, Kharga and Selima Oases could have had less than 300 mm, since they do not support the moisture-loving genera that reached Siwa Oasis and Tripolitania. Bahriya Oasis must have had a wet connection with Fayum and the Nile, since it has the aquatic *Alma nilotica* (probably from a Tertiary connection). Khartoum could have had 400 mm above present rainfall of 163 mm. Butzer has estimated that, during the Early Würm Pluvial, there may have been an annual precipitation of over 300 mm for Kurkur Oasis (70 km south-west of Asswan) during that Pluvial (Butzer, 1964).

Only those oases near enough to the Nile could receive new species during the Neolithic wet phase. Thus Kharga acquired its *Nannodrilus* and Selima its *Gordiodrilus*. Therefore, at various periods during the past half million years, the dryness of Egypt's climate, the submergence of the Delta, and basin irrigation, formed barriers against the migration of earthworms across or along the Sahara.

8.2.5. Adaptations of earthworms in the Sahara

The adaptations of earthworms living in the Sahara must be efficient since they are able to exist there, in spite of their known indispensable need of moisture. They have attracted attention for this reason, the common rather than the rare and restricted species, which are difficult to use as experimental animals. *Allolobophora caliginosa*, the most successful species in the Sahara, actually shows greater adaptations than other species involved, including *Pheretima californica*, which is successful in gardens.

When both species are placed in cans covered by wire gauze to slow down evaporation and are left to be air-dried, *Allolobophora* loses 7% of its body water during a day, against a loss of 27% by *Pheretima*. By the second day, *Allolobophora* has lost 17% compared with 44% by *Pheretima*. *Allolobophora* dies within 7 days after a loss of 52–62% body water, but *Pheretima* dies within 3–4 days after a loss of 62–68% body water (El-Duweini and Ghabbour, 1968c).

There is a marked effect of soil water content in two types of soil, clay and sand. It is clear that *Allolobophora* can survive for longer periods in both soils and at a lower moisture content than can *Pheretima* (El-Duweini and Ghabbour, 1968c).

The effect of temperature on the two species has also been investigated (El-Duweini and Ghabbour, 1965a). It was found that *Allolobophora* has a wide temperature preference range (2 to 37°C), against 26–35°C for *Pheretima*. Of 571 *Allolobophora* individuals 1.2% chose to remain at 37°C, without ill effects, but none of 128 *Pheretima* individuals. One *Pheretima*, however, chose 36°C. *Allolobophora* also has quite a high lethal temperature. Two individuals survived a maximum of 42°C, while the maximum for *Pheretima* was only 39°C. These upper tolerance limits are significant when soil temperature rises in the hot dry summer. Soil temperature at Inchass (in the Delta) in farmland, at 10 cm depth, reached a maximum of

39.2°C in the period 1976–1980 at 1200 h in August 1979 (unpublished data, Meteorological Dept., Cairo). Tolerance to high temperatures can be further improved in earthworms by acclimation; that is, it increases with gradual seasonal increase of ambient temperature.

Tolerance to an alkaline pH is high in both species; both can live for more than 24 hours at pH 9.16: this is important in the alkaline soils of the Sahara. *Allolobophora* can tolerate 1.3% sodium chloride for 24 hours, sodium carbonate at 0.3%, and sodium bicarbonate at 1.0% for the same period. The figures for *Pheretima* are 0.8–1.1, 0, and 0.1% for the same salts respectively (El-Duweini and Ghabbour, 1964). These high tolerance limits go far to explain the ubiquity of *Allolobophora caliginosa* in the Sahara. Furthermore, this species undergoes diapause when the soil dries up (El-Duweini and Ghabbour, 1968c).

8.2.6. Numbers and biomass of earthworms in the Sahara

The population density and biomass of earthworms in Saharan localities, where favourable conditions prevail, are found to be comparable with those in other countries, as shown by El-Duweini and Ghabbour (1965b) in the Beheira Province (western Delta), Ghabbour and Shakir (1982) in Mariut (Mediterranean coastal desert west of Alexandria) and by Kollmannsperger (quoted by Niethammer, 1971). In Beheira Province, density varied from 4 to 264 ind./m² in desert fringe localities with moderately moist soil. In Mariut, density varied from 0.2 to 28.6/m² in reclaimed soils, 4–12 years under cultivation. Biomass in Beheira soils varied from 4 to 88 g/m², and at Mariut from 0.1 to 1.2 g/m². The lower figures of population and biomass of earthworm populations in Mariut than in Beheira, although the former are derived by invasion from the latter, show that, in spite of adaptations, desert conditions are still a major problem of survival and establishment. The soil has to undergo a longer period of cultivation until normal populations can be maintained in a healthy condition. Kollmannsperger found 1000 ind./m² *Allolobophora caliginosa* in Tazeruk in the Hoggar, a remarkable density of dwarf worms which must represent a very old colony.

8.2.7. References

BRINKHURST, R. O. and JAMIESON, B. G. M. (1971). *Aquatic Oligochaeta of the World*. Oliver & Boyd, London. 860 pp.

BUTZER, K. W. (1958) Quaternary stratigraphy and climate in the Near East. *Bonner Geogr. Abbs.* **24**, 1–157.

BUTZER, K. W. (1964) Pleistocene palaeoclimates of the Kurkur Oasis, Egypt. *Canad. Geogr.* **8**, 125–140.

CERNOSVITOV, L. (1938) Oligochaeta. *Mission Scientifique de l'Omo. Mus. Natl. d'Hist. nat.* **4**, 255–318.

EL-DUWEINI, A. K. and GHABBOUR, S. I. (1964) Effect of pH and of electrolytes on earthworms. *Bull. Zool. Soc. Egypt.* **19**, 89–100.

EL-DUWEINI, A. K. and GHABBOUR, S. I. (1965a) Temperature relations of three Egyptian oligochaete species. *Oikos*, **16**, 9–15.

EL-DUWEINI, A. K. and GHABBOUR, S. I. (1965b) Population density and biomass of earthworms in different types of Egyptian soils. *J. appl. Ecol.* **2**, 271–287.

EL-DUWEINI, A. K. and GHABBOUR, S. I. (1968a) The zoogeography of oligochaetes in north-east Africa. *Zool. Jb. Syst.* **95**, 189–212.

EL-DUWEINI, A. K. and GHABBOUR, S. I. (1968b) The geographical speciation of north-east African oligochaetes. *Pedobiologia*, **7**, 371–374.

EL-DUWEINI, A. K. and GHABBOUR, S. I. (1968c) Nephridial systems and water balance of three oligochaete genera. *Oikos*, **19**, 61–70.

GHABBOUR, S. I. (1976) The faunal relations of Oligochaeta in the Nile Basin. In *The Nile, Biology of an Ancient River*, edited by J. Rzòska and Dr W. JUNK. B.V. Publ., The Hague, 117–125.

GHABBOUR, S. I. and SHAKIR, S. H. (1982) Population density and biomass of earthworms in agro-ecosystems of the Mariut coastal desert region in Egypt. *Pedobiologia*, **23**, 189–198.

IMAM, M. and GHABBOUR, S. I. (1972) A contribution to the moss flora of Egypt. *Bot. Notiser.* **125**, 518–522.

JAMIESON, B. G. M. (1957) Some species of *Pygmaeodrilus* (Oligochaeta) from East Africa. *Ann. Mag. nat. Hist.* (12), **10**, 449–470.

JAMIESON, B. G. M. (1969) A new Egyptian species of *Chuniodrilus* (Eudrilidae, Oligochaeta) with observations on internal fertilization and parallelism with the genus *Stuhlmannia*. *J. nat. Hist.* **3**, 41–51.

JAMIESON, B. G. M. and GHABBOUR, S. I. (1969) The genus *Alma* (Microchaetidae: Oligochaeta) in Egypt and the Sudan. *J. nat. Hist.* **3**, 471–484.

LEE, K. E. (1959. *The Earthworm Fauna of New Zealand*. D.S.I.R. Bull. **130**, 486 pp.

LÖVE, D (1963) Dispersal and survival of plants. In *North American Biota and Their History*, edited by A. LÖVE and D. LÖVE, pp. 189–205. NATO Publ., Oxford.

NIETHAMMER, G. (1971) Die Fauna der Sahara. In *Die Sahara und ihre Randgebiete*, edited by H. SCHIFFERS, pp. 499–587. Weltforum Verlag, München.

OMODEO, P. (1952) Particolarita della zoogeografia dei lombrichi. *Boll. Zool.* **19** (4–6), 349–369, 5 pls.

OMODEO, P. (1954) Problemi faunistici riguardanti gli Oligocheti terricoli della Sardegna. *Atti Soc. Toscana Sci. Nat., Suppl.* **61**, 1–15.

OMODEO, P. (1961) Le peuplement des Iles méditerranéennes et le problème du l'insularité. *Colloque CNRS, Banyuls-sur-Mer, 21-27/9/1959*, pp. 127–133, éd. CNRS, paris.

OMODEO, P. (1963) Distribution of the terricolous oligochaetes on the two shores of the Atlantic. In *North Atlantic Biota and Their History*, edited by A. LÖVE and D. LÖVE, pp. 127–151. NATO Publ., Oxford.

PICKFORD, G. E. (1937) *A Monograph of the Acanthodriline Earthworms of South Africa.* Cambridge Univ. Press, 612 pp.

SAID, R. (1981) *The Geological Evolution of the River Nile.* Springer-Verlag, New York, Heidelberg, Berlin. 151 pp.

TIMM, T (1980) Distribution of aquatic oligochaetes. In *Aquatic Oligochaete Biology*, edited by R. O. BRINKHURST and D. G. COOK, pp. 55–77. Plenum Press, New York and London.

CHAPTER 9

Temporary and Other Waters

J. RZÒSKA

6 Blakesley Avenue, London W5

CONTENTS

TEMPORARY waters are derived directly or indirectly from rain, and their existence, extent and duration are governed by climatic factors. Their life phenomena have attracted the attention of scientists in all continents. In the Sahara, they are found from Mauritania to the Nile Valley. A map of rainfall zones of the Sahara (Fauré, 1966, 1969) shows winter rains on the Mediterranean coast. In the south, a dense configuration of rainfall zones is caused by the summer and autumn monsoon. Figure 9.1 has been redrawn from a paper by Fauré (1967) describing the Quaternary lakes which existed in the middle of the Sahara, and these are indicated by numbers.

9.1. PAST AND PRESENT

Geologically speaking, the Sahara is a young desert. As the last phase of the Palaeoclimatic changes and fluctuations, a state of aridity had set in by the third millenium B.P. Before that, the Sahara was an area with rich populations of wild animals and thriving life. A magnificent human documentation is contained in a book by Henri Lotte (1958) who has reproduced many of the rock drawings and frescoes made by Man especially in his Neolithic phase (4000–3000 B.C.). These frescoes show a great variety of people and illustrate their ways of life: swimming, and hunting African game, including hippos, and herding large numbers of cattle—a veritable fountain of ecological information. Subsequently, Hugot (1974) assembled all the data about human colonization of the Sahara in a book with the evocative title 'The Sahara before the desert'. By about 2000 B.C. the country began to dry up (Wickens, 1975). The vegetation disappeared during the following centuries, and the process took place right across the whole of northern Africa as documented by rock drawings in the Nile Valley, by traces of Neolithic agriculture, and later by historical

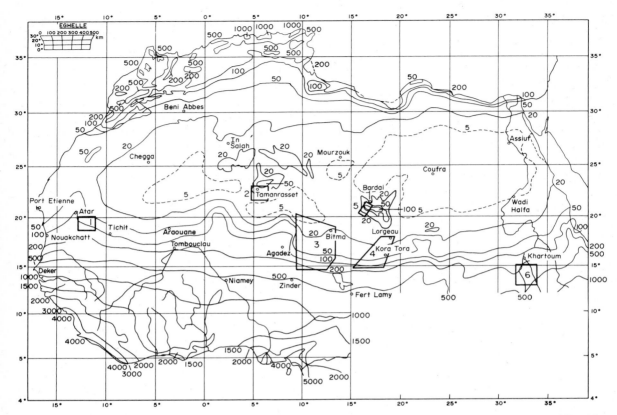

FIG. 9.1. Rain chart of Northern Africa after H. Fauré (preprint 1967). Note the location of Quarternary lakes in the middle of the Sahara.

documentation in Egyptian monuments. Today, rainfall varies between 50–30 mm per year in the middle of the Sahara, slightly increasing in the mountains of the western Sahara, the Atlas, Adrar, Hogar, Tassili, Ennedi and also on Jebel Marra in the Sudan.

In addition to the rain data (isohyets) shown in the map (Fig. 9.1) there occur irregular outbursts of rain, the freak storms which have been reported by French observers (see also Cloudsley-Thompson, 1977). These may cause parts of the desert to become green as millions of dormant seeds, especially of grasses, begin to sprout. I have observed such an occurrence personally in the Nubian desert north of Khartoum: a great area was green, and this was commented upon by the local population. On the other hand, there are places in the Sahara where rain has not fallen for up to 13 years, adding to the severity and complexity of biological conditions.

Of the former richness of waters, as indicated by the locations of lakes in Fauré's map, very little remains, and what does has been the object of investigation—even Lake Chad has recently undergone drastic changes, diminishing in size and becoming overgrown with vegetation in large parts.

Our knowledge of the waters and especially of rainpools in the western Sahara comes from collections made during the last decades by passing biologists. The most important of these are the work of that doyen of Saharan exploration, Theodore Monod. Other collectors include Dussart, Brehm, Blanc and d'Auberton. From their collections no comprehensive picture of the remaining Saharan waters emerges but a characteristic fauna from rainpools has been found in Mauritania and other places by Monod. The most extensive investigations have been made by H. Gauthier who, from 1928 until 1954, published a series of important papers. In the paper of 1951 on the Senegal, Gauthier concluded that there are two types of waters in the Sahara, permanent waters in sheltered positions like rocky crevices, and *thalwegs*. The latter harbour a ubiquitous and circumtropical fauna without a trace of the characteristic fauna of rainpools. In 1954 Gauthier devoted a monograph to one of the inhabitants of both permanent and temporary waters of Africa, the Cladoceran *Moina dubia* which forms races with different physiology.

Temporary waters are present in all continents and impose a selective influence on the life in them, depending on the factors mentioned above. Thus, in temperate climates, temporary waters last much longer, but further south become shortlived lasting, when observed, for 7 to 20 days. These data come from detailed observations in the Nubian desert, in the eastern part of the Sahara, as will be related later. No detailed observation exists for the western Sahara where the limited data is derived from collections by numerous observers. These collectors did not stop to observe the sequence of events and, consequently, we have to rely only on the evidence of the appearance of the particular animals which are characteristic of those transient waters.

The most conspicuous components of the fauna of temporary waters are the euphyllopods because of their appearance, which is different from that of other Branchiopoda. According to various sources, these animals are of ancient origin and fossils have been found in Cambrian and Silurian deposits. They seem to have been a conservative group, preserving their shapes, structure and habits for thousands of years.

The eastern part of the Sahara in Egypt and Sudan is bordered by the Nile Valley and seems to be an old part of the desert. Palaeoclimatic investigations have been extensive here, and are partly mentioned in Rzòska (1976, chapters 4a, 4b and 5). Work on the past history of the Sudan is continuing, and will doubtless reinforce the conclusion that the Nubian desert shares the history of desiccation with the western Sahara. Shifts of the rainfall zones have been collated by Wickens (1975). The present state of the eastern Sahara is biologically documented by its vegetation zones (Fig. 9.2). Whereas the western desert is delimited in the south by the presence of the ocean and shows a dense configuration of vegetational zones, aridity and rainless months reach further south in the Sudan. It is, therefore, not surprising that the characteristic fauna of rainpools has been found from Egypt westwards to Kordofan.

The present state of this region and, indeed, of the whole northern Sudan is a result of long-term changes since 20,000 B.P. described by Wickens (1975) and partly summarized in Rzòska (1976, pp. 33–45). It seems that, as in the central Sahara, desert conditions began to set in about 2000 B.C., and are recorded in ancient Egyptian inscriptions. As in the central Sahara, the great African fauna disappeared from the drying savannah and retreated southwards.

The most important documentation for these waters comes from the numerous publications of Gauthier (1928–1954) already referred to. These were mentioned in a paper on rainpools round Khartoum (Rzòska, 1961). More recently Dumont (1978–9) published the results of collections in numerous localities

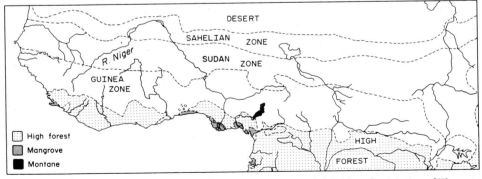

Fig. 9.2. Map of vegetation zones of the southern Sahara, reflecting the present rainfall. Note the situation of Khartoum where the Sahel merges with the Nubian desert.

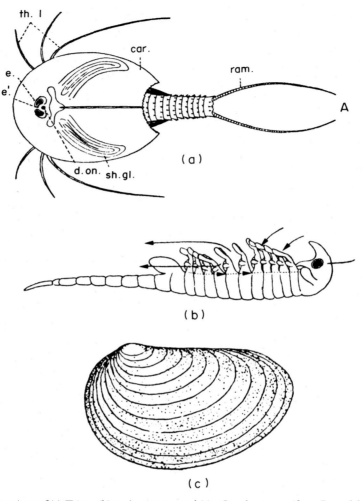

FIG. 9.3. Drawings of (a) *Triops*, (b) an Anostracan and (c) a Conchostracan (from Borradale, 1951).

from Mauritania to Egypt. From these publications it emerges that two types of waters exist in the Sahara. Gauthier has defined them as (a) permanent waters and (b) temporary pools. Dumont's work confirms this: most of the waters that he studied were of the permanent type, with a fauna of either widely distributed or circumtropical species. In a few cases, however, Dumont found indications of the second type of faunal assembly.

It has been said already that animal life in temporary waters is restricted by their short duration. Their outstanding inhabitants are euphyllopods composed of three groups of Crustacea all exhibiting numerous leaflike appendages which are used for breeding. The three groups are the Anostraca, Conchostraca and the Notostraca. Figure 9.3 shows the appearance of these animals which may not be familiar to readers in Britain where they are either rare or appear only in unusual circumstances. The dorsal shield covering these crustaceans is most developed in the Notostraca. The Anostraca are the most widespread and the many species exhibit adaptation to a larger range of conditions than do the other two groups. G. Fryer (*in lit.*) has drawn my attention to carnivorous forms amongst the more numerous filter feeders. The Conchostraca, enclosed by mussel-like extensions of their bodies, are probably mainly filter feeders, whereas the Notostraca are scavengers and carnivores. This last group shows zoogeographical and climatic distribution. The genus *Lepidurus* is northern and may even occur in Arctic conditions, whereas the genus *Triops* is confined to warmer waters, especially in arid regions and hot climates. Many references to the distribution of euphyllopods are contained in the paper by Rzòska (1961) and will not be repeated here. With the extension of hydrobiological research in tropical countries knowledge of their biology has been enlarged. In fact, the observations made in rainpools of hot climates have revealed an extraordinary degree of adaptation. All these animals are drought resistant in their resting stages.

The distribution of euphyllopods and probably of the other groups is worldwide. They have been found not only in the Sahara but in Arabia, Jordan, Palestine and Iraq. Some of the euphyllopods, especially Anostraca, appear in saline waters in some parts of the world. Such a distribution can only be achieved by resting stages—eggs enclosed in hard shells, cysts, and other adaptations suitable for distribution by wind. This is the only explanation possible for their enormous distribution. The photograph (Fig. 9.4) illustrates the hard-baked mud residue of a dried-up rainpool in the Sudan. The cracked surface of this pool withstood 8 months of total drought and a soil temperature reaching 60–80°C. One can well imagine the hot winds of arid regions sweeping over these remains and carrying, amongst the dust, the resistant stages of the fauna. Even more astonishing is the rich development of life in these short-lived pools. In addition to euphyllopods, other animals participate in the life-cycle of these waters, especially the cladocerans *Moina* spp. and the copepod, *Metacyclops minutus*, which appears in great masses. In the Azraq Oasis of the Jordanian desert, a much richer fauna has been found by myself, and by Loffler and Bonomi (1966). Their study describes a comprehensive collection of the animals which appear in large pools after the winter rains.

Henri Dumont (1979) has recently published an account of his exploration of the waters in and round the Sahara. This is written in Flemish. It confirms Gauthier's opinion that two types of waters exist, but Dumont does not distinguish strictly between the fauna of temporary waters and that of permanent waters (of which he gives many photographs). Temporary waters are usually formed by rain, which is scanty and irregular in the Sahel and the Sahara, and only on a few days after the rain has fallen can the fauna so characteristic of temporary waters be found. Most of Dumont's results, therefore, reflect the ordinary freshwater fauna: he has recorded twenty-eight species of fishes, thirteen species of amphibia and six of *Daphnia* amongst the 220 species of freshwater animals collected. Judging from the Nubian Sahel and Sahara there must be many temporary waters in the Nigerian Sahel and Sahara. On pp. 148–149 of his thesis, a number of localities are listed in which the typical fauna of temporary waters is mentioned. On request, Dr. Dumont sent me the following explanation.

> As to your question concerning the Euphyllopods. Not all samples have been processed, but I can tell you in Mauretania and Rio de Oro, we found *Streptocephalus torvicornus* in semi-permanent gueltas (i.e. waterbodies that dry up on the average only once per several years, but evidently have no fish predators). In the Hoggar and Tassili-n-Ajjer, *Branchippus schaeferi* was frequent, but it was found before all in more ephemeral waters, often accompanied by *Triops granarius* and some, not yet identified, Conchostraca. Remarkably, Theodore Monod reports *B. schaeferi* from similar (ephemeral) biotopes in Mauretania as well, but I do not recall having found it there myself. Evidently, Monod has spent much more time in Mauretania than I and found water in places that I have seen only completely dry.

9.2. TEMPORARY WATERS IN THE SUDAN

At 35–36°N and 32°E, Khartoum lies on the fringe of the Sahel Savanna and the Nubian desert. These form part of the Sahara but, in the Sudan, the aridity is not influenced by the Atlantic as in West Africa, and arid lands extend from southern Egypt to Kordofan. Temporary waters in this region are subjected to high temperatures, violent winds with distinct names (*Hamsin, Habboub, Hamrattan*) and scanty irregular rainfall. Khartoum was a most convenient place in which to investigate temporary waters, because they were just outside the town and therefore allowed a day by day record of events to be noted. The climate

FIG. 9.4. Photographs of Khartoum rainpools. (a) Size of rainpools with figures in front. (b) Dried-up rainpool at Khartoum 1957, the cracked mud represents a depository of dead organisms and dust: 10 months will pass before the pool is filled again.

of Khartoum is characterized by a spell of 8 months aridity and a rainy season which may last from July to October, with four to ten precipitations. From 1900 to 1957 records show a variation in annual rainfall from 48 to 340 mm, but this figure only underlines the great fluctuations in certain years. Normally one would expect an average rainfall of 160–180 mm. Only precipitations above 15 mm create rainpools; lesser showers have no effect and evaporate quickly. Investigations (Rzòska, 1961) were conducted casually from 1954 to 1957—in the last year with more concentrated effort. There are no other standing waters in the whole of the northern Sudan and southern Egypt, other than those created as rainpools in depressions of the flat expanses of the Nubian desert as far as the dry savannah of Kordofan. The rainpools observed had dimensions of several hundred square metres and a maximal depth of 50 cm. Once formed, winds, dry air and an enormous rate of evaporation set in to restrict their duration to 7–15 days. Added to this, the soil heated to 60°C or even more and had a consistency, at least round Khartoum, of clay with a cover of sand. Air temperatures in the season of rain vary from 28 to 41°C, but may reach even higher. One can recognize last year's rainpools by their depressions, filled with cracked sediments. These fill every year. In 1957 work allowed me to follow events more closely. In that year the first substantial rain fell on the 11 August and the rainpools were established. This first series lasted from 11–24 August (14 days). On 25 August the pools were dry and remained so until 29 August (5 days). On 29 August rain fell from 0400 h and lasted until 0700 h, filling the pools once more. This second series of pools lasted until 8 September, followed by a short dry period of half a day. The third series lasted from 9 September until 17 September. Further light showers failed to re-create any pools and the rainy season was over, having lasted altogether 37 days.

The proximity of pools to my residence allowed easy inspection. Two methods were used to investigate the appearance of life: net samples, and a 1-litre jar dipped into the free water. This last method gave results only for two organisms, which live in the free water and are numerous enough to be recorded: *Moina dubia* De Guerne & Richard and *Metacyclops minutus* (Claus). Their astonishing numerical development and rapid breeding has been recorded (Rzòska, 1961).

The fauna of these Khartoum rainpools was found to consist of:

Anostraca:	*Streptocephalus proboscideus* Frauenfeldt
	Streptocephalus vitreus Brauer
	Branchipus stagnalis L.
Conchostraca:	*Eocyzicus Klunzingeri* (E. Wolf)
	Eocyzicus irritans (E. Wolf)
	Leptestheria aegyptiaca Daday
	Limnadia spec.
Notostraca:	*Triops granarius* (Lucas)
	Triops cancriformis (Bosc)
Cladocera:	*Moina dubia* De Guerne & Richard
Copepoda:	*Metacyclops minutus* (Claus)
	Metadiaptomus mauretanicus Kiefer & Roy

In addition, two rotifers *Asplanchna* and *Pedalion* appeared in masses, nematodes, Protozoa swam round, midge larvae appeared and other insects were frequent visitors.

9.2.1. Water characteristics

In 1955 J. F. Talling provided me with an analysis of waters in some of the rainpools (Rzòska, 1961) and here is a summary of this:

pH varied from 8.2 to 9.3;
alkalinity (10^{-4} N) 21.6–66.0;
phosphates (PO_4P) in mg/l 0.11–2.4; silica (SiO_2) in mg/l 12–24;
calcium in mg/l 4.8–18.7; nitrates and Cl were represented in traces.
oxygen as mg/l was 4.3–7.1, and as saturation ranged from 51–99%.
Conductivity and temperatures were also noted but most extensively in 1957 (Rzòska, 1961) (Fig. 9.5).

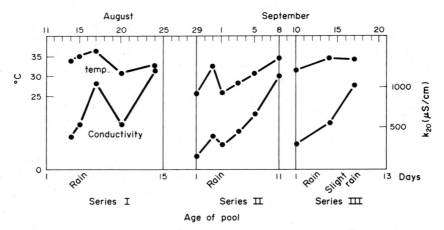

FIG. 9.5. Graph of temperature and conductivity in Khartoum rainpools. The rainwater collected had a temperature of 24°C and conductivity of 35.

The most spectacular animals of rainpools are undoubtedly members of the genus *Triops*, and it is no wonder that its appearance has elicited observations on its biology (Fig. 9.6). Munro Fox (1948) regarded the finding of *Triops* in Britain as an event worthy of a special publication. He determined the presence of haemoglobin which gives the animal a brownish–reddish colour. In Khartoum, observations on its lethal temperature and orientation responses were made by Cloudsley-Thompson (1965, 1966). In these experiments, adult *Triops* died at 34°C, and the animals were found to be photonegative. An important discovery was made by Carlisle (1968) on the resistance to heat of the permanent eggs (ephyphia). These were subjected to temperatures near boiling point and only then failed to develop. The soils of the Sudan reach 80°C so the survival of the species depends on the resistance of its eggs to drought. The spectacular appearance of *Triops* has even produced sensational headlines like those in the *Gulf Weekly Mirror* (Northern Gulf Edition, 25 April 1976)—TRIOPS INVADE! This sudden appearance was recorded in Bahrain—much to the astonishment of the local inhabitants.

9.2.2. Biology

A considerable amount of biological information is given in Rzòska (1961), especially on the biology of animals of temporary waters in Europe. Some few contrasts of data between Khartoum and European conditions may be mentioned. *Triops* appeared in Khartoum on the seventh day of pool existence, in Europe after 2–3 weeks; its life span at Khartoum was limited to 20 days of maximal pool existence: in Europe it lasted 2–3 months. Conchostraca appeared at Khartoum in 3 days, were mature and egg bearing in 5 days. In Europe they appeared in 20 days and lasted for many weeks. *Moina* appeared full grown in 72 hours and bred on the third and fourth day, first parthenogenetically and also by epphyppia. *Metacyclops*

FIG. 9.6. Drying desert pool with stranded *Triops granarius* and tadpoles (photo: J. L. Cloudsley-Thompson).

was full grown in 48 hours and breeding immediately. Numbers of organisms developed rapidly. Up to 460 *Moina* per litre were counted and up to 800 *Metacyclops*. To this must be added 1700 juvenile stages, specially naupliae.

9.2.3. Survival

If one looks at the photograph of the dried-out rainpool, and recalls that the soil may heat up to 60–80°C, the question presents itself as to by what means rainpools are repopulated each year, very often in the same place. It is obvious that all of the regular inhabitants must form drought-resisting resting stages which are embedded with the fine cracking mud visible in the photograph. These resting stages must have a much higher heat tolerance than their living representatives. At Khartoum the lethal temperature of adult *Triops* was assessed by Cloudsley-Thompson at 34°C—the resistance of eggs was found experimentally by Carlisle (1968) as nearing 100°C. I here found Conchostraca swimming freely in a pool on tarmac with a water temperature of 41°C, but the eggs must be much more resistant.

9.2.4. Food relations

Only once was there a bloom of algae (*Oscyllatoria*), and the question arises where does the food supply for so many animals come from? *Triops* is a scavenger and carnivore: the Conchostraca are filter feeders and so are the Anostraca. *Asplancha* seems to be a carnivore and *Metacyclops* probably grasps particles. Speculations can only be made as to the food chain in rainpools with bacteria and Protozoa forming the basic supply. The question remains open and is worth close examination.

9.3. GENERAL REMARKS

It is not only the food-supply problem, but also other extraordinary phenomena in rainpools which require an answer. Why are only few selected groups of animals attracted to and able to persist in rainpools? And why do they not appear in permanent waters? What is the inner drive or 'trigger' mechanism which releases the cycle of appearance, breeding and survival each year? What governs the acceleration of growth and reproduction in these conditions? What induces the race of *Moina* inhabiting the pools to produce strong resting eggs whereas, in permanent waters, as far as observed in Africa, these are produced very rarely? The extraordinary dynamism of the populations of *Moina* and *Metacyclops* are illustrated by the following data from a large table in Rzòska (1961). *Moina* appeared after 72 hours as 208 specimens per litre of which 58 females were breeding. *Metacyclops*, in the same 72 hours, reached 205 specimens. The maximum numbers recorded per litre were 460 *Moina* and 838 *Metacyclops*.

ACKNOWLEDGEMENTS

I am grateful to the following persons for help: J. F. Talling, FRS, J. Fryer, FRS, Professor J. Green and H. Dumont. Miss J. Williams has patiently recorded my dictation.

REFERENCES

Numerous references have been given in Rzòska (1961) which are not repeated here.

CARLISLE, D. B. (1968) *Triops* (Entomostraca) eggs killed only by boiling. *Science*, **161**, 279–280.
CLOUDSLEY-THOMPSON, J. L. (1965) The lethal temperature of *Triops granarius* (Lucas) (Branchiopoda: Notostraca). *Hydrobiologia*, **26**, 424–425.
CLOUDSLEY-THOMPSON, J. L. (1966) Orientation responses of *Triops granarius* (Lucas) (Branchiopoda: Notostraca) and *Streptocephalus* spp. (Branchiopoda: Anostraca). *Hydrobiologia*, **27**, 33–37.
CLOUDSLEY-THOMPSON, J. L. (1977) *Man and the Biology of Arid Zones*. Arnold, London.
DUMONT, H. (1979) *Limnologie van Sahara en Sahel*. Thesis Rijksuniversiteit Gent Fakultiet der Wetenschappen, pp. 1–557.
FAURÉ, H. (1967) *Lacs Quarternaires du Sahara*. International Association for Theoretical & Applied Limnology (SIL), Communications No. 17, pp. 131–146.
FOX, H. MUNRO (1948) Apus and a rare cladoceran in Britain. *Nature*, **162**, 116.
GAUTHIER, H. (1951) Contribution à 18 Etude de la Faune et des Eaux Douces. Imprimerie Minerva, Algerie.
GAUTHIER, H. (1954) Essai sur la variabilite . . . de . . . *Moina* . . . recoltés en Afrique et à Madagascar. Imprimerie Minerva, Alger.
HUGOT, H. J. (1974) *Le Sahara avant le Desert*. Editions Hesperides, Paris.
LOFFLER, H. and BONOMI. G. (1966) in *International Jordan Expedition 1966:* Report Compiled by John Morton Boyd, pp. 25–41. IBP/CT Section: Limnology.
LHOTE, H. (1958) *A la découverte des freque du Tassili*. Arthaud, Paris.
RZÒSKA, J. (1961) Observations on tropical rainpools and general remarks on temporary waters. *Hydrobiologia*, **17**, 265–286.
RZÒSKA, J. (Ed.) (1976) *The Nile: Biology of an Ancient River*. Monographia Biologicae, The Hague.
WICKENS, G. E. (1975) Changes in the climate and vegetation of the Sudan since 20,000 BP. *Boissiera*, **24**, 43–65.

CHAPTER 10

Woodlice and Myriapods

J. G. E. LEWIS

Taunton School, Taunton, Somerset

CONTENTS

10.1. INTRODUCTION

Of the five groups of extant arthropods, the primitive, soft-bodied Onychophora with about twenty pairs of stubby conical legs are confined to tropical rain forests. The insects and arachnids (spiders, scorpions, mites and ticks) have a waterproofed cuticle and are distributed in a wide variety of habitats, including deserts. The remaining two groups, namely the Crustacea, represented on land by the woodlice and the Myriapoda (centipedes and millipedes) are generally thought to lack a waterproof layer in the cuticle and would therefore appear to be poorly adapted to desert life.

10.2. WOODLICE

10.2.1. General characteristics

The Crustacea are arthropods with two pairs of antennae, three pairs of mouthparts and many legs. They are mainly aquatic, breathing by gills and the cuticle, which forms the exoskeleton, is frequently massive, being impregnated with calcium carbonate. Familiar examples of the Crustacea are crabs, lobsters, shrimps, prawns and water-fleas. Related to these aquatic groups is the order Isopoda, a group containing marine, freshwater and terrestrial Crustacea, the latter are commonly called woodlice. A good general account of their biology is given by Sutton (1972).

The majority of woodlice are between 0.5 and 2.0 cm in length, dorso-ventrally flattened with a somewhat oval shape and an arched back (Fig. 10.1). In most species the head bears a pair of large compound eyes. Of the two pairs of antennae, the first pair is vestigial but the second is well developed, each antenna consisting of many segments.

The mouth is situated on the underside of the head. It is bounded in front by the upper lip (labrum). Of the three pairs of mouthparts, the first (mandibles) bear heavily sclerotized teeth. There follow two pairs of maxillae and a pair of legs (maxillipedes) modified to form mouthparts. The maxillae and maxillipedes hold and abrade the food which is cut up by the mandibles and normally consists of dead plant material.

Behind the head is the thorax (pereion) of seven segments and the abdomen (pleon) of six. The abdomen ends in a pointed telson. Each of the thoracic segments bears a pair of legs. In the breeding season females have a brood pouch formed of overlapping plates (oostegites) attached to the underside of segments two to five at the bases of the legs. The eggs and young are carried in the cavity so formed.

The appendages of the abdomen (pleopods) retain the two-branched (biramous) form characteristic of the class Crustacea. The outer branch (exopodite) of each limb forms a cover, which can be raised and lowered, for the inner branch (endopodite) which is modified to form a gill. In males, the first two pairs of pleopods are modified to form the external genitalia which effect sperm transfer during copulation.

In the more terrestrial species of woodlouse, the pleopods contain pseudotracheae, masses of air channels in each exopodite opening to the exterior by a small pore. The pseudotracheae appear to be associated with the reduction of water loss. They resemble the breathing tubes (tracheae) of insects.

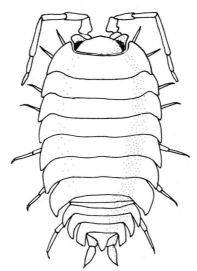

Fig. 10.1. A woodlouse, *Porcellio* sp., length 10 mm. (After Sutton.)

10.2.2. Saharan species

The following species would appear to have some claim to be regarded as Saharan.

Armadillo sp. Found under boulders and the canopy of shrubs in desert, Burg El-Arab west of Alexandria, Egypt (Kheirallah, 1980).

Porcellio albinus Budde-Lunde. Distribution: Biskra and Ouargla in northern Algeria (Seurat, 1944) and dry situations in the oasis of Tozeur in southern Tunisia (Cloudsley-Thompson, 1956)

Porcellio olivieri Audouin & Savigny. Distribution: Egypt and the Algero-Tunisian Sahara (Seurat, 1944). Cloudsley-Thompson found this species in dry situations in the oasis of Tozeur.

Porcellio laevis Latreille. Distribution: Assuan in Egypt (Strouhal, 1965), in litter on sodden soil in oasis of Tozeur (Cloudsley-Thompson, 1956).

Porcellio carthaginensis Silvestri. Distribution: Tunisia, ruins of Carthage and Ain-el-Hadjadj on the Tademait plateau in Algeria (26.55°N, 7.25°E) (Seurat, 1944).

Porcellio simulator Budde-Lund. Distribution: Biskra region of northern Algeria and several localities in the Hoggar mountains in southern Algeria.

Fossoniscus nubicus Strouhal. Described from the Nile oasis at Wadi Halfa, northern Sudan (Strouhal, 1965).

Agabiformius obtusus (Budde-Lund). A small species only 3–4 mm long found with the above at Wadi Halfa. It is widely distributed: Malta; the oases of Fezzan and Giarabub, Cyrenaica and Kufra oasis in Libya; Egypt, Cairo; Israel and Lebanon.

Philoscia sp. Found under boulders and the canopy of shrubs in desert west of Alexandria (Kheirallah, 1980).

Metaponorthus pruinosus (Brandt). Distribution: world-wide. Found under cow dung, stones and palm litter on the Nile bank at Wadi Halfa and in the desert near Abd el Quadir it occurs as far south as Khartoum in the Sudan (Strouhal, 1965).

Leptotrichus panzerii (Audouin & Savigny). Found under boulders and the canopy of shrubs in desert west of Alexandria (Kheirallah, 1980).

Periscyphis albescens (Budde-Lund). Distribution: Egypt; desert near Abd el Quadir and Khartoum in the Sudan (Strouhal, 1965). It occurs on the Nile bank at Khartoum (Lewis, 1966).

Periscyphis convexus (Budde-Lund). Distribution: Asyut, Karnak, Luxor, Assuan and Philae in Egypt. The desert near Abd el Quadir, second Nile Cataract and Sarra Ost in the Sudan. East Africa (Strouhal, 1965).

Hemilepistus reaumuri (Audouin & Savigny). A large species about 2 cm long with tubercles on the head and first four or five thoracic segments (Fig. 10.2). It is unusual in that a number of individuals co-operate to dig vertical burrows. It is abundant on the high Algerian plateaux, the north of the Sahara, Tunisia, Egypt, Syria and Israel where it is found in very arid localities (Verrier, 1932). Cloudsley-Thompson (1956) found the species in localities in Tunisia where the sand was fine and firmly packed.

10.2.3. Ecological distribution

Armadillo sp., *Philoscia* sp., *Leptotrichus panzerii*, *Porcellio albinus*, *P. olivieri* and *P. laevis* appear to be limited to the desert edge. *Fossoniscus nubicus*, *Agabiformius obtusus*, *Metaponorthus pruinosus*, *Periscyphis albescens* and *P. convexus* occur along the Nile and some also at oases. *A. obtusus*, *M. pruinosus* and *P. albescens* are recorded from the desert near Abd el Quadir but no habitat data was given and I have been unable to locate this place. *Porcellio simulator* is the only woodlouse recorded from the Hoggar mountains.

The two species that venture farthest into the desert would seem to be *Porcellio carthaginensis* and *Hemilepistus reaumuri*, both of which are North African forms.

FIG. 10.2. *Hemilepistus reaumuri*. (Photo: J. L. Cloudsley-Thompson.)

10.2.4. Behavioural and physiological adaptations

There is little information on the physiology and behaviour of the woodlice of the Sahara. Most of the data refer to *Hemilepistus reaumuri*.

H. reaumuri is generally described as a crepuscular species. Although specimens may be seen running on the surface of the desert at all times of the day and night in southern Tunisia, experiments in the laboratory showed that the activity was confined to dusk and dawn (Cloudsley-Thompson, 1956).

The species is very resistant to desiccation and markedly unresponsive to unfavourable environmental conditions when compared with woodlice from temperate regions. The rate of water loss of the woodlouse in dry air averages 0.39 mg/cm²/h but gradually diminishes as desiccation continues. There is no evidence of a waterproofing wax layer on the cuticle, nor of any means of controlling water loss from the respiratory organs in dry conditions.

Edney (1960) showed that short periods of water loss could lead to a lowering of the body temperature of *Hemilepistus* of some 3° from 37°C. Such evaporative cooling would allow animals time to seek the shelter of their burrows when suddenly exposed to bright sunlight.

Much of the work on the biology of *H. reaumuri* has been carried out in Israel and has been summarized by Crawford (1981). Following a winter of hibernation the animals appear in early spring. A shallow burrow is dug, usually by the female and the mating takes place. After mating, both partners extend the burrow. Later, as the young begin to forage for themselves, the parental digging diminishes. It is some 6–8 weeks before the young isopods become nutritionally independent. After this, the main role of the parents is to guard the burrow against intruders. This they do with their hunched bodies which flick rapidly when they are disturbed.

Crawford points out that the social existence of *Hemilepistus reaumuri* is required to allow the creation of a burrow deep enough to provide favourable climatic conditions. In firmly packed sand the burrows are over 30 cm deep. A hundred or so individuals can together produce a burrow of sufficient depth to enable them to survive the dry season.

Cloudsley-Thompson (1969) showed that specimens of *Metaponorthus pruinosus* collected from the banks of the Blue Nile and gardens in Khartoum have rates of water loss almost as low as those recorded for *Hemilepistus reaumuri*.

10.2.5. Discussion and conclusion

Although woodlice from arid habitats have a lower rate of water loss than those from damper habitats, none appear to have developed an impermeable cuticle. Several suggestions have been made to account for this.

Woodlice excrete the waste nitrogen resulting from the breakdown of proteins in the form of ammonia which diffuses through the cuticle. Ammonia is very poisonous so the tissues of woodlice have a high tolerance to it. It has been suggested that the possession of a permeable cuticle saves energy because it does not have to be used to convert ammonia into relatively non-toxic substances such as urea or uric acid.

A permeable cuticle also allows evaporative cooling which enables woodlice to withstand high temperatures for brief periods. Cloudsley-Thompson (1969) suggested that the selective advantage of a permeable cuticle might be that it allows a woodlouse to take up and lose water rapidly from a habitat subjected both to drought and flooding. The ability to take up water rapidly would enable a desert woodlouse to take advantage of dew and fog.

10.3. MYRIAPODS

10.3.1. Introduction

Myriapods may be defined as arthropods with one pair of antennae, two or three pairs of mouthparts and numerous pairs of legs. Four classes of arthropods have been grouped together as the myriapods. The centipedes (Class Chilopoda) are soft-bodied, dorso-ventrally flattened carnivores with three pairs of mouthparts and a pair of poison claws and one pair of legs on each trunk segment. The millipedes (Class Diplopoda) are slow-moving decomposers or herbivores with two pairs of legs per segment with the exception of the first four segments. They lack poison claws and have only two pairs of mouthparts. The millipede cuticle is usually impregnated with calcium carbonate and is therefore rigid.

The minute Pauropoda have nine or ten pairs of legs and biramous (forked antennae) and the Symphyla have twelve pairs of legs and a stout pair of anal cerci. These latter two classes are soil dwellers and do not appear to have been recorded from the Saharan region. If they do occur, they are likely to be found in oases or along the Nile.

10.3.2. Centipedes

10.3.2.1. *General characteristics*

Like woodlice, centipedes appear to lack a wax layer on the cuticle and thus lose water relatively rapidly in dry conditions.

There are five orders of centipedes. The Craterostigmomorpha are found only in Australasia. The Geophilomorpha are elongated, worm-like, soil-dwelling centipedes with from 31 to 181 pairs of legs (Fig. 10.3). The Scolopendromorpha are mostly large tropical species, usually with 21 pairs of legs (Figs. 10.4, 10.5). The Lithobiomorpha are short-bodied, relatively fleet centipedes with 15 pairs of legs. The largely tropical order Scutigeromorpha have 15 pairs of very elongated legs and are extremely fast-moving (Fig. 10.6).

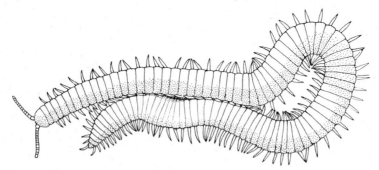

FIG. 10.3. The geophilomorph centipede *Orya barbarica*, length 160 mm. (From a photograph by S. M. Manton.)

The head of centipedes bears a pair of moniliform antennae. The Geophilomorpha are blind but the Scolopendromorpha and Lithobiomorpha have a group of simple eyes (ocelli) on each side of the head and the Scutigeromorpha have well-developed compound eyes. The mouthparts are on the underside of the head. The mouth is bounded in front by the upper lip (labrum), there follows a pair of mandibles, two pairs of maxillae and a pair of large poison claws. The poison claws are highly modified walking legs which are used to catch and inject poison into prey. Some species given an extremely painful bite. In all orders but the Scutigeromorpha some or all the trunk segments bear a single pair of lateral pores (spiracles) which open into the respiratory system which consists of branching chitinous tubes (tracheae) through which oxygen diffuses to the tissues. There appear to be no well-developed spiracle closing mechanisms in centipedes (see Lewis, 1981 for discussion).

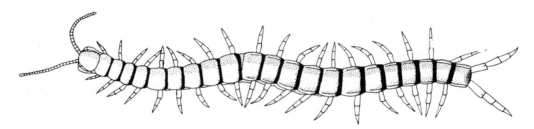

FIG. 10.4. The scolopendromorph centipede *Scolopendra morsitans*, length 88 mm. (From a photograph by S. M. Manton.)

Fig. 10.5. *Scolopendra cingulata*. (Photo: J. L. Cloudsley-Thompson.)

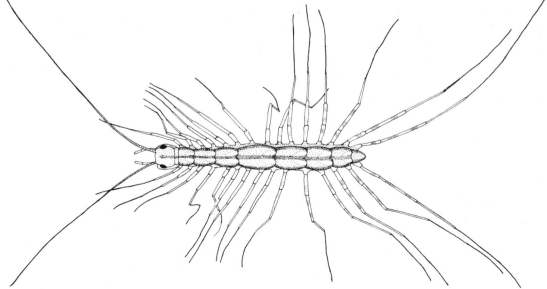

Fig. 10.6. The scutigeromorph centipede *Scutigera coleoptrata*, length 30 mm. (From a photograph by S. M. Manton.)

The spiracles of the scutigeromorphs are median dorsal slits on segments 1, 3, 5, 8, 10, 12 and 14, each leading to a cavity into which open some hundreds of small tracheae, forming lung-like structures.

Sperm transfer in centipedes is accomplished by means of spermatophores, packets of sperm that are passed from male to female during a courtship that is normally a lengthy process.

Geophilomorphs and scolopendromorphs lay their eggs in clutches which are brooded by the female as are the newly-hatched young. Brooding sites may be in soil, under stones, in rotten wood or under bark. Scutigeromorphs and lithobiomorphs lay their eggs singly.

Centipedes seem, for the most part, to be general carnivores feeding on a variety of invertebrates. Large scolopendromorphs have been known to feed on nestling birds, snakes, lizards and mice. Plant material may be important in the diet of some geophilomorphs.

10.3.2.2. *Saharan species*

It is not easy to decide which species to include and which to leave out of the following list which is, therefore, somewhat subjective.

Order Geophilomorpha

Bothriogaster egyptiaca Attems. A very long species with 93–129 pairs of legs recorded from Syria, Palestine, Cyprus, Egypt, Tripoli and Tunis. The localities in Egypt are Mokattan hills near Cairo and desert near pyramids (Attems, 1910).

Mesocanthus albus Meinert. Distribution: Algeria, Sudan, Eritrea, Kenya and India. The Algerian localities are Ideles and In Ameri in the Hoggar mountains, altitudes 1560 and 2450 m (Brolemann, 1930).

Orya barbarica (Gervais). This is a large species reaching 22 cm in length with 107–125 pairs of legs. Distribution: Southern Spain, Morocco, Algeria, Tunisia and Libya (Tripoli).

Pachymerium ferrugineum C. L. Koch. Distribution: North America, Europe, Central Asia, North Africa (Tunis, Algeria, a number of localities in Libya). Brolemann (1930) recorded it from Asekrem (2810 m), In Ameri (1560 m), Ilaman (2050 m) in the Hoggar mountains.

Order Scolopendromorpha

Scolopendra canidens Newport. A widely distributed species from Samarkand in the USSR, through Persia, Syria and Israel and across North Africa from Egypt to Morocco. It also occurs on some Greek and Italian Islands (Würmli, 1980). The North African localities are mostly coastal but Ghadamis (30.10°N, 9.70°E) and Ghat (24.59°N 10.11°E) are desert localities in Libya.

Würmli considers that Turk's (1955) record of *Scolopendra clavipes* C. L. Koch from Jebel Cherichera in Tunisia refers in fact to *S. canidens*.

Scolopendra cretica Attems. Distribution: Turkey, Crete, Barka and Ghat in Libya. Formally regarded as a subspecies of *S. canidens*, this form was given specific status by Würmli (1980). *S. cretica* is very closely related to *S. canidens*.

Scolopendra cingulata Latreille (Fig. 10.5). Distribution: The Mediterranean region extending into the Balkans and Syria. Recorded from Morocco and Libya in North Africa.

Scolopendra valida Lucas is a widely scattered species being recorded from the Canaries, Cameroons, Sudan and Syria. In the Sudan it occurs as far north as Atbara on the Nile (17.30°N, 34.45°E).

Scolopendra morsitans Linn. This is virtually cosmopolitan occurring in South America, Africa and Asia. It is a very common species south of the Sahara. It is limited to the high Nile bank at Khartoum. It has been recorded from Giza and Assuan in Egypt. Specimens from Morocco, Algeria and Tunis are sometimes referred to the variety *scopoliana*. The British Museum (Natural History) possesses a specimen from Guelt es Stel on the Central Plateau of Algeria.

Asanada sokotrana Pocock. Recorded from Libya: Brach e Sebha by Manfredi (1935). Sebha is a desert locality (27.04°N, 14.25°E).

Otostigmus spinicaudus (Newport). Distribution: North Africa from Morocco to Tripoli.

Order Lithobiomorpha

A large number of species of Lithobiomorpha have been recorded from North Africa, many of them European species of *Lithobius*. They do not appear to occur in the Sahara.

Order Scutigeromorpha

Thereuonema syriaca Verhoeff. Distribution: Turkey, Syria, Israel, Iraq, Iran, Egypt, Yemen, Sudan (Mongalla) and Kenya (Würmli, 1975). Lewis (1966) recorded a *Thereuonema* sp. from crevices in mud in the Nile bank at Khartoum. It was presumably *T. syriaca*.

10.3.2.3. *Ecological distribution*

Orya barbarica, *Scolopendra cingulata* and *Otostigmus spinosus* appear to be species of the desert edge. *Mesocanthus albus*, *Bothriogaster egyptiaca*, *Scolopendra canidens*, *S. morsitans* and *Asanada sokotrana* occur in more arid localities. *Pachymerium* has reached the Hoggar mountains and *Thereuonema syriaca* seems to be a Nile bank species.

10.3.2.4. *Behavioural and physiological adaptations*

The lithobiomorphs and scutigeromorphs do not appear to occur in arid environments, whereas the geophilomorphs and scolopendromorphs are relatively more successful.

The Geophilomorpha may owe their success, in part, to their burrowing ability. Burrowing efficiency is increased by a shortening in segment length and an increase in their number. Both *Orya* and *Bothriogaster* have very high segment numbers. The scolopendromorphs are not as specialized as geophilomorphs but are better able to penetrate the soil than are lithobiomorphs or scutigeromorphs. *Scolopendra cingulata* is known to burrow actively.

The fact that geophilomorphs and scolopendromorphs brood their eggs and young probably gives the latter protection from desiccation for the brood cavities may well be deep in the soil.

The lithobiomorphs and scutigeromorphs which are not adapted for burrowing are limited to more superficial habitats as are the eggs which will be laid in the surface soil or in superficial crevices. The young larval stages are very susceptible to desiccation.

Brolemann (1930) pointed out that the centipedes of the Hoggar are amongst the most resistant to dry conditions. This is certainly true of *Pachymerium ferrugineum* which survives desiccation longer than any other geophilomorph so far studied. Specimens survive for 38–109 h at 35 per cent relative humidity whereas the maximum survival time for four British species ranges from 16–32 h.

Cloudsley-Thompson (1956) showed that *Scolopendra clavipes* (probably *Scolopendra canidens*, see above) is active only at night. Its rate of water loss is very high in dry air. He found no evidence that the centipede could control water loss neither was there any indication of a waterproofing wax layer on the cuticle. He concluded that the species showed no particular adaptation to the desert climate and survival must depend on its nocturnal habits.

10.3.3. Millipedes

10.3.3.1. *General characteristics*

There are two subclasses of millipedes, the Penicillata which are very small animals with 11–13 tergal plates covered with tufts of serrated bristles (Fig. 10.7) and the Chilognatha, the more familiar millipedes (Figs. 10.8, 10.9, 10.10), lacking serrated bristles but with a rigid calcified cuticle and 11 to many tergal plates. The subclass Penicillata contains one order, the Polyxenida, the subclass Chilognatha contains ten.

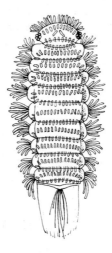

FIG. 10.7. The millipede *Polyxenus lagurus*, length 3 mm. (After various authors.)

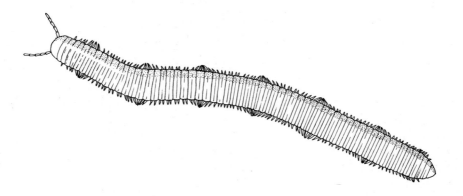

FIG. 10.8. A spirostreptid millipede, length 130 mm. (After various authors.)

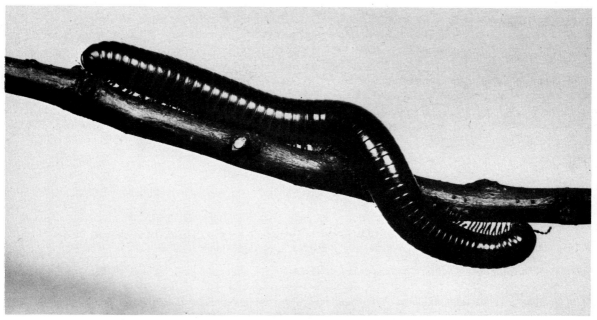

FIG. 10.9. *Spirostreptus* sp. (Shell photograph.)

FIG. 10.10. *Spirobolus* sp. (Photo: J. L. Cloudsley-Thompson.)

The millipede head bears a pair of antennae and normally a number of ocelli on each side. The mouthparts consist of a pair of mandibles and a gnathochilarium, a broad plate formed by the fusion of the second pair of mouthparts. In the Chilognatha the body is typically circular in cross-section, the body rings are formed largely of the tergal plates. In the trunk region the rings are diplosegments, each bearing two pairs of legs. The spiracles open just above the basal segment of the eggs. The genital openings of millipedes are situated behind the second pair of legs and in the males usually one or both of the leg pairs of segment seven are modified to form gonopods which are used during copulation.

Millipedes feed largely on dead plant material. Some attack seedlings and are pests of cotton in West Africa.

10.3.3.2. *Saharan species*

Subclass Penicillata: Order Polyxenida

Polyxenus lagurus (Linn.). A European species recorded from the plateau of Asekrem in the Hoggar mountains of Algeria (2810 m) (Brolemann, 1930).

Subclass Chilognatha: Order Spirostreptida

Archispirostreptus lugubris villiersi and an immature odontopygid of uncertain generic status were recorded from Irabellaben, Baguezans Mtns., in the Aïr region of Niger at an altitude of 1200–1300 m by Schubart (1951). Professor Hoffman believes that the odontopygid might be an *Omopyge* or *Peridontopyge* or even a new genus but guesses the first.

10.3.3.3. *Behavioural and physiological adaptations*

Millipedes appear to be poorly adapted to dry conditions: although woodlice and centipedes are relatively common along the Nile at Khartoum, south of the Sahara, millipedes do not occur there. Crawford (1979) points out that in general there are few records of millipedes inhabiting very arid places. He believes that it is among the Spirostreptida that most of the well-adapted desert species will probably be found, one reason being their large body size, which gives them an advantage over small-bodied forms in that they lose water less rapidly. The only 'desert' species to have been investigated is *Orthoporus ornatus* (Girard) which is common in the deserts of the south-western United States. It has lower rates of water loss than other millipedes and may have an epicuticular wax layer. It may be able to take up water from an unsaturated atmosphere.

10.4. DISCUSSION

With the current state of our knowledge any discussion of woodlice and myriapods in the Sahara must be unsatisfactory. There are a number of reasons for this. The region has been poorly collected and even when place names are given it is often difficult to pinpoint the site of many localities. This is in part due to variations in spelling and the fact that several places may have the same name. Habitat data is seldom provided so that it is difficult to make generalizations about the ecology of a species.

The available information suggests that two species of woodlice, perhaps two geophilomorph and three scolopendromorph centipedes, can survive in the desert. It seems that the ability to burrow and parental care of the young are important factors in this. The character of the soil may be important, it would have to be suitable for burrowing or contain deep crevices offering suitable refuges. Woodlice require soils with a relatively high calcium carbonate content.

Far more information is required on the animals' behaviour, food, life histories and physiology, especially their water relations.

A final difficulty is that of correct identification. This is certainly difficult for centipedes where there is still confusion as to the status of and correct names for certain forms.

ACKNOWLEDGEMENTS

I am grateful to Professor J. L. Cloudsley-Thompson, Professor R. L. Hoffman, Dr O. Kraus and D. Macfarlane for their help with the literature.

REFERENCES

ATTEMS, K. G. (1910) Ergebnisse der mit Subvention aus der Erbschaft Treitl unternommenen zoologische Forschungsreise Dr. Franz Werner's nach dem ägyptischen Sudan und Nord-Uganda. *Sber. Akad. Wiss. Wien*, **119**, 355–360.

BROLEMANN, H. W. (1930) Myriapodes du Sahara central recueillis par L. G. Seurat au cours de la mission du Hoggar (février–avril 1928). *Bull. Soc. Hist. nat. Alger*, **21**, 6–8.

CLOUDSLEY-THOMPSON, J. L. (1956) Studies on diurnal rhythms. VI. Bioclimatic observations in Tunisia and their significance in relation to the physiology of the fauna, especially woodlice, centipedes, scorpions and beetles. *Ann. Mag. nat. Hist.* Ser. 12, **9**, 305–329

CLOUDSLEY-THOMPSON, J. L. (1969) Acclimation, water and temperature relations of woodlice *Metoponorthus pruinosus* and *Periscyphis jannonei* in the Sudan. *J. Zool., London.* **158**, 267–276.

CRAWFORD, C. S. (1979) Desert millipedes: A rationale for their distribution, in *Myriapod Biology*, Edited by M. CAMATINI, pp. 171–181. Academic Press, London.

CRAWFORD, C. S. (1981) *Biology of Desert Invertebrates*. Springer-Verlag, Berlin.

EDNEY, E. B. (1960) The survival of animals in hot deserts. *Smithsonian Rep. 1959:* 407–425.

KHEIRALLAH, A. M. (1980) Aspects of the distribution and community structure of isopods in the Mediterranean coastal desert of Egypt. *J. Arid Env.* **2**, 51–59.

LEWIS, J. G. E. (1966) Seasonal fluctuations in the riverain invertebrate fauna of the Blue Nile near Khartoum. *J. Zool., Lond.* **149**, 1–14.

LEWIS, J. G. E. (1981) *The Biology of Centipedes*. Cambridge University Press, Cambridge.

MANFREDI, P. (1935) Alcuni Chilopodi della Tripolitania. *Atti Soc. ital. Milano*, **74**, 419–422.

SCHUBART, O. (1951) Ueber einige. Diplopoden von Bergmassiv Aïr in der Süd Sudan, gesammelt von L. Chopard und A. Villiers in 1947. *Bull. Inst. fr. Afrique noire*, **13**, 116–125.

SEURAT, L. G. (1944) *Zoologie Saharienne*. Publications du Centre National de la Recherche Scientifique, Alger.

STROUHAL, H. (1965) Ergebnisse der zoologischen Nubien-Expedition 1962. *Ann. Naturhistor. Mus. Wien*, **68**, 609–629.

SUTTON, S. L. (1972) *Woodlice*. Ginn, London.

TURK, F. A. (1955) The myriapoda of Dr. Cloudsley-Thompson's expedition to the Tunisian desert. *Ann. Mag. nat. Hist.* Ser. 12, **8**, 277–284.

VERRIER, M.-L. (1932) Etude des rapports de la form de l'habitat et du compartement de quequ
es crustacés isopodes. *Bull. biol. Fr. Belg.* **66**, 200–231.

WÜRMLI, M. (1975) Revision der Hundertfusser-Gattung *Thereuonema* (Chilopoda: Scutigeridae). *Entomologica Germanica*, **2**, 189–196.

WÜRMLI, M. (1980) Statistische Untersuchungen zur Systematic und postembryonalen Entwicklung der Scolopendra-canidens-Gruppe (Chilopoda: Scolopendromorpha: Scolopendridae). *Sber. Akad. Wiss. Wien*, **189**, 315–353.

CHAPTER 11

Epigeal Insects

F. T. ABUSHAMA

Department of Zoology, University of Khartoum, Sudan

CONTENTS

11.1. INTRODUCTION

Insects comprise the largest class of Arthropoda and are divided into some thirty orders including the Hemiptera (bugs), Coleoptera (beetles), Diptera (flies) and so on. Adult insects are characterized by having the body divided into three regions or 'tagmata': head, thorax and abdomen. The thorax is three-segmented, with each segment typically bearing a pair of legs and, in many insect groups, the second and third thoracic segments possess a pair of wings. The head bears a pair of segmented antennae, mouthparts and, nearly always in adult insects, a pair of multifaceted compound eyes. The abdomen consists of as many as eleven segments: usually the eighth, ninth and tenth segments have appendages modified for mating or egg-laying.

In common with other arthropods, insects have an exterior integument of exoskeleton, composed mainly of the nitrogenous substance *chitin*. This provides protection for the vital organs and the support which maintains the body shape. Insects normally lay eggs, and the young moult several times in their development to the mature or adult stage. With few exceptions, immature forms do not have wings. Insects respire through apertures along the sides of the body, known as 'spiracles' or 'stigmata'.

Insects are remarkably successful organisms; they attain the largest diversity of any animal class, with an estimated 900,000 species. In competition with other animals, and due to their way of life and extreme resistance to adverse conditions, they have become adapted to every kind of environment except, perhaps, to the depths of the oceans. They are to be encountered in the air and in fresh water even when it is frozen in the form of ice or snow. Mosquito larvae develop well in geysers with average water temperatures reaching 80°C; they are able to detect and remain in small microhabitats of cooler water.

Insects can also live in liquids other than water; the fly *Psilopa petrolei* inhabits seepage pools of crude oil in California, and feeds on the dead bodies of insects that fall in. Furthermore, its intestine contains symbiotic bacteria which transform petroleum into protein that can be digested by the insect (Chauvin, 1967). Insects are common in the soil, where they may burrow to a depth, which extends as far as the decomposing vegetable matter upon which they feed. In desert regions, most insects are nocturnal, hiding in deep holes during the hottest part of the day when the soil surface temperature may reach 80°C. They come out at night to chew dried vegetable fragments, especially when these are moist with dew that condenses during the hours of darkness.

Cloudsley-Thompson and Chadwick (1964) commented that one might gain the impression that the desert is sterile and uninhabited. The true picture varies considerably, as the distribution of animals in arid regions is extremely uneven. A large rock or a small bush may shelter a surprising number of insects compared with the surrounding desert; but the number is trifling when compared with similar habitats in temperate or wet tropical regions.

As described by Leopold (1972), there are two phases of life in the desert; one during the time of rainfall, when leaves, flowers, seeds and fruits are in good supply, the other the long period of drought when the only available plant food consists of stems, roots and drought-resistant leaves. The critical time for insects is the drought period, and species that live in the desert have found ways to adjust to or avoid its long dry spells. Following the first good rains, plants begin growing and the desert crawls and buzzes with an enormous number and variety of insects; beetles, ants, wasps, moths and butterflies, which feed greedily on the flourishing plants and reproduce. Then most of them die, but leave behind a rich supply of eggs and also of pupae. These eggs and pupae serve as their main food for many birds and reptiles.

A vivid picture of the vernal rain fauna which appears at the time of inflorescence in the desert was given by Cloudsley-Thompson (1977), who wrote that the flowers are visited by butterflies and moths, bees, wasps, hoverflies, beeflies and other Diptera. The droppings of camels and goats are rolled away by dung beetles and grass-seeds are harvested by ants. Termites extend their subterranean galleries to the soil surface and indulge in nuptial flights, while predators such as scorpions, camel-spiders, ant-lions, bugs, wasps, robber-flies and predatory beetles, glut themselves on an abundance of food. With the rains, too, come swarms of desert locusts which breed in the damp sand. The ephemeral vegetation is devoured by hordes of caterpillars, and the air buzzes with an abundance of flies, wasps and bees, rarely seen at other times of the year.

11.2. INSECTS OF THE SAHARA

The insects of the Sahara can be divided according to their habitat into four groups: the steppe type, the fauna of the true desert, the fauna of oases and the fauna of mountains. The species encountered in each type of habitat are of different origin. The oasis fauna, for instance, includes numerous species which are primarily Mediterranean, while, in the real desert area, Mediterranean species become uncommon and progressively disappear. The mountain fauna is somewhat comparable to that of the oases but, on the whole, the Saharan insect fauna is a mixture of real desert types, Mediterranean species, and species of tropical origin. Nevertheless, the Palaearctic element seems to dominate (Dekeyser and Derivat, 1959).

Twenty-six insect orders are represented in the north-west Sahara; fourteen are typical inhabitants of extreme desert, viz. Collembola, Thysanura, Dictyoptera, Orthoptera, Heteroptera, Homoptera, Thysanoptera, Neuroptera, Coleoptera, Lepidoptera, Diptera, Hymenoptera and Strepsiptera. Of these, Thysanura, Orthoptera, Neuroptera, Coleoptera, Lepidoptera, Diptera and Hymenoptera are the most numerous (Cloudsley-Thompson and Chadwick, 1964).

11.2.1. Thysanura

Members of the family Lepismatidae are common sand dwellers, found under rocks and other débris. They probably form the basis of a food chain upon which many other animals depend.

11.2.2. Orthoptera

Comprises crickets, locusts and grasshoppers. The latter are mainly of the family Acrididae. In the Saharan steppe, the Acrididae are dominated by species of large size, with limited flight range and poor jumping ability. More mobile species, capable of rapid but not long sustained flight, are also encountered. This leads them from one tuft of Graminaceae to another; for example, members of the genus *Platypterna*. Most crickets (Gryllidae) which are common in the oasis are generally hygrophilic, living in the galleries or burrows in the soil. They are omnivorous and produce songs and chirrups.

Apart from its swarms, which extend beyond ecological barriers, the desert locust *Schistocerca gregaria* (Fig. 11.1(c)) is normally confined to sand dunes, coastal plains, scrub belts along the beds of *wadis* (seasonal rivers) and similar habitats which represent ecological 'islands' in the desert (Uvarov, 1954). The desert locust is not well adapted physiologically to desert conditions and its existence is closely bound up with its migratory or nomadic habits.

In the northern Sudanese desert, the climatic conditions and food supply along the banks of the Nile are most favourable to most insects. Acridid grasshoppers are a successful group in riverain localities. However, the problem they face is the annual rise and fall of the river, and thus the flooding of their habitat for about three months. To overcome this, some species migrate to the desert during the rainy season where climatically suitable retreats are created. *Truxalis grandis grandis* (Fig. 11.1(a)) breeds in desert localities and the nymphs produced move back to the riverside during the dry season (Abushama and El Hag, 1971).

North African desert carnivores, seeking a nutrient nibble at the expense of the desert grasshopper *Poecilocerus hieroglyphicus* (Fig. 11.1(b)), are likely to be greeted by two well-aimed jets of irritating and toxic fluid. The insect's production of this fluid is a remarkable example of physiological economy, for its active ingredients are obtained intact from the food plant *Calotropis procera* (Abushama, 1973).

Cockroaches and mantids (order Dictyoptra), close relatives of the Orthoptera, are encountered in some Saharan habitats. Cockroaches of the genus *Heterogamodes*, originally a steppe element, are adapted to survive arid conditions. The females dig holes in the sand while the males fly during the night. Mantids live on bushes; but the common *Eremiaphila*, which has a flat body and reduced wings (Fig. 11.2), runs on the ground especially on the rocky area of the Sahara desert.

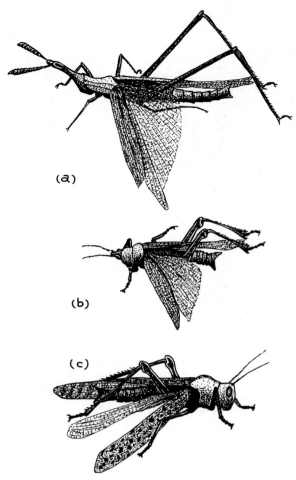

FIG. 11.1. Some acridids of the desert. (a) *Truxalis grandis*, (b) *Poecilocerus hieroglyphicus*, (c) *Schistocerca gregaria* (drawings not to scale).

FIG. 11.2. Desert mantid, *Ermiaphila* sp. (after Dekeyser and Derivot, 1959).

FIG. 11.3. Desert mantid, *Blepharopsris mendica* (nymph) (photo: J. L. Cloudsley-Thompson).

11.2.3. Earwigs and ant-lions

Earwigs (Dermaptera) are hygrophilic, common in the Saharan oasis. They hide under stones and cracks in the soil and are generally nocturnal. The ant-lion larvae (Fig. 11.3(a)), called 'demons of the dust' by Wheeler (1930), are characteristic insects of the summer season. Their cone-shaped pits can be seen especially well in fine sand. At the bottom of each pit lies buried an ant-lion, in ambush for passing insects that may fall into the trap (Cloudsley-Thompson and Chadwick, 1964). Green (1955) has shown that the larvae of the ant-lion *Myrmeleon immaculatus* orient themselves within their pits so that their bodies are in the coolest region available to them. They remain active at all temperatures of the sand at the base of the pit ranging up to about 48°C; but may be inactivated by high midday temperatures.

11.2.4. Termites

Termites (Isoptera) of the families Hodotermitidae and Psammotermitidae are found in the mountains as well as the dry areas of the Sahara desert. It has been observed (Dekeyser and Derivat, 1959) that, in open bare *reg*, without any conspicuous plants around, *Psammotermes* was sometimes found a few centimetres below the ground. They survived there mostly by virtue of subfossil trunks of the ancient flora that disappeared after the humid Pleistocene, especially in the wood of big *Tamarix* trees covered by alluvium from past floods.

Untreated *Acacia nilotica* and *Isoberlina doca* fence-posts, around desert reserves established in the northern Sudan to help in stopping desert encroachment, are reduced to mere wood and soil sandwiches in less than five years, while *Eucalyptus* telephone poles in the Khartoum–Cairo line were completely destroyed. This was due to infestation by the sand termite *Psammotermes hybostoma*, an inhabitant of the fringes of the Sahara and Nubian desert (Abushama and Abdel Nur, 1973). This termite manages to maintain a high humidity in its nest even though there is little water in the food it eats. Worker termites actually go down to the water table, which may lie at a depth of up to 40 metres, and bring up moist particles with which they moisten the nest (Howse, 1964).

11.2.5. Hemiptera

Many species of Heteroptera and Homoptera are xerophylic, feeding on the sparse vegetation of the desert. Generally, Heteroptera tend to be diurnal in habit and withstand considerable temperatures. Many of them are brightly coloured; for example, the lygaeid *Spilostethus pandurus*, which is found in appreciable numbers in the desert of the northern Sudan. Its life is closely bound up with the asclepiadaceous plant *Calotropis procera*. It feeds on the leaves, and lays its eggs on the fibres inside the fruits. The bug possesses two dorsolateral prothoracic glands, the secretion of which is repugnatorial and unpalatable to predators (Abushama and Ahmed, 1976).

11.2.6. Thrips (Thysanoptera)

These are small insects, only about 2 mm in length, usually having wings with limited innervation and long marginal area, but are sometimes apterous. Some have been found in the contents of the alimentary canal of the *Ermiaphila* mantids collected in the Sahara desert (Dekeyser and Derivat, 1959). However, the life of thrips is normally related to vegetation, and their presence in the desert looks surprising. They are encountered in limited numbers in places where Graminaceae grow.

11.2.7. Lepidoptera

In the Sahara, Lepidoptera (butterflies and moths) are represented by species common to other parts of Africa. They make use of their characteristic pupal diapause to pass the unfavourable winter climate. In response to high desert wind, certain small blue butterflies which inhabit the Great Palaearctic Desert possess the power of continued flight within one small bush, from the shelter of which they seldom issue (Cloudsley-Thompson and Chadwick, 1964). Buxton (1924) pointed out that the butterflies of the genus *Tarucus* can be observed fluttering up and down continuously inside a bush of *Zizyphus*. The minute *Freyeria galba* is able to limit its flight within a plant of *Ononis* only a foot in diameter and remain on the wing in this little bush when such a wind is raging outside as to prevent the flight of all other butterflies.

Some butterfly larvae of the families Arctiidae, Lymantriidae and Lasiocampidae make use of the dispersive effect of the wind for purpose of distribution. These caterpillars are abundant in North Africa and Middle East in spring. They can be seen rolled up, blown along with the wind. Thus they are able to benefit by the desert winds without modifying either their habits or their structure.

11.2.8. Hymenoptera (bees, wasps and ants)

Ants are common in nearly all the Saharan region, while bees are found in mountain areas, their activity being restricted to the rainy season since they require pollen and nectar. Wasps, on the other hand, live both at higher and lower elevations. The giant, steel-blue, carpenter bees, *Xylocopa* spp. (Fig. 11.5), build their nests in old wood and are very active especially during the morning. Species of *Andrena, Eucera* and *Tetralonia* nest deep in the soil, while the small *Ceratina* construct their nests in twigs.

The number of ant species known in the Sahara is as high as sixty-six, most of which live in rather humid areas. Only ten species are xerophiles. Many ant species are carnivorous but some, for example, *Messor, Monomorium* and *Pheidole* spp., live mainly on the seeds of grasses and other plants which they harvest and feed to their brood. The honey-pot ants store the excretion of aphids and coccids, the exudation of oak galls and the nectar of flowers, in the crops of certain workers which are known as 'repletes'. These are incapable of walking, but remain suspended from the ceiling of underground chambers in the nest. They become filled with food collected and regurgitated by foragers, until their abdomens are completely spherical and several times the normal size (Fig. 11.6). In the dry season, they provide for the entire colony

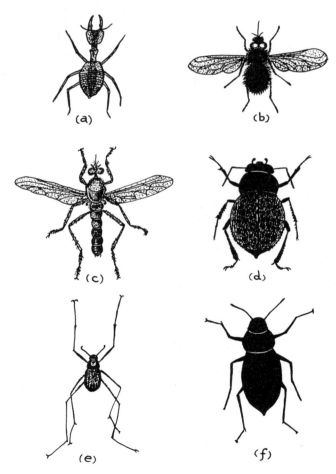

FIG. 11.4. Some insects of the desert: (a) Nemopterid lava (1 cm), (b) Bombyliid fly (1 cm), (c) Asilid fly (1 cm), (d) *Pimelia* sp. (3 cm), (e) *Sternocara phalangium* (2 cm) (from the Namib desert), (f) *Blaps* sp. (after Cloudsley-Thompson, 1965).

FIG. 11.5. *Xylocopa* sp. (photo: J. L. Cloudsley-Thompson).

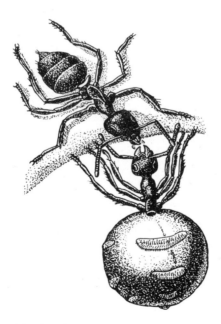

FIG. 11.6. Worker ant pumping sweet juices into the 'honey-pot' whose abdomen expands and stores the juice until it is needed (after Leopold, 1967). Honey-pot ants have been recorded from North America, South Africa and Australia!

from their capacious bellies. The forms which adopt this behaviour belong to several unrelated genera, for example, *Pheidole, Melophorus, Myrmecocystus, Componotus* and *Prenolepis* (Wheeler, 1930).

Wasps of the family Scoliidae are large insects (up to 5 cm long), with short strong legs well suited for digging in the soil. The females can be seen digging for the larvae of scrabaeid beetles, which they paralyse with their stings. They lay a single egg on their prey; this hatches into a larva which eats the beetle's grub and then spins a cocoon in its empty skin (Cloudsley-Thompson and Chadwick, 1964). Eumenid wasps are conspicuous in the desert. They are solitary and construct their larval nests of loam and saliva often on the walls of the houses. The social wasps (Vespidae) prepare their nest from chewed wood converted into paper like sheets utilized for the formation of regular hexagonal cells. *Polistes* spp. are conspicuous in the oases of North Africa and the Middle East. They are abundant in the vicinity of water, and hundreds of nests can be seen in the oases of southern Tunisia (Cloudsley-Thompson, 1956).

11.2.9. Diptera

Flies and mosquitoes in the Sahara are represented by species common to other parts of Africa. The housefly *Musca domestica* is common everywhere, being abundant around human settlements. Other common flies include small sand-flies (Psychodidae), crane-flies (Tipulidae), long-snouted flies (Nemestrinidae), and bee-flies (Bombyliidae) (Fig. 11.4(b)). Predacious flies of the families Asilidae (Fig. 11.4(c)), Therevidae and Empididae are also common. Blood-sucking horseflies (Tabanidae) (Fig. 11.7) constantly attack camels, donkeys and horses.

One of the oestrid flies develops in the nose of the camel and, when mature, is sneezed out onto the sand where it pupates. Another species causes myiasis of the dorcas gazelle, appearing as lesions or swellings on the back of the animal where the larva develops. It drops to the ground when mature, mostly at the end of the rainy season. In the Dinder area of the northern Sudan, the oestrid fly *Strobiloestrus* sp. is common, causing maiysis to the Bohor reed-buck *Redunca redunca cottoni* (Abushama, 1971).

FIG. 11.7. *Tabanus* sp. (photo: J. L. Cloudsley-Thompson).

11.2.10. Coleoptera

Beetles comprise an important group of insects in the Sahara. Most common are members of the family Tenebrionidae; for example, *Pimelia* (Fig. 11.4(d)), *Blaps* (Fig. 11.4(f)), *Prionotheca* (Fig. 11.8) and *Adesmia* spp. In the Sudan *Pimelia grandis*, *Ocnera hispida* and *Adesmia antiqua* are the most common species (Figs. 11.9, 11.10). In the winter and spring they tend to be diurnal or crepuscular in activity while, in hot weather they become nocturnal. Their wings are generally absent, and the elytra are fused with the body,

FIG. 11.8. The desert tenebrionid beetle, *Priontheca coronata* (after Dekeyser and Derivot, 1959).

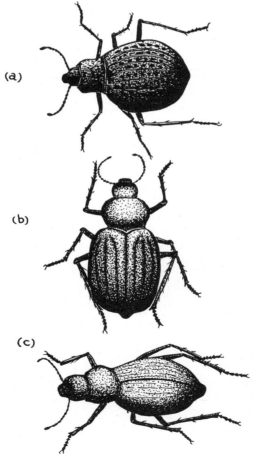

(a)

(b)

(c)

FIG. 11.9. Desert Tenebrionid beetles, common in the northern Sudan: (a) *Adesmia antiqua*, (b) *Pimelia grandis*, (c) *Ocnera hispida*.

FIG. 11.10. *Adesmia antiqua* and *Zophosis plana* feeding on dung (photo: J. L. Cloudsley-Thompson).

leaving a sub-elytral air space which serves the purpose of insulation, and of reducing the amount of water lost through transpiration (Cloudsley-Thompson, 1965a). Beetles of the family Scarabaeidae are also common. *Scarabaeus sacer* (the 'sacred scarab') (Figs. 11.11, 11.12) breaks up the dung of camels, goats and other animals, rolls it into balls and buries them in subterranean chambers. The beetle then remains with the balls until they are consumed. In autumn a larger ball is stored in a bigger chamber where an egg is deposited.

Blister beetles (Meloidae) are common in the desert areas, normally possessing conspicuous warning coloration; vivid black and red, green, or brown. *Epicauta aethiops* is common in the Sudanese northern desert appearing in large numbers during the rainy season in August and September. It does not usually fly and, if disturbed, secretes a yellow toxic substance (cantharidin), exuded from glands in the leg-joints. If this reaches the human skin and is not immediately washed off, it causes painful blisters.

Carabid beetles, which are predators, spend the day beneath rocks and stones. Many of the desert species are wingless. *Anthia venator* (Fig. 11.13) and *A. sexmaculata* are gigantic nocturnal predators, with conspicuous black and white warning coloration.

FIG. 11.11. Scarab beetle, *Scarabaeus sacer*, with her ball of dung (after Cloudsley-Thompson, 1960).

FIG. 11.12. *Scarabaeus sacer* (photo: J. L. Cloudsley-Thompson).

FIG. 11.13. *Anthia venator* (photo: J. L. Cloudsley-Thompson).

11.3. ADAPTATION TO DESERT CONDITIONS

Climatic conditions in the desert are characterized by extremes of temperature and humidity. The rainfall is low and irregular. Strong winds carrying fine particles of sand often blow and increase the drying effect of the air. The soil surface temperature reaches a very high level, while a few centimetres beneath the surface the climate becomes more stable. Under such conditions, survival depends upon avoiding desiccation and keeping cool. Thus, there must be a conflict between the requirements of conserving water for vital purposes and of transpiring it for cooling. Insects are too small to withstand transpiration for long, and their adaptation to desert conditions involves change in degree rather than the evolution of new behavioural or physiological mechanisms.

11.3.1. Burrowing

Many insects in the desert escape the unfavourable surface conditions during the day by burrowing, making use of favourable microclimatic conditions a few centimetres below the surface. Insects associated with dunes and ergs in the Sahara show morphological adaptation to burrowing; for example, the female of the cockroach *Heterogamodes* sp. Some are modified to swim through a loose substratum without making a hole, while some excavate pits and others mine tunnels in the sand (Pierre, 1958). Thysanura take refuge under tussocks of grass where they wriggle or almost swim with a fish-like movement in the sand. The earthen galleries which termites build along their path can also be considered as a protective measure against excessive heat and evaporation. Most of the ant species found in the Sahara desert live in comparatively damp places. A few live in extremely arid areas. The most adapted species to desert conditions is *Acantholepis* which brings water to its nest from the salt–damp sand of water-bearing strata (Bernard, 1951).

11.3.2. Rhythmic activity

Many desert insects confine their activity to the hours of darkness when the temperature is low and the relative humidity of the air is comparatively high. The desert beetle *P. grandis* is active by night; its locomotory activity increases markedly during the first hours of the evening, and then gradually decreases, and by morning the activity stops completely (El Rayah, 1969). Certain desert insects, such as grasshoppers and some beetles, may be active during the hottest part of the day. The adult stage of the successful desert and semi–desert grasshopper *P. hieroglyphicus* shows a marked diurnal rhythm of locomotory activity with a fairly regular afternoon peak (Abushama, 1968). This grasshopper, as well as its North African relative *Poecilocerus bufonius* (Klug), if attacked or irritated, emits a repugnatorial secretion which drives the predators away (Abushama, 1972; Fishelson, 1960).

Desert beetles which may be seen wandering around in broad daylight, are usually distasteful to predators and have extremely hard integuments. Their black coloration has a warning function (Cloudsley-Thompson, 1977).

Diurnal desert insects tend to leave the sand when its temperature reaches about 50°C (112°F). Some climb grasses, some dive into holes, while others fly about above the ground, making hurried landings to enter their burrows. Many of the beetles which can withstand the high soil temperature have long legs which raise their bodies above the scorching sand; for example, *Sternocara phalangium* (Fig. 11.4(e)). The daily rhythm of activity is constantly modified both by endogenous seasonal rhythm and by environmental changes in day–length, ambient temperature and so on. Day-active desert beetles often show two peaks of activity when the weather is very hot. These peaks occur at dawn and dusk. In spring and autumn the insects show a single peak around midday. The winter is spent in hibernation (Cloudsley-Thompson, 1977).

The numbers and stages of development of desert insects are a marked reflection of the desert environment, characterized by seasonal precipitation and periods of cooler weather. Populations of adult insects reach their peak at the time of the rains. Adult *Adesmia bicarinata* begin to appear in small numbers in the Egyptian desert in October. Their population then gradually increases, reaching a peak in March. By the end of May, the beetles disappear completely and only dead bodies can be found. During the hot season their life is continued by the larval and pupal stages (Hafez and Makky, 1959).

Reproduction in some desert insects needs to be triggered by some immediate stimulus from the environment such as rainfall or the appearance of green vegetation. Maturation of the desert locust *S. gregaria* occurs in response to terpenoids and other aromatic chemicals produced seasonally by certain desert shrubs at the time of the annual rains (Carlisle *et al.*, 1965).

11.3.3. Water conservation and heat resistance

Desert insects may sometimes obtain moisture from dew or hygroscopic vegetation, but are usually able to exist independently of this. Insects have a relatively enormous area in proportion to their mass when compared to larger animals. The latter can suffer a given rate of water loss for a longer time than insects, before their water contents are reduced to a lethal minimum (Fig. 11.14). Thus, the prevention of water loss by evaporation in desert insects becomes of great importance. One way of tackling the problem is by developing an impervious intergument. This impermeability is due to a lipid cuticular layer that prevents desiccation, and brings the rate of transpiration to a low level (Wigglesworth, 1945; Lees, 1947; Beament 1958, 1961).

The rate of water loss by transpiration at various temperatures in the desert grasshopper *P. hieroglyphicus* is slower than that of the tree locust *Anacridium melanorhodon melanorhodon* an inhabitant of the savanna region of the Sudan (Abushama, 1970). Ants inhabiting drier situations usually survive experimental drying better than other species of the same genera that normally inhabit more humid environments.

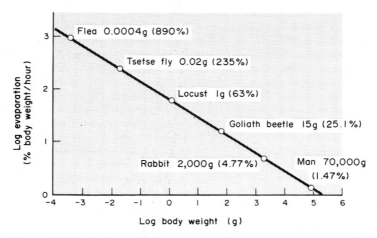

FIG. 11.14. The relation between body weight and the amount of water which must evaporate per hour from the body to preserve a constant temperature in desert conditions. Such a rate is tolerable for some hours by a large animal because it represents only a small percentage of total weight, but a small animal would have to lose several times its own weight per hour. Figures in parenthesis show the amount of water, expressed as a percentage of total weight, to be evaporated each hour by animals whose weights are also shown. (From Cloudsley-Thompson, 1977, after Edney, 1974.)

Moreover, there is evidence that the epidermal cells of the migratory locust (*Locusta migratoria*) expend energy continuously in regulating the water balance of the cuticle which to be in equilibrium with blood, would contain 60% more water than it actually does. This may result from active transport or from cuticular water pump (Shaw and Stobbart, 1972).

Cuticular transpiration, although low in absolute rate, is a greater source of water loss from desert beetles than respiratory transpiration, and an adaptive feature of long-lived Tenebrionidae is the ability to produce secondary sclerotization if the elytra are damaged. This is of particular importance since the main function of the subelytral cavity is to reduce water loss through the spiracles (Cloudsley-Thompson, 1965a, 1977).

The lipoid cuticular layer which prevents desiccation in insects is also impermeable to oxygen and carbon dioxide. Thus, insects have to develop a respiratory mechanism which permits gaseous exchange, whilst restricting water loss to a minimum. The insect's spiracles are kept shut by special muscles until the amount of carbon dioxide in the respiratory system exceeds about 5%. Only then are they released to permit ventilation.

Uric acid, excreted by insects, is extremely insoluble and can be eliminated from the body in a dry state. Water is also reabsorbed from the faeces through the rectal glands. In addition to water obtained from food, and metabolic water conserved, some desert insects are probably capable of absorbing water vapour from unsaturated air through their integument or rectum. This has been shown to take place in the Californian desert cockroach, *Arenivaga* sp. (Edney, 1966), and in the thysanuran, *Ctenolepsima terebrans* (Edney, 1971b).

Desert insects can survive exposure to extremely high temperatures. *P. grandis* and *A. antiqua* died at 43°C and 46°C respectively, on exposure to those temperatures for 24 hours (Cloudsley-Thompson, 1962). The colour of desert beetles affects, to some extent, their body temperature. In direct sunshine, the abdomen of the living tenebrionid beetle *Onymacris brincki* of the Namib, which is covered by white elytra, was 3 to 4°C cooler than its black thorax. Under identical conditions, the abdomen and thorax of *O. rugatipennis*, which is an entirely black beetle from the same region, had the same temperature as each other and the thorax of *O. brincki* (Edney, 1971a). Since similar effects were observed in dead beetles, the warmer thorax was not due to muscular activity. When the white elytra of *O. brincki* were covered with carbon black, its abdominal temperature rose to equal that of *O. rugatipennis*. Due to lack of easily obtainable food in the desert, insects show surprisingly prolonged resistance to starvation; the beetle *Pimelia bajula* has survived 137 days; and *Blaps requini* from Tunisia lived for five years without food (Cloudsley-Thompson, 1965).

REFERENCES

ABUSHAMA, F. T. (1968) The rhythmic activity of the grasshopper *Poecilocerus hieroglyphicus* (Klug) (Acridiae, Pyrogomorphinae). *Ent. exp. Appl.* **11**, 341–347.

ABUSHAMA, F. T. (1970) Loss of water from the grasshopper, *Poecilocerus hieroglyphicus* (Klug) compared with the tree locust, *Anacridium melanorhodon melanorhodon* (Walker). *Zeit. ang. Ent.* **75**, 124–134.

ABUSHAMA, F. T. (1971) *Strobiloestus* sp. causing myiasis in the Sudan Bohor reed-buck *Redunca redunca cottoni* (W. Rothschild). *The Entomologist*. December.

ABUSHAMA, F. T. (1972) The repugnatorial gland of the grasshopper *Poecilocerus hieroglyphicus* (Klug). *J. Ent.* (A) **47** (1), 95–100.

ABUSHAMA, F. T. (1973) Grasshoppers borrow plants' protection. *New Scientist*, **56** (835), 472.

ABUSHAMA, F. T. and ABDEL NUR, H. O. (1973) Damage inflicted on wood by the termite, *Psammotermes hybostoma* Desneux. *Zeit. ang. Ent.* **73**, 216–223.

ABUSHAMA, F. T. and AHMED, A. A. (1976) Food plant preference and defence mechanism in the lygaeid bug *Spilostethus pandurus* (Scop.). *Zeit. ang. Ent.* **75**, 206–213.

ABUSHAMA, F. T. and EL HAG, A. G. (1970) Distribution and food-plant selection of the acridids near Khartoum. *Zeit. ang. Ent.* **69**, 212–221.

BEAMENT, J. W. L. (1958) The effect of temperature on the water-proofing mechanism of an insect. *J. exp. Biol.* **35**, 494–519.

BEAMENT, J. W. L. (1961) The water relation of insect cuticle. *Biol. Rev.* **36**, 281–320.

BERNARD, F. (1951) Les insects Sociaux du Fezzan. Comportement et biogeographie. *Mission Scientifique du Fezzan* (1944–45). Ch. 5, Zoologie (Arthropods 1). Inst. Richerches Sahariennes de l'Univ. de Alger.

BUXTON, P. A. (1923) *Animal Life in Deserts.* Arnold, London.

CARLISLE, D. B., ELLIS, P. E. and BETTS, E. (1965) The influence of aromatic shrubs in sexual maturation in the desert locust, *Schistocerca gregaria. J. Insect Physiol.* **11**, 1541–1558.

CHAUVIN, R. (1967) *The World of an Insect* (translated by H. OLDROYD). World University Library, London.

CLOUDSLEY-THOMPSON, J. L. (1956) Studies in diurnal rhythms. VI. Bioclimatic observations in Tunisia and their significance in relation to the physiology of the fauna, especially woodlice, centipedes, scorpions and beetles. *Ann. Mag. Nat. Hist.* Ser. 12 **9**, 305–329.

CLOUDSLEY-THOMPSON, J. L. (1960) *Animal Behaviour.* Oliver & Boyd, Edinburgh and London.

CLOUDSLEY-THOMPSON, J. L. (1962) The lethal temperatures of some desert arthropods and the mechanisms of heat death. *Ent. Exp. Appl.* **5**, 270–280.

CLOUDSLEY-THOMPSON, J. L. (1965a) On the function of the sub-elytral cavity in desert Tenebrionidae (Col.). *Ent. monthly Mag.* **100**, 148–151.

CLOUDSLEY-THOMPSON, J. L. (1965b) *Desert Life.* Pergamon Press, London.

CLOUDSLEY-THOMPSON, J. L. (1977) *Man and the Biology of Arid Zones.* Edward Arnold, London.

CLOUDSLEY-THOMPSON, J. L. and CHADWICK, M. J. (1964) *Life in Deserts.* Foulis, London.

DEKEYSER, P. L. and DERIVOT, J. (1959) *La Vie animale au Sahara.* Libraire Armand, Cohin. Paris.

EDNEY, E. B. (1966) Absorption of water vapour from unsaturated air by *Arenivaga* sp. (Polyphagidae, Dictyoptera). *Comp. Biochem. Physiol.* **19**, 386–408.

EDNEY, E. B. (1971a) The body temperature of tenebrionid beetles in the Namib Desert of Southern Africa. *J. exp. Biol.* **55**, 253–272.

EDNEY, E. B. (1971b) Some aspects of water balance in tenebrionid beetles and a thysanuran from the Namib Desert of Southern Africa. *Physiol. Zoöl.* **44**, 61–76.

EL RAYAH, E. (1969) Behaviour and sensory physilology of some desert beetles. M.Sc. Thesis, Zoology Department, University of Khartoum, Sudan.

FISHELSON, L. (1960) The biology and behaviour of *Poekilocerus bufonius* (Klug), with special reference to the repellant gland. *Eos. Mad.* **41**, 602–658.

GREEN, G. W. (1955) Temperature relation of ant-lion larvae (Neuroptera: Myrmeleontidae). *Can. Ent.* **87**, 441–425.

HAFEZ, M. and MAKKY, A. M. M. (1959) Studies on desert insects in Egypt. III. On bionomics of *Adesmia bicarinata* (Klug). *Bull. Soc. Ent. Egypte,* **43**, 89–113.

HOWSE, P. E. (1964) *Termites: A Study in Social Behaviour.* Hutchinson Univ. Lib., London.

LEES, A. D. (1947) Transpiration and the structure of the epicuticle in ticks. *J. exp. Biol.* **23**, 390–410.

LEOPOLD, A. S. (1967) *The Desert.* Time Life Books, Time Inc., New York.

PIERRE, F. (1958) *Écologie et Peuplement Entomologique des Sables vifs du Sahara Nord-Occidental.* CNRS, Paris.

SHAW, J. and STOBBART, R. H. (1972) The water-balance and osmoregulatory physiology of the desert locust, *Schistocerca gregaria* and other desert and xeric arthropods. *Symp. Zool. Soc. London,* **31**, 15–38.

UVAROV, B. P. (1954) The desert-locust and its environment. In *Biology of Deserts,* edited by J. L. CLOUDSLEY-THOMPSON, pp. 85–89. Institute of Biology, London.

WHEELER, W. M. (1930) *Demons of the Dust.* Norton, New York.

WIGGLESWORTH, V. B. (1945) Transpiration through the cuticle of insects. *J. exp. Biol.* **21**, 97–114.

CHAPTER 12

Insect Pests of the Sahạra

G. B. POPOV, M.B.E., T. G. WOOD and M. J. HAGGIS

Tropical Development and Research Institute
(Formerly Centre for Overseas Pest Research, London)
College House, Wrights Lane, London W8 5SJ

CONTENTS

12.1. FOREWORD

This chapter is dedicated to Graham Mitchell, our late colleague at the Tropical Development and Research Institute, London, who was to have written it, but his untimely, much regretted death after a brief illness, intervened before he had had time to begin the work. We are grateful for the privilege of writing the chapter in his place as a tribute to his memory. The section on Termites is by Tom Wood, the remainder by George Popov, with a contribution on *Heliothis armigera* (Hb.) by Margaret Haggis. We are grateful to the Director of the Institute for permission to reproduce Figs. 12.1, 12.2, 12.3 and 12.4 from the *Desert Locust Forecasting Manual*.

12.2. INTRODUCTION

During the Pleistocene the Sahara, in common with the rest of Africa, has known great ecological and climatic vicissitudes, brought about by alternating wet and dry periods, known as pluvials and interpluvials (see Monod, 1963; Moreau, 1966). Although on the evidence of chains of fossil dunes stretching from Senegal to northern Nigeria (Grove, 1958), at one time the Sahara extended some 500 km farther south than it does at present, it has since, and on the last occasion no more than 6000 years ago, known much wetter conditions than it does today. As the pollen evidence indicates, during the Mousterian–Aterian times some 20,000–30,000 years ago, the Ahaggar and Aïr in central Sahara bore a vegetation of Mediterranean scrub and dry woodland type (Quezel, 1960). Some relics of this vegetation survive to this day. Of them the olive (*Olea laperrinei*) in the Ahaggar and the magnificent, centuries-old cedars (*Cupressus dupreziana*) of Tassili des Ajjer are the most outstanding examples. There is also fossil and other evidence to show that genera and species characteristic of tropical Africa had been able to get across the Sahara and reach northern Africa. Evidence of this tropical ancestry is particularly evident in the mammals and grasshoppers (Uvarov, 1953). Furthermore, artefacts show that as recently as 5000 years ago, Neolithic man and his stocks could wander around widely in the Sahara (see especially Clarke, 1962, 1964; and review by Monod, 1963). In short, there is ample evidence that the Sahara then interposed no great barrier to a north/south exchange by many species of animals and plants. Rapid desertfication occurred since and today the Sahara is too severe a desert for plants or terrestrial animals to traverse; only some migrant birds and insects may do so.

Agriculture in the Sahara depends on irrigation and is thus centred at water points, the oases. The Saharan oases are few and scattered, in the nature of islands, or perhaps archipelagos in a vast ocean desert. Nevertheless the desert is not quite the alien environment that is the sea; it has its own terrestrial fauna and flora, albeit an impoverished and specialized one. Moreover, after the sporadic and rare rains, the desert may for a while bloom and become a less hostile environment, even for some of the more delicate species, which may then penetrate, spread and even multiply for a time and reach the refuge of an oasis, there to survive. The desert is thus in the nature of a filter allowing ingress to some species and keeping out others. Compared with the pest faunas that are found on similar crops on both sides of the Sahara, that of the Saharan oases is a depauperate one; its composition varies from one oasis to another and becomes progressively poorer the more they are remote and isolated. A notable exception is the Nile valley that, like a broad highway, traverses eastern Sahara allowing the penetration of various life forms from either side. Indeed the Nile valley is so special a case, that it deserves recognition as a key environment in its own rights and a more or less comprehensive account of its pest fauna is outside the scope of this chapter.

In keeping with its relatively recent history as a desert, the Sahara, in contrast to the older deserts like the Namib, is low in endemics; most of them are on the lower rungs of the classification ladder, while some others, like the above-mentioned olive of the Ahaggar, or the cedar of Tassili are not so much Saharan, as relics of the wetter epochs, that still survive in their mountain refuges. Similarly outside the Nile valley, agriculture in the Sahara has a relatively short history and there is no evidence that the Neolithic man, who left such abundant traces of his presence all over the desert, was more than a hunter-gatherer, or a pastoralist (Balachowsky, 1954). In any case, all the crops cultivated today in the Sahara have been introduced from the outside. This is also true of the date palm, the principal food crop—'le pain des sahariens' (A. Chevalier, 1932), which probably arrived from Mesopotamia.

12.3. INSECT PESTS OF THE SAHARA

The insect pest fauna of the Sahara reflects well this environmental and climatic background and evolutionary history. Balachowsky (1954), one of our major sources of information, divides the Saharan pests into two main categories: the *introduced*, for the most part cosmopolitan species, and the *adapted* species, that also occur outside the oases, but which have found the artifical, man-made environment to be an acceptable, not to say a preferable alternative to their natural one. Many of the latter are migrant species, that may from time to time invade the oases, to leave more or less persistent populations.

12.3.1. The introduced species

The most characteristic of the introduced species are the stored products pests which may be found the world over. The most important among them in the Sahara are several species of beetles (Coleoptera), notably the saw-toothed grain beetle (*Oryzaephilus surinamensis* L.) and the dried fruit beetle (*Carpophilus hemipterus* L.) on dates; the common flour beetle (*Tribolium confusum* Duv.) and the rice weevil (*Sitophilus orizae* L.) in cereal grain, flour and starchy products and the *Dermestes* spp. in animal produce and organic matter. Among the Lepidoptera, flour moths (*Ephestia* spp.) and date moths (*Myelois* spp.) are the most common. There are also some others, but the complete inventory of stored products pests is a modest one. This paucity could be ascribed to the adverse Saharan conditions, notably the low humidities and the very high summer temperatures that prevail even in the oases, while outside them the pests need to survive the hazards and the ardours of transport across the desert. These were particularly adverse in the days of the camel caravans and were more or less effective in limiting the spread of the pests. Today with the advent of wheel transport and the aeroplane, the spread of infestations is more rapid and the importance of this group of pests is on the increase.

Predictably, the introduced pests are more numerous and varied along the border on both sides of the Sahara, decreasing towards the middle, where many of the species have so far failed to penetrate. As an example one can cite the Mediterranean fruit-fly (*Ceratitis capitata* Wied. (Diptera)) that has reached Biskra and the cottony cushion scale *Icerya purchasi* Mask. (Coccoidea) Biskra and Laghouat, while the black bean aphid *Aphis fabae* Scop. (Aphidoidea) is found in most northern and pre-Saharan oases, but does not penetrate the Sahara.

Another important group of pests are the scales (Coccidae), notably the date-palm scales. Three species occur in the Sahara, of them the date scale (*Parlatoria blanchardi* Targ) is by far the most important. It appears to be a truly Saharan species, for its distribution coincides with the Saharan climatic conditions where the dates ripen and it is absent from the peripheral areas such as the Mediterranean coast, the Algerian highlands, or northern Cameroon, where the date-palm may grow, frequently as an ornamental plant, but where the dates fail to ripen. The theory of an eastern origin of the date-palm is borne out by the fact that the appearance of *P. blanchardi* in western Sahara is a recent one. Thus it is only since 1949 that it has become a pest in Mauritania, where it has since acquired the status of the principal pest of the principal food crop outside the Senegal river valley. As many as 180 million scales were at times recorded on a single 10–15-year-old palm, completely covering all the aerial vegetative parts. Such infestations would kill a younger palm and seriously weaken an older one, reducing its yield. Thus at Haroun the crop yield decreased by 50–60% (Tourneur, 1975). Biological control trials were initiated under the aegis of the Institut Français de Recherches Fruitières d'Outremer (IFAC). Eventually good results were obtained with the coccinellid ladybird *Chilocorus bipustulatus* var. *iranensis* introduced from Iran, which became well adapted to local climatic conditions. In 1970 experimental work was succeeded by large-scale field trials, when 32,000 ladybirds were released in the Atar and Tidjikdja areas. The experiment proved a great success and in 1971 a mass-breeding centre of *C. bipustulatus* var. *iranensis* was established near Nouakchott. Subsequently bioecological campaigns were conducted in the Adrar, Tagant and Assaba areas from operational bases in these provinces. More recently IFAC extended its biological control operations against *P. blanchardi* to the Adrar des Iforas area of Mali and the Aïr area of Niger (Tourneur, 1975).

For operational purposes, the predators were packed in special boxes and provided with a supply of scale-infested plants, in which conditions they could be kept for up to one week. They were then taken to the field by road, or sometimes broadcast from aircraft, the boxes so designed that they opened on impact, with the minimum damage to the ladybirds. Although the predators proved to be more sensitive to the climatic conditions than the scales, so that their breeding cycle was 4 to 6 weeks shorter, they were nevertheless effective in substantially decreasing the levels of scale infestations. Thus in many areas where in 1966 less than 30% of the palms were free from attack, by 1972 the percentage of the healthy palms increased to 90–100%.

12.3.2. The adapted species

These are insect species that occur in the Sahara under natural conditions, but which have accepted, and indeed have prospered under the artificial, man-made conditions in oases. They cover a wide range, from polyphagous species such as locusts, grasshoppers and noctuid moths, which accept a wide range of food plants including many kinds of crops, to highly specific, stenophagous ones with a very narrow range of food tolerance, but which have also accepted some, often a closely related, cultivated variety.

12.3.2.1. *The polyphagous species*

These form a large and very important group of insect pests comprising many orders of insects. The Orthoptera contain the locusts and grasshoppers, discussed in a separate section of this chapter, and also the mole crickets (*Gryllotalpa africana* Beauv.) and several species of crickets, the most spectacular amongst them being *Brachytrypes megacephalus* Lef. of central and eastern Sahara. These are relatively minor pests which are well adapted to the Saharan conditions, living in burrows and emerging only at night. The mole crickets are found both in oases and in the sandy beds of wadis well away from them.

Coleoptera comprise the scarabeid beetles, chafers whose root-feeding terrestrial larvae, particularly those of *Rhizotrogus*, can cause considerable crop damage. They are important pests of cereal crops in northern Algeria, but do not penetrate south beyond the northernmost Saharan oases and east beyond Cyrenaica. Some Cetoniinae such as *Phyllognatus excavatus* and *Oxythyrea pantherina* are frequently found feeding on the flowers of wild *Acacia* spp., but would also attack the flowers of date-palms.

Among the Lepidoptera various Noctuid moths are common in oases and their caterpillars sometimes inflict serious damage on a wide range of crops, particularly legumes and vegetables. The most common species belong to the common African or cosmopolitan genera *Autographa* (*Plusia*), *Prodenia*, *Spodoptera* (*Laphygma*) and *Heliothis* (various leafworms, bollworms, armyworms, etc.). These pests are commonly found both on the crops and in natural habitats outside them, and it is probable that the crops are frequently invaded from the outside source areas possibly from some distance. Indeed many of the species are known to be migrant. Little is known of migrations under Saharan conditions, but a summary of recent observations on *Heliothis armigera* (Hb) in the Sudan Gezira, prepared by Miss Margaret Haggis, is given below as an example of the range of migration of these moths and its association with the weather and ecological conditions.

12.3.2.1.1. *Heliothis armigera* (Hb) in the Sudan Gezira

In the Sudan Gezira *Heliothis armigera* (Hb) has been considered a major pest of cotton since 1965, although for several decades before that it was recorded as responsible for serious, sporadic damage, particularly in areas sprayed with DDT. *H. armigera* breeds not only on all the important crops grown in the Gezira, such as cotton, Sorghum, groundnuts, Dolichos lablab and many vegetables, but also in large numbers on the dominant grain, *Sorghum purpureo-sericeum*, outside the Gezira, particularly to the east and south, where it has been seen in far greater numbers than on the Gezira crops themselves (Joyce, 1973, 1976 and pers. comm.).

Intensive studies were made of the biology and migration potential of *H. armigera* in the Gezira area in the 1970s—a period that coincided with extensive changes in the agronomy and crop rotation of the Scheme. Insect pest surveys, including counts of *H. armigera* eggs, were routinely made on cotton twice weekly throughout each season (late August–December) throughout the 800,000 hectares of irrigated crops, to assess the need for and the success of chemical control. The spray threshold was determined as 10 eggs or larvae per 100 plants of *Gossypium barbadense* that was planted most widely up to 1975. *H. armigera* oviposition was not related to bud formation or flowering, which began in mid–October, but over six seasons peak numbers of eggs occurred variously between late September and late October. Egg numbers rose and fell synchronously over areas extending from a few hundred to some thousands of square kilometres, and the boundaries of such areas changed continuously (Joyce, 1976; Haggis, 1981).

This period corresponds with the seasonal southward movement of the Intertopical Convergence Zone (ITCZ). Radar studies showed that while the converging winds at low levels did indeed concentrate flying insects, including *H. armigera*, in a narrow belt at the main discontinuity (ITD) between the south-west Monsoon and the north-east Trades, the wind systems most strongly affecting insect distribution were the outflows from storms within the ITCZ system. In these outflows insects could be 60 times more dense than in the air ahead of the storm (Schaefer, 1976). Detailed examination of the distribution of *H. armigera* eggs immediately after such storms showed that eggs were found extensively at increased density, but that the highest counts were generally outside, or on the edge of the rain, with very small numbers of eggs recorded within the rain area itself (Haggis, 1982). Increased oviposition also occurred at times of rapid extensive southward movement of the ITD, and close beneath the windshift line when it lay over the Gezira (Haggis, 1982). In each season *H. armigera* oviposition on cotton dropped to economically negligible levels in early November, when the ITCZ had moved south of the Gezira.

After 1975, when the earlier flowering 'Acala' cotton, *G. hirsutum*, became the dominant variety sown, distinct generations of *H. armigera* became more readily identifiable and could be related to the early build-up of the population, initially on weeds and then on the groundnuts that are sown in July (Topper, 1978 and Ph.D. Thesis). Biological studies on diapause in *H. armigera* established that few pupae formed under field conditions in the Sudan had an extended development period. In the laboratory, increased incidence of dispause could be induced by rearing under low-temperature, short-day conditions and the diapause was broken by simulating the rising temperatures and then drop in soil temperature at the start of the rainy season in the field. Studies on the migratory potential of *H. armigera* moths concluded that long-duration flight could occur under conditions such as the onset of cool or excessively hot weather, that caused delay in reproductive maturity (Hackett and Gatehouse, 1982; Hackett, Ph.D. Thesis).

12.3.2.2. *The specialized species*

These are the pest species that have a much narrower range of food plants, that are often closely related, or belong to the same family of plants. As examples of these Balachowsky (1954) cites the Pierid butterflies, the Whites, notably *Pieris rapae* L. and *P. napi* L. that attack cruciferous crops—radishes, cabbages, etc., in oases and outside them live on wild crucifers. Some aphids, e.g. *Eurydemna*, follow the same rules. The perennial cotton cultivated in the oases is attacked by a diverse fauna of insects that occur naturally on wild species of *Malva* and *Althea*. The most important of them are the cotton seed bugs (*Oxycarenus* spp., Lygaeidae), while stainer bugs (*Dysdercus* spp., Pyrrocoridae) are common in southern Sahara.

The melon ladybird (*Epilachna chrysomelina* F. Coccinellidae), which is a serious pest of melons, watermelons and other cucurbits in many parts of Algeria, finds an acceptable natural host plant in the colocynth (*Colocynthis vulgaris* Schred.) which grows throughout the Sahara. It is thus able to use it as a relay plant and thus penetrate the remotest corners of the desert and to cross it to the Sudan. Balachowsky (1954) quotes examples of other insect species, not all of them pests, utilizing such Saharan relay plants, sometimes a succession of them, to span the desert. Conversely, sometimes the absence of particular pests in certain parts of the desert can be explained by an absence of suitable relay plants and be used as an argument that in their case at any rate transport by wind, or by Man alone may not be sufficient.

On the whole the importance of specialized pests in the Sahara is small and their numbers very restricted. Thus only one insect species has been recorded on the Saharan olive (*Olea laperrinei*), a scale *Aspidapsis laperrinei* Balachw., while its Mediterranean counterpart (*Olea europea* L.) supports nearly 100. The same is true on the southern side; the relic *Acacia* and *Balanites* trees in the Sahara have very few insect pests; the same species in their original Sahel and Sudan savannas support a rich insect fauna. It is evident that these specialized insects have greater difficulty in becoming adapted to the rigours of the Saharan environment than the polyphagous, often migrant species, which with their wider tolerance and choice of environmental conditions have a greater capacity for adaptation and survival.

12.3.3. Locusts and grasshoppers

12.3.3.1 *The desert locust (Schistocerca gregaria (Forsk.))*

The desert locust deserves pride of place among the Saharan pests not only because of its importance as a pest, but also because of its success of survival under desert conditions as a nomad *par excellence*.

Swarms of the desert locust have plagued agriculture from the earliest recorded times; the eighth plague of Egypt recorded in the Book of Exodus (about 1300 B.C.) was of this species. Pliny records that in 125 B.C. swarms from the Sahara caused 800,000 people to perish in Cyrenaica and another 300,000 in Tunisia. Medieval writings make it clear that plagues have occurred intermittently since those early records and the more detailed reports during the last century testify to the damage wrought. Four main factors contribute to its status as a major pest: the food intake per individual, which on average equals, or exceeds its own weight per day; the great diversity of plants eaten; the large size and density of its swarms, so that even a moderate one would consume some 1000 tonnes of fresh vegetation daily and the mobility of its population (see particularly Uvarov, 1966, 1977; Pedgley, 1981; COPR, 1983).

There is great variation in the amount of damage caused seasonally from country to country and from region to region, but taking the 1949–57 nine-year period as an example, the FAO estimate of damage in twelve most affected countries out of some forty subject to invasion was £15 million (FAO, 1958). Little damage has occurred since the end of the last major plague in 1963, largely due to the efforts of national and regional locust-control organizations, albeit at very considerable cost of input in manpower, equipment and materials.

The invasion area of the desert locust, i.e. the area over which its populations, usually as *gregarious swarms* and *bands*, occur during *plagues* is vast. It covers some 30 million square kilometres and stretches from the Atlantic Ocean to eastern India and from the southern republics of the USSR to southern Tanzania, comprising whole or parts of fifty-seven countries. Moreover, desert locusts have also been recorded reaching south-west Europe and some of the islands off the coasts of Africa and south-west Asia and have been seen at sea, both flying and in the water, as far as 2400 km from land (Waloff, 1960, 1966) (Fig. 12.1).

The recession area, where the desert locust occurs during *recessions*—i.e. the intervals between plagues, usually at much lower densities—is about half the size of the invasion area and stretches in a broad belt of arid and semi-arid land through its middle from the Atlantic Ocean to north-west India. It corresponds fairly closely to the hot Palaearctic deserts, or what biogeographers term the Sindo-Saharan zone. Thus the Sahara and its borderlands correspond to the western half of the recession area of the desert locust (Fig. 12.1).

As was pointed out by Uvarov (1954) (also cited by Cloudsley-Thompson and Chadwick, 1964), in some of its physiological requirements, the desert locust does not exhibit a high degree of adaptation to what is usually understood by desert conditions. For instance, for successful development, the eggs of the desert locust have a very high humidity requirement, since they need to absorb their own weight of water (Shulov, 1952). Therefore the soil surrounding the eggs must preserve sufficient water during the period of egg development—12–15 days under hot and up to 70 days under cold conditions. To some extent the behaviour of the female locust at the time of oviposition provides some safeguard for the eggs, for while the locust prefers to lay in sand that is dry on the surface, oviposition does not take place unless it can reach moist sand with its ovipositor during digging (i.e. moisture should be at a depth of not more than 10 cm

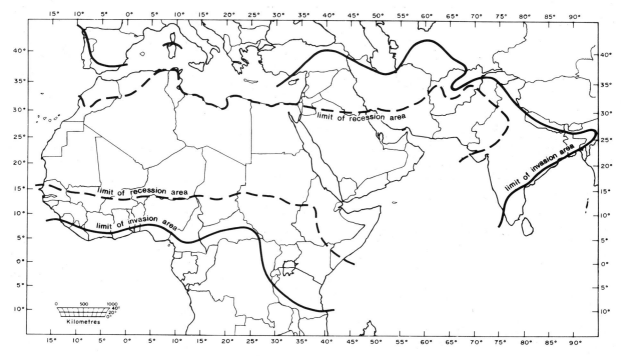

FIG. 12.1. Invasion and recession areas of the desert locust. Figures 12.1–4 reproduced from the *Desert Locust Forecasting Manual* (Pedgley (ed.), 1981), with permission from the Centre for Overseas Pest Research. The remainder are G. Popov originals.

from the surface). Several successive attempts may then be made at several more or less distant sites, but if suitable conditions are not found within a period of about 3 days, the eggs are abandoned on the surface of the ground and the vegetation and perish (Popov, 1958).

The nymphs (hoppers) are no less demanding of high moisture conditions, for they devour their own weight of fresh vegetation each day during their period of development of some 30 to 70 days (Pedgley, 1981). Excessive heat and dryness during the period of development has been known to cause wholesale mortality of hoppers (Ashall and Ellis, 1962). For these reasons, successful breeding is generally associated with significant quantities of direct falls of rain—some 20–25 mm (Magor, 1962)—a rare event in the desert, where in many parts such a quantity would represent half the annual rainfall.

Even during the adult life water requirements are high, for while adult desert locusts are known to be able to survive dry conditions for many months, for their sexual maturation they need either succulent green food or high air humidity and preferably both (Hamilton, 1950; Norris, 1952; Pedgley, 1981).

However, despite these severe handicaps, the desert locust survives in the desert, and even at times multiplies to fantastic numbers and starts a new plague. The key to this remarkable success lies in two characteristics of the desert locust—its *life-cycle strategy* and its behaviour, particularly its *flight performance*. These are well adapted to the opportunities sporadically provided by the weather systems, which provide both the transport for the flying locust and create the ephemeral and temporary habitats, thus enabling it to lead the life of a true nomad.

The life-cycle. Among locusts and grasshoppers, the desert locust belongs to the group that may be termed opportunists—species that so long as the environmental conditions are not limiting, develop and reproduce continuously without any built-in diapause or interruption during any of its life stages—the egg, the nymph (hopper), or the adult. The environmental conditions, and particularly water (usually in the form of rain) and temperature play the controlling and regulating role in determining the duration of the development and indeed whether any development can occur at all. The importance of water directly and indirectly and the dependence of the desert locust on it has already been mentioned. Temperature has a regulating effect on the level of activity and physiological functions. The thresholds are situated around 20°C at the lower, and 35–40°C at the higher end of the range (Pedgley, 1981; COPR, 1983). At lower temperatures such activity as flight in adults and marching in hoppers is reduced, or stops altogether, and embryonic development is suspended. Increase of temperature from the low threshold brings about an increase of activity and speeds up development, but beyond the high threshold there is no longer an increase, but high mortality occurs. The levels of temperature are associated with marked regional differences, which are of importance in forecasting (Pedgley, 1981).

The flight and migrations of the desert locust. Weather systems. The Sahara and the lands on both sides of it are under the influence of two climatic regimes, Mediterranean to the north and tropical to the south. The Mediterranean system is dominated by subtropical anticyclones and eastward-moving waves and cyclones, which are active and bring the bulk of the rain in winter and spring, with summer largely dry. South of the Sahara weather is dominated by the Inter-Tropical Convergence Zone (ITCZ), which in winter is at its southernmost position; it begins to move north from March to April, sometimes advancing in surges, to reach its northernmost position at about 20°N in July. It retreats in September-October, usually more rapidly. South of the ITCZ are the monsoon west winds, disturbed by westward-moving waves and cyclones on the ITCZ. It is convection in the moist monsoon winds modified by westward-moving waves that bring the rain which falls in summer. Annual rainfall amounts, and duration of the rainy season, vary with latitude; the winter–spring rains in the north decrease south and more or less disappear south of 30°N. Conversely summer rains in the south strengthen southwards starting at about 20°N. Between 20° and 30°N rains are rare and erratic, average annual rainfall is below 50 mm and often below 10 mm. Some places have no rain for a year or more, but heavy rainstorms may occur (Pedgley, 1981).

Thus as a result of the distribution and seasonal chronology of rainfall the two sides of the Sahara form complementary breeding grounds for the desert locust—during winter and spring in the north and during summer in the south.

The close associations of migrations of the desert locust with weather dynamics were first put forward in a well-documented hypothesis by Rainey (1951). The essence of the hypothesis is, that major swarm movements are with prevailing winds towards zones of convergence of air masses, which are often associated with precipitation. The result is that locusts and rain are likely to arrive in an area together, an occurrence which had been noticed before its mechanism was properly understood. Considerable advances have since been made towards better understanding of the dynamics of locust migrations and breeding in relation to weather conditions. A major contribution was the biogeographical studies conducted at the Anti-Locust Research Centre (later Centre for Overseas Pest Research and now Tropical Development Research Institute). The principal conclusions together with background information on the biology and ecology of the desert locust and numerous case studies representative of the diverse situations that are likely to arise at different times in different areas and in different weather conditions, are embodied in the *Desert Locust Forecasting Manual* recently published by the Centre (Pedgley, 1981). The information is very complete for the plague periods, rather less so for recession periods.

Figure 12.2, reproduced from the *Manual*, shows the major seasonal breeding and migrations of swarms during plagues. The two main seasonal breeding areas—the spring breeding area north, and the summer breeding area south of the Sahara—are defined very clearly; both are well clear of the desert and for the major part situated in areas receiving over 250 mm of rainfall.

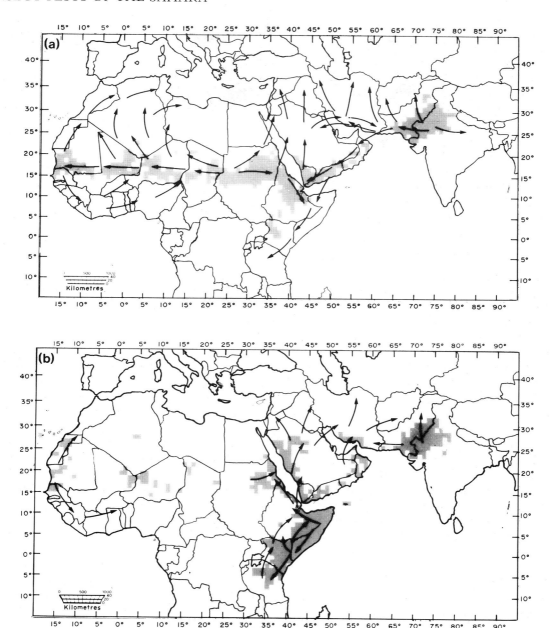

FIG. 12.2. Breeding areas and major movements of resulting swarms.

(a) Summer. Degree squares with hopper band infestations in at least *two* years during the period 1939 to 1975 from August to October, but only August and September on the Somali Peninsula. Degree squares with infestations in at least *one* year are shown in Arabia east of 50°E. Arrows show major movements of summer swarms in the following winter and spring.

(b) Winter. Degree squares with hopper band infestations in at least *one* year during the period 1939 to 1975 from November to January, but October to January on the Somali Peninsula. Arrows show major movements of winter swarms in winter and spring.

[*continued overleaf*

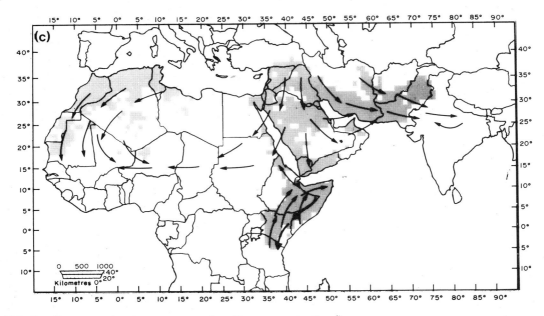

FIG. 12.2. Breeding areas and major movements of resulting swarms (*continued*)

    (c) Spring. Degree squares with hopper band infestations in at least *two* years during the period 1939 to 1975 from March to June. Degree squares with infestations in at least *one* year are shown in West Africa west of 15°E and in Arabia east of 50°E. Arrows show major movements of spring swarms in spring and summer. Heavy arrows in East Africa show the movements in the following winter.

Summer breeding takes place mainly between latitudes of 20° and 14°N in a belt extending from Mauritania and northern Senegal, through Mali, Niger and Chad, eastwards through the Sudan, the bulk of it coincides with the Sahelian and the northern part of the Sudanian vegetation zones. Maturation of swarms invading the summer breeding belt begins from late June and July with the bulk of breeding in August and September. As a rule only one generation is produced and most of the new generation swarms appear from late September to late October. They begin to leave their source areas and migrate towards the countries in which they will breed in winter or spring, or both (Fig. 12.2(a)). The initial emigrations are often to north and north-west, or north-north-west, with swarms moving down the advancing southerlies during northward surges of ITCZ, or more commonly with the south-easterlies associated with the approach of depressions moving in from the Atlantic and eastwards over the Sahara. Such movements result in the spread of swarms into the Algerian Sahara, northern Mauritania and Western Sahara, all of which may be invaded from September, though more frequently from October and even from November. Swarms reaching the barrier of the Atlas Mountains in Morocco and Algeria tend initially to move ENE to the south of the ranges; in Morocco they may become trapped between the ranges in the funnel-shaped Sous Valley (as notably in 1954 and 1960). Eventually swarms spread across the mountains on to the northern plains, though sometimes not till early spring, when they may reach Tunisia and north-west Libya. These long-distance movements may carry the swarms eastwards across northern Sahara through Tunisia and Libya to Egypt, or westwards to the Atlantic Ocean. There are examples, as in October 1954, when strong southerly and south-easterly winds associated with depressions over the Atlantic carried several swarms to the Canaries and dispersed some over the Atlantic as far as the Scilly Isles (Rainey, 1954, 1963).

During plagues swarms tend to overfly the Sahara and reach the spring breeding grounds, but occasionally some may remain and breed during the winter, November to February, in countries bordering the Atlantic, or the central Saharan highlands, all of which may receive winter rain (Fig. 12.2(b)). But in general breeding in these areas appears to be more frequent during recessions.

Spring breeding associated with rains, brought mostly by the eastward-moving disturbances, may occur throughout an extensive belt from Morocco through northern Algeria to Tunisia and north-western Libya, and more locally in Mauritania, Western Sahara, Algerian Sahara and central Libya (Fig. 12.2(c)). The earlier breeding with hatching in February and fledging in March–early April may occur in Mauritania and Algeria, but in Morocco due to lower spring temperatures, it is more protracted and fledging may not begin until May. The main spring breeding, in which hatching and band formation may begin any month between March and May or June, occurs over most of the spring belt; the fledging of hoppers may start in the second half of April, become general in May or June, and continue into July and early August. In some years two successive generations may be produced in spring, when swarms of the early generation, forming in north-western Africa, in March and April, move to the north and east over the spring breeding belt, rapidly reach sexual maturity and breed together with the swarms of their parent swarm generation during the later stages of spring breeding.

From May onwards, the spring swarms forming in north-western and northern Africa begin to move southward and to appear to south of the Sahara in any month between April and July. The southward migration of young swarms can usually be traced clearly only through Mauritania; elsewhere in the Sahara, their sighting has been rare. This is because they move across very sparsely inhabited country, while another reason could be the very high day temperatures causing the swarms to fly by night and remain unseen. In May and June, when the summer belt is being invaded from the north, it is also being reached from the east by swarms originating in Arabia and parts of East Africa. In general the summer breeding grounds in the Sudan are complementary not so much with the spring breeding grounds in north-west Africa, as with winter and spring breeding grounds of the Red Sea and Arabian Peninsula. Thus swarms produced at the end of summer breeding in the Sudan tend to move eastwards into spring breeding grounds of Arabia, rather than westwards into North Africa. Conversely swarms from spring breeding grounds in Arabia and the adjacent countries emigrate (partly) to the Sudan, although they may then continue to move westwards to reach the more westerly parts of the summer breeding belt south of the Sahara (Fig. 12.2(c)).

Recessions and outbreaks. Compared with the volume of data on swarm movements, the information on displacement of non-swarming solitarious populations during recessions is very scanty. This is because solitarious individuals fly singly by night and are considerably less conspicuous than day–flying swarms and are thus hardly ever seen and reported. Nevertheless all available evidence indicates that seasonal movements of recession populations occur on substantial scale and on the whole follow the general trend of plague populations.

The most direct evidence on night flight in individual locusts (and some other insects) comes from radar studies. Of greatest relevance here are those conducted by a joint COPR/Loughborough University Project led by Professor Glen Schaefer, in southern Sahara in Niger, during September–October 1968 (Roffey, 1969; Schaefer, 1972, 1976). These studies established *inter alia* that:

— Desert locusts flew on every night when the radar was in operation.
— The average area densities were in the order of 1 locust/hectare.
— The maximum height of flight was 1.8 km, but the majority flew below 400 m above ground level.
— Flight was generally in the low-level jet stream giving ground speeds from 15 to 65 km/hr, with an average of 35 km/hr.
— Orientation was almost invariably down-wind, even on darkest, moonless nights.
— Typically no locusts were flying at sunset, but numbers appeared about 20–30 minutes later, reaching a peak about 40–50 minutes after sunset. Numbers then tailed off, but occasionally this was interrupted by passage of waves of locusts at higher densities; often such increases were associated with wind change. On some nights locusts continued to fly well after dawn.

Although individual locusts could be detected by radar only up to about 8 km, it was felt from the general performance and characteristics of the flying locusts, that their displacements were frequently over considerable distances, in terms of tens or even hundreds of kilometres. Further evidence of this is circumstantial, arising from changes in the seasonal distribution of locusts. For instance, when the populations in the Niger and Mali Sahara decline at the end of the breeding season in autumn, there is usually a concomitant increase of locust numbers in parts of central Sahara in habitats brought into existence by recent rains. Moreover, the populations involved in both areas tend to have similar colour and morphological characteristics. Also significantly in the event of gregarization, the resultant swarms exhibit similar northward displacement and are often accompanied by individual locusts of similar appearance to those previously recorded in the likely source areas in the south.

As a result of differences in the pattern of behaviour, the distribution of recession populations is considerably more restricted than that of plague populations. In particular, in the northern part of the recession area low night temperatures inhibit flight between autumn and spring, when northward movements are most likely to occur, whereas in the south and east the limit appears to be set by the greater ecological selectivity of low-density adults whose responses to their physical environment are not swamped by gregarious behaviour (Popov, 1965).

Nevertheless the area over which the desert locust is found during recessions is still very large, covering nearly 16 million square kilometres and including all or parts of thirty countries. Figure 12.3(a–e), reproduced from the *Forecasting Manual*, shows the degree squares where adults and hoppers of the desert locust have been recorded during recession periods. These show that breeding by recession populations is almost entirely confined to the drier central part, where the average annual rainfall is under 200 mm. In fact much of the breeding occurs in areas with an average annual rainfall of less than 100 mm (Waloff, 1972), i.e. within the confines of Sahara proper. However, the complimentarity is maintained and the main summer breeding of recession populations takes place on the rains associated with the ITCZ and

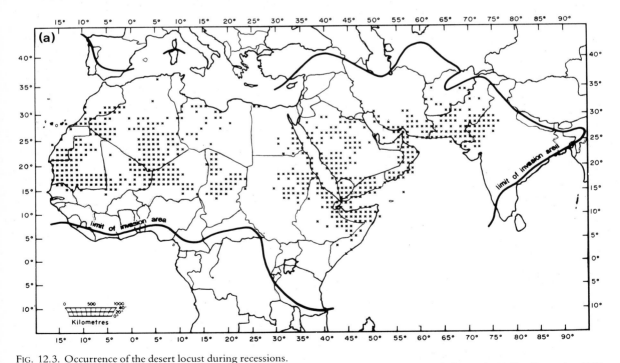

FIG. 12.3. Occurrence of the desert locust during recessions.
(a) Adults. Degree squares where adults have been recorded during recession periods from July 1963 to December 1967, and from 1969 to 1976.

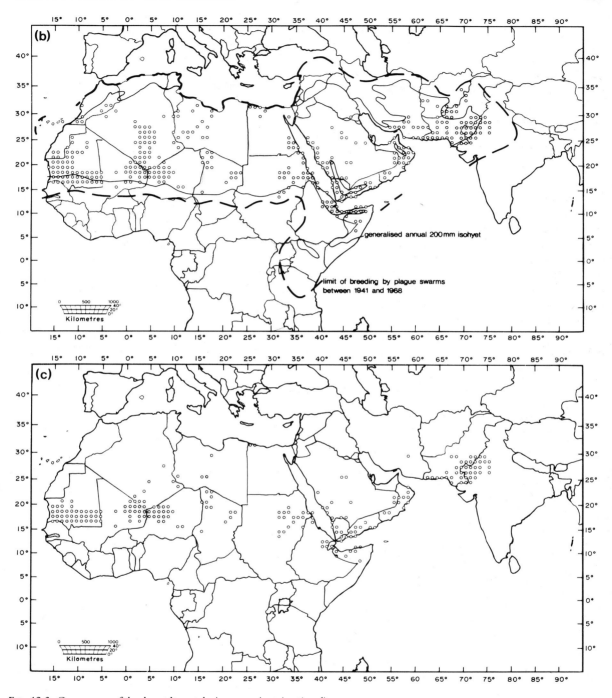

FIG. 12.3. Occurrence of the desert locust during recessions (*continued*)
   (b) Hoppers. Degree squares where hoppers have been recorded during recession periods from 1920 to 1948 (after Waloff 1966) and from 1963 to 1976.
   (c) Summer generation hoppers. Degree squares where hoppers have been recorded from July to October (July to September on the Somali Peninsula) during recession periods from 1920 to 1948 (after Waloff 1966) and from 1963 to 1976. [*continued overleaf*]

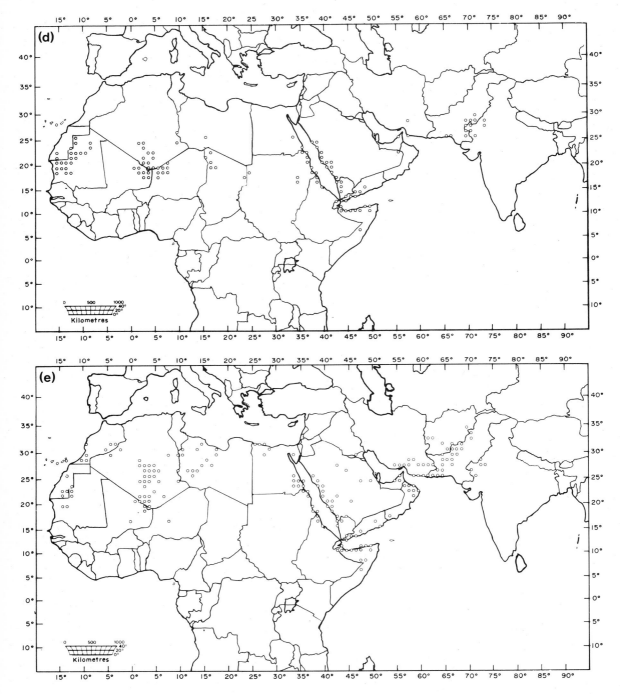

FIG. 12.3. Occurrence of the desert locust during recessions (*continued*)
(d) Winter generation hoppers. Degree squares where hoppers have been recorded from November to January (October to January on the Somali Peninsula) during recession periods from 1920 to 1948 (after Waloff 1966) and from 1963 to 1976.
(e) Spring generation hoppers. Degree squares where hoppers have been recorded from February to June during recession periods from 1920 to 1948 (after Waloff 1966), and from 1963 to 1976.

south-west monsoon, in a belt running through southern Sahara and its southern borderlands eastwards
to the Red Sea and through southern Arabia to India and Pakistan. The resulting young adults, unless
retained by late summer rains, emigrate towards winter and spring breeding areas, generally northwards
in western Africa and towards the Red Sea in the Sudan, where breeding then follows in areas favoured by
rain during winter and spring.

From Fig. 12.3(b–e) and Fig. 12.4 it is evident that areas where the principal breeding and particularly
the outbreaks of the desert locust occur during recessions are generally clustered around the areas of
marked relief, where the incidence of rainfall is greater and where the amount of direct rainfall is enhanced
by drainage. Under Saharan conditions, where the flow of surface water is unimpeded by vegetation, the
supplement of water brought by run-off can be considerable despite the high evaporation; in particular
flooding down the major wadis can occasionally be very great. Thus the August–September 1966 rains on
the south-west flank of the Tibesti mountains in central Sahara resulted in the flooding of all the major
wadis (enneris), reaching points up to 150 km away from the mountains and creating locust habitats in the
desert (that itself had had no rain), which then remained suitable for locust breeding until the following
January (GP pers. obs., Fig. 12.5).

The area of Tamesna in northern Niger is selected as an example to illustrate the desert locust habitats
in southern Sahara.

Tamesna is a vast plain situated between the Aïr mountains in northern Niger in the east and the Adrar
des Iforas mountains in northern Mali in the west. Extensive network of largely fossil drainage has, during
the pluvials, led to deposition of hydromorphic clays, whose cracks and fissures have, during the ensuing
dry periods, become progressively filled by eolian sands, with some of the sand accumulating as dunes
(Fig. 12.6). This process has created soils at once permeable and retentive of any rain water they receive.
Nowadays the average annual rainfall over the Tamesna is only within the range of 20 to 75 mm and is
little more than this in the mountains. The network of wadis is therefore functional only in the vicinity of

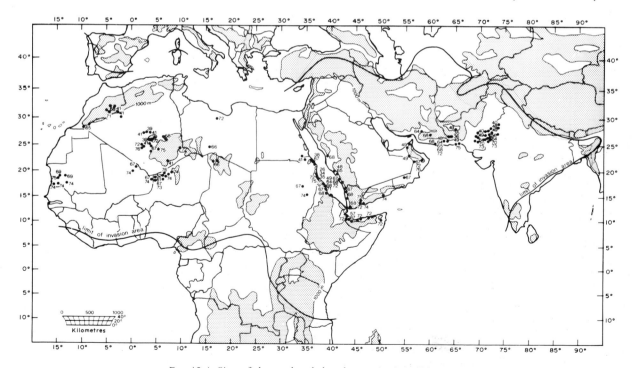

FIG. 12.4. Sites of observed or deduced gregarization, 1926–1976.

FIG. 12.5. The delta of enneri Damar, where extensive locust habitats developed following August 1966 floods, from rains in Tibesti mountains 100–150 km away. Some of these habitats remained suitable for locust breeding until the following January.

FIG. 12.6. An aerial view of Aghlan Niklen, in Tamesna in northern Niger: This is part of the now fossil drainage network flowing westwards from the Aïr mountains. Today there is no flow, but rain-water drains from the surrounding higher ground into such shallow depressions, where the combination of clayey soils and surface sand help the preservation of water and further the formation of *Schouwia* and *Tribulus* desert locust habitats.

FIG. 12.7. An annual community of *Schouwia purpurea* (upright, large-leaved species) and *Tribulus ochroleucus*, which constitutes the principal breeding and gregarization habitat of the desert locust in Tamesna.

the mountains and their floods do not normally reach Tamesna proper. However, after local rain, the sheet flow from the surrounding ground into the barely perceptible depressions that correspond to the fossil drainage is as much as 5–6 times the amount of the direct rain. Here develop communities of *Tribulus ochroleucus, Schouwia purpurea, Aerva persica, Colocynthis vulgaris* and some other plants that form the principal habitats of the desert locust. Some of these plants, notably *Schouwia*, are very demanding of water and seem as ill adapted to desert conditions as are the eggs of the desert locust, yet both develop well thanks to the properties of the Tamesna soil, provided they receive adequate rain (Figs. 12.7–9).

The principal locust habitats recorded during aerial and ground reconnaissance in the course of four consecutive years (1964–1967) are shown in Figs. 12.11(a)–(d). The rainfall and the habitat conditions varied widely during these years, and so did the locust situation.

1964. This was a year of good rainfall, with 55 mm recorded at In Abangharit, the locust control base, during late August and early September. This led to abundant development of *Schouwia* and *Tribulus* habitats; locust population remained low and breeding in August–October did not lead to gregarization.

1965. A year of moderate rainfall and more restricted habitats than in 1964; locust population densities somewhat higher with early stages of gregarization noted among hoppers, but without band or swarm formation (Fig. 12.10).

1966. A year of poor rainfall, total at In Abangharit for May–September 60 mm. Habitats small, highly localized and poorly developed. Severe competition between locusts and grazing camels, leading in many cases to complete destruction of habitats and failure of locust breeding. Several local small-scale, largely abortive gregarizations.

1967 (described in Roffey and Popov, 1968; Pedgley, 1981). A year of widespread, abundant rains with 123 mm recorded at In Abangharit in August alone (about twice the annual total), leading to extensive development of locust habitats. Increasing numbers of adults migrating as isolated individuals at night, reach Tamesna during September and October. These probably originated from July–August breeding in

FIG. 12.8. *Schouwia-Tribulus* habitat in the valleys of Adrar des Iforas in northern Mali.

FIG. 12.9. **An egg-laying site of the desert locust within the *Schouwia-Tribulus* habitat, individual egg-pods are marked by matchsticks.**

Fig. 12.10. Hoppers of the desert locust aggregating and gregarizing on *Schouwia* plants in Tamesna. Some hoppers have gregarized more completely than others and the more gregarious ones have a darker pattern.

the Sahel and preceding spring breeding in Tibesti, Fezzan and south-east Algeria. In Tamesna locusts became concentrated in the *Tribulus* and *Schouwia* habitats, which, while more extensive than during the preceding years, still constituted less than 10% of the total area. Surveys and counts established that less than 5% of the locusts occurred outside the habitats, which suggests strongly that locusts flying singly at night were settling in them selectively. Further counts revealed that while in the area as a whole the average densities were only 0.04–0.08/ha, in the green areas they reached 210/ha, so that concentration due to selective settling was some thousands of times greater than the average density. Further concentration occurred at the time of laying in restricted sites, where moist soil was still within reach of the digging females. These sometimes led to onset of gregarious behaviour and laying in dense groups; this in due course resulted in synchronous emergence of large numbers of hoppers, which gregarized at an early stage and formed groups, which fused to form larger groups, then bands, which in time began to march and ultimately produced small swarms. This explosive situation prevailed across the whole of Tamesna, over much of the Adrar des Iforas and parts of Aïr, a total of some 15 degrees square. Although control operations greatly reduced the locust numbers, many escaped and subsequently bred during spring in parts of the Algerian Sahara. Here again there was multiplication and gregarization and some escapes despite control operations, so that at the beginning of summer 1968 the situation was worse than in the

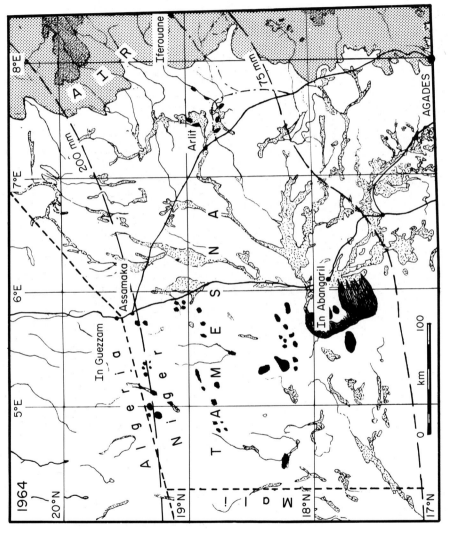

Fig. 12.11. (a) The development of the desert locust habitats in Tamesna during four consecutive years, 1964, 1965, 1966 and 1967, as recorded during aerial and ground reconnaissance. The size and limits of the habitats, which appear as black spots, are very approximate. Tamesna is situated roughly between the 20- and 75-mm isohyets. [*continued opposite*

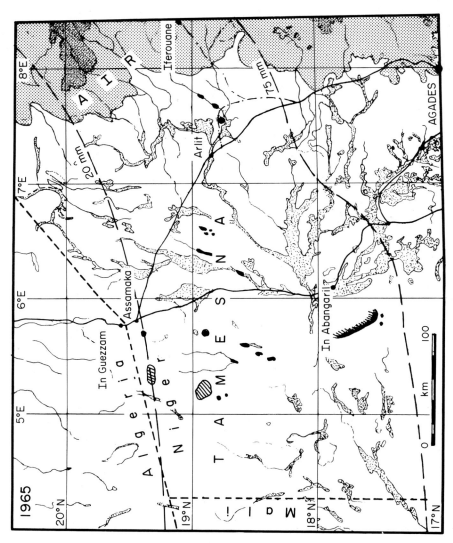

Fig. 12.11 (continued) (b). The development of the desert locust habitats in Tamesna during four consecutive years, 1964, 1965, 1966 and 1967, as recorded during aerial and ground reconnaissance. The size and limits of the habitats, which appear as black spots, are very approximate. Tamesna is situated roughly between the 20- and 75-mm isohyets.　　　　[continued overleaf

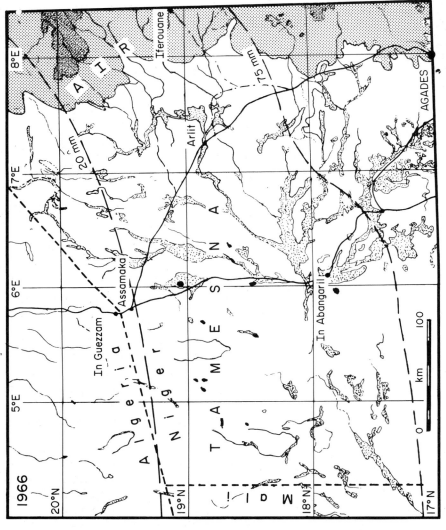

FIG. 12.11 (continued) (c). The development of the desert locust habitats in Tamesna during four consecutive years, 1964, 1965, 1966 and 1967, as recorded during aerial and ground reconnaissance. The size and limits of the habitats, which appear as black spots, are very approximate. Tamesna is situated roughly between the 20- and 75-mm isohyets.   [continued opposite

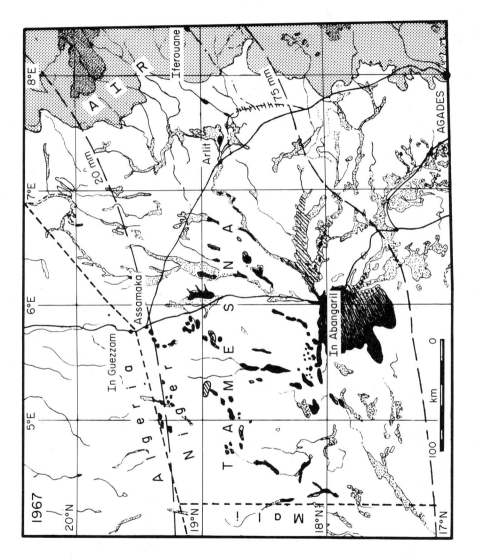

Fig. 12.11 (*continued*) (d). The development of the desert locust habitats in Tamesna during four consecutive years, 1964, 1965, 1966 and 1967, as recorded during aerial and ground reconnaissance. The size and limits of the habitats, which appear as black spots, are very approximate. Tamesna is situated roughly between the 20- and 75-mm isohyets.

preceding year. It was then further aggravated by an influx of swarms and scattered locusts escaping from the Arabian spring breeding grounds in June, some of them traversing the Red Sea, the Sudan, and Chad to reach Niger and Mali (Pedgley, 1981). Further and more serious invasions of western Africa occurred in October, when several large and highly mobile swarms, produced during summer breeding in the Sudan, rapidly traversed the Sahara to reach southern Morocco in the space of one week. It was only thanks to the concerted all-out campaigns waged by all the national and regional locust-control organizations that the upsurge of a new plague was averted. Even so locust populations remained high, requiring preventive control until 1970–71, when they eventually declined at the time of the Sahelian drought. Since then there were further outbreaks in the Sahara in 1974, 1976, 1978 and 1980–81, the last one being particularly serious.

To conclude. During recessions between plagues, the desert locust nomadizes in the Sahara on pastures created by seasonal rains; during the summer in southern and during winter and spring, in central and western Sahara. When the rains are average or poor, locust numbers may remain low for several consecutive years, but locusts survive even such extreme conditions as the Sahelian drought. On the other hand, in the event of widespread, prolonged rains, Saharan habitats may provide conditions allowing prolonged multiplication ultimately leading to outbreaks. Exceptionally, when a succession of such good rains in several successive seasonal breeding areas ensures that conditions remain favourable for two years or more, plagues may develop.

The vagaries and the unpredictability of the Saharan rainfall are such, that locust-control organizations must maintain continuous vigilance over their domain at considerable drain on their resources. However, modern technology offers some perspectives for improvement, with the greatest promise in the field of the application of remote sensing. Meteosat weather satellites now scan the surface of the globe, permitting detailed monitoring of the development, performance and displacements of important weather systems and their associated cloud masses over the desert locust recession area. In addition, the techniques and methodologies, developed notably by Dr. J. U. Hielkema of FAO (Hielkema, 1980), allow rapid processing of Landsat data to extract information on vegetation biomass density, which is then used as an indicator of availability of soil moisture, food and shelter for the locust. Monitoring the environmental conditions in the potential habitats of the desert locust in the Sahara and elsewhere in the recession area by remote sensing techniques is therefore a technical possibility, although this has yet some way to go before becoming a routine monitoring service.

12.3.3.2. *Other locusts and grasshoppers*

The true home of the African migratory locust (*Locusta migratoria migratorioides* R. & F.), where its plagues are known to have originated in the past, is in the marshes of the central delta of the river Niger. Outbreaks of greater or lesser magnitude, some on the scale of regional plagues, have also occurred from time to time in the Chad basin, the Gash delta of the Sudan, in Angola and some other parts of Africa. Although individual migratory locusts probably originating from the Niger delta spread out during the rains and some may reach parts of the Adrar des Iforas and possibly penetrate even deeper into southern Sahara, these migrations remain closely linked to the dynamics of the ITCZ and unlike the desert locust, the migratory locust does not become involved in regular movements into or across the Sahara either during plagues, or during recessions.

However, individual migratory locusts of undetermined race and original source (it could also be from the Mediterranean basin, where the nominate race and *L.m. cinerascens* F. occur) are found in many Saharan oases, where together with some grasshopper species such as *Aiolopus thalassinus* (F.), *A. simulatrix* (Wlk.), *Eyprepocnemis plorans* (Charp.), *Heteracris* spp., *Pyrgomorpha* sp., *Chrotogonus* sp. and *Ochrilidia* spp. they exist as polyphagous pests of relatively minor importance, occasionally inflicting some damage, notably to vegetables and wheat. It is probable that in the older oases, where these pest species have been established for many years, the conditions are relatively stable and the numbers tend to be low under the constraint of the environmental and biotic factors, so that outbreaks occur relatively rarely.

However, recent agricultural developments in parts of the Sahara tell a very different story. The following is an extract from Duranton *et al.* (1983). Recently Libya, stimulated by its great wealth and natural resources, notably the presence of aquifers under its oil fields and ready availability of modern technology, decided to realize an old dream and turn the desert sands green with fields of wheat, sorghum, lucerne, onions and potatoes. The Sarir project of agricultural and forage production in southern Cyrenaica is one of the largest in Libya. It consists of 239 production units, each of 80 hectares—a complete circle 1 km in diameter. Irrigation is by means of a horizontal boom mounted on a central pivot, which contains the pump providing the water from the underground supply; the boom is supported and rotated slowly by eight to twelve tractors. These circular plots are spaced at about 2.5 km from each other, in lines of 10 km. The current total of 19,120 ha of cultivations is spread over some 4800 sq km of desert.

The project started about 3 years ago over almost the whole area, initially with a single planting of wheat in winter and more recently with the addition of sorghum in summer. A few locusts were noted from the beginning; they were inobtrusive and remained unimportant, until the single winter crop of wheat was allowed to stand and was coupled with experimental planting of sorghum and forage millet during the hot season. Among the five species of acridids that prospered under these conditions, the migratory locust, as a strict grass feeder, did particularly well on the abundant favourite food and shelter, appearing in just the right rotation, together with permanent laying sites in the sand kept moist by irrigation. These conditions ensured non-stop reproduction and to make them really perfect, there were also the bright pilot lamps mounted on each central pivot that helped to attract the locusts to the right site. Moreover, to begin with, there was an almost complete absence of natural enemies in this initially sterile desert environment. Breeding conditions being optimal and constraints minimal, there was an inevitable population explosion followed by gregarization, formation of bands and swarms and a spread of the infestation.

Although the hazards were envisaged at an early stage, the danger signs were at first not recognized by the management, then recognized, but for a while wilfully ignored, so that the situation was allowed to deteriorate and lead to serious crop losses, before remedial action was taken. Given the great resources and modern technology, the situation was brought under control by aerial and ground spraying but at very considerable cost, both in crop losses and control expenditure. The perspective now involves regular monitoring of the situation and the necessary prophylactic control operations. These are costly and the alternative of longer-term improvement by cultural and biological control methods, notably through the introduction of insect natural enemies, is well worth consideration. Moreover, it should be borne in mind that the Sarir and similar agricultural developments could become more attractive to the desert locust, in the event of an introduction of leguminous and other non-gramineous crops, and possibly even more through the advent of weeds, which could transform the marginal crop areas and fallows into acceptable food/shelter and breeding habitats. Even if these do not themselves develop into new outbreak areas, they could contribute towards improving the survival conditions for the desert locust and thus an increase in its numbers.

The Sarir is a perfect example of the dangers inherent in desert reclamation pointed out by Uvarov almost 30 years ago (Uvarov, 1954); his words of advice are a fitting conclusion to this section.

> The main general conclusion which may be suggested by considering the Desert Locust [and other pests] in relation to desert reclamation, is that while reclamation may certainly be able to increase crop producing areas, it would also increase the risks of losing the crops, unless repercussions of reclamation on certain members of desert fauna, such as the Desert Locust, are realised clearly and before it is too late.
>
> It should not be concluded, of course, that desert reclamation is undesirable because it may encourage the [pests,] but this danger must be borne in mind when desert development schemes are considered. It should be possible to provide safeguards against undesirable consequences of irrigation and cultivation, but the need for such safeguards must be realised in time.

REFERENCES

ASHALL, C. and ELLIS, P. E. (1962) Studies on numbers and mortality in field populations of the Desert Locust (*Schistocerca gregaria* Forskål). *Anti-Locust Bull.*, London, No. 38: 59 pp.

BALACHOWSKY, A. S. (1954) Sur l'origine et le développement des insectes nuisibles aux plantes cultivées dans les oasis du Sahara français. In *Biology of Deserts*, edited by J. L. CLOUDSLEY-THOMPSON, Institute of Biology, London.

C.O.P.R. (1983) *The Locust and Grasshopper Agricultural Manual*. Centre for Overseas Pest Research, London.

CHEVALIER, A. (1932) *Resources végétales du Sahara et ses confins Nord et Sud*. Museum d'Histoire Naturelle, Laboratoire d'Agronomie Coloniale, Paris. 256 pp.

CLARK, J. D. (1962) The spread of food production in sub-Saharan Africa. *J. Afr. Hist.* **3**, 211–228.

CLARK, J. D. (1964) The prehistoric origins of African culture. *J. Afr. Hist.* **5**, 161–183.

CLOUDSLEY-THOMPSON, J. L. and CHADWICK, M. L. (1964) *Life in Deserts*. G. T. Foulis & Co., London.

DURANTON, J. F., LAUNOIS, M., LAUNOIS-LUONG, M. H. and LECOQ, M. (1981) De l'acridologie à l'ecologie operationnelle. *Pour la Science* no. 63, January 1983.

GROVE, A. T. (1958) The ancient erg of Hausaland, and similar formations on the south side of the Sahara. *Georgr. J.* **124**, 528–533.

HACKETT, D. S. and GATEHOUSE, A. G. (1982) Studies on the biology of *Heliothis* spp. in Sudan. *Proc. Int. Workshop on Heliothis Management*. ICRISAT, November 1981, pp. 29–38.

HAGGIS, M. J. (1981) Spatial and temporal changes in the distribution of eggs of *Heliothis armiger* on cotton in the Sudan Gezira. *Bull ent. Res.* **71**, 183–193.

HAGGIS, M. J. (1982) Distribution of *Heliothis armigera* eggs on cotton in the Sudan Gezira: spatial and temporal changes and their possible relation to weather. *Proc. Ioc. Int. Workshop on Heliothis Management*. ICRISAT, November 1981, pp. 87–99.

HAMILTON, A. G. (1950) Further studies on the relation of humidity and temperature to the development of two species of African locusts—*Locusta migratoria migratorioides* (R. & F.) and *Schistocerca gregaria* (Forsk.). *Trans. R. ent. Soc. Lond.* **101**, 1–58.

HIELKEMA, J. U. (1980) Remote sensing techniques and methodologies for monitoring ecological conditions for Desert Locust population development. Final Technical Report FAO/CGP/INT/349/USA.

JOYCE, R. J. V. (1973) Insect mobility and the philosophy of crop protection with reference to the Sudan Gezira. *PANS* **19**, (1), 62–70.

JOYCE, R. J. V. (1976) Insect flight in relation to problems of pest control. In *Insect Flight*, edited by R. C. RAINEY, pp. 135–155. 7th Symp. Royal Entomological Society Lond., Oxford, Blackwell Scientific.

MAGOR, J. I. (1962) Rainfall as a factor in the geographical distribution of the Desert Locust breeding areas, with particular reference to summer breeding areas of India and Pakistan. Ph.D. Thesis, University of Edinburgh: 118 pp. Unpublished typescript.

MONOD, T. (1963) The late Tertiary and Pleistocene in the Sahara. In *African Ecology and Human Evolution*, edited by F. C. HOWELL and F. BOURLIÈRE, pp. 117–229. Viking Fund Publications in Anthropology.

MOREAU, R. E. (1966) *The Bird Faunas of Africa and its Islands*. Academic Press, New York, London.

NORRIS, M. J. (1952) Reproduction in the Desert Locust (*Schistocerca gregaria* Forsk.) in relation to density and phase. *Anti-Locust Bull.* London, No. 13, 49 pp.

QUEZEL, P. (1960) Flore et palynologie sahariennes, leur signification bioclimatique et paléoclimatique. *Bull. Inst. fr. Afr. noire* (A), **22**, 253–360.

PEDGLEY, D. (ed.) (1981) *Desert Locust Forecasting Manual*, Vols 1 and 2. Centre for Overseas Pest Research, London.

POPOV, G. B. (1958) Ecological studies on oviposition by swarms of the Desert Locust (*Schistocerca gregaria* Forskål) in eastern Africa. *Anti-Locust Bull.* London, No. 31, 70 pp.

POPOV, G. B. (1965) Review of the work of the Desert Locust Ecological Survey June 1958–March 1964. *F.A.O. Progress Report* No. UNSF/DL/ES/8, 80 pp.

RAINEY, R. C. (1951) Weather and the movements of locusts swarms: a new hypothesis. *Nature*, London, **168**, 1057–1060.

RAINEY, R. C. (1954) Recent arrival of Desert Locust in the British Isles. *Proc. R. ent. Soc. Lond.* **19**, 45–46.

RAINEY, R. C. (1963) Meteorology and the migration of Desert Locusts. *Anti-Locust Mem.* No. 7, 115 pp. also *Wld. Met. Org. Tech. Note* No. 54.

ROFFEY, J. (1969) Radar studies on the Desert Locust, *Schistocerca gregaria* (Forskål) in the Niger Republic September–October 1968. *Anti-Locust Research Centre Occ. Report*, No. 17, 14 pp.

ROFFEY, J. and POPOV, G. B. (1968) Environmental and behavioural processes in a Desert Locust outbreak. *Nature*, London, **219**, 446–450.

SCHAEFER, G. W. (1972) Radar detection of individual locusts and swarms. In *Proc. Int. Study Conf. Current and Future Problems of Acridology, London 1970*, edited by C. F. HEMMING and T. H. C. TAYLOR, pp. 379–380. Centre for Overseas Pest Research, London.

SCHAEFER, G. W. (1976) Radar observations of insect flight. In *Insect Flight*, edited by R. C. RAINEY, pp. 157–197. 7th Symp. Royal Entomological Society Lond., Oxford, Blackwell Scientific.

SHULOV, A. (1952) The development of eggs of *Schistocerca gregaria* (Forskål) in relation to water. *Bull. ent. Res.* **43**, 469–476.

TOPPER, C. (1978) The incidence of *Holiothis armigera* larvae and adults on groundnuts and sorghum and the prediction of oviposition on cotton. *Proc. Seminar on the Strategy for Cotton Pest Control in the Sudan*, May 1978, Basle, CIBA-GEIGY, pp. 17–33.

TOURNEUR, J. C. (1975) La lutte biologique menée par l'Institut Français de Recherches Fruitières Outre-Mer (I.F.A.C) contre *Parlatoria blanchardi* Targ. ravageur du palmier dattier en Mauritanie (text of lecture given at the FAO/UNDP Training Course, Dakar, 17 February–21 March, 1975).

UVAROV, B. P. (1954) The Desert Locust and its environment. In *Biology of Deserts*, edited by J. L. CLOUDSLEY-THOMPSON, pp. 85–89. Institute of Biology, London.

UVAROV, B. P. (1966) *Grasshoppers and Locusts. A handbook of general acridology*, Vol. I. Cambridge Univ. Press.

UVAROV, B. P. (1977) *Grasshoppers and Locusts. A handbook of general acridology*, Vol. II. Centre for Overseas Pest Research, London.

WALOFF, Z. (1960) Some notes on the Desert Locust and its occurrence at sea. *Mar. Obsr.* London, **30**, 40–45.

WALOFF, Z. (1966) The upsurges and recessions of the Desert Locust plague: an historical survey. *Anti-Locust Mem.*, London, No. 8, 118 pp.

WALOFF, Z. (1972) The plague dynamics of the Desert Locust, *Scistocerca gregaria* (Forsk.). In *Proc. Int. Study Conf. Current and Future Problems of Acridology, London 1970*, edited by C. F. HEMMING and T. H. C. TAYLOR, pp. 343–349. Centre for Overseas Pest Research, London.

12.4. TERMITES

Termites are social insects inhabiting most tropical and sub-tropical regions. Their basic food is plant material: living, recently dead, dead in various stages of decomposition and organic-rich soil containing humified plant remains (Wood, 1977). Most are harmless detritivores and only 10% of the world's 2000 species are of economic importance in causing damage to crops, plantation and forestry trees and buildings (Johnson, 1981). Ecologically they are notable not so much for their diversity of species as for their numerical abundance. They are particularly abundant in tropical forests and savannas with maximum populations of approximately 4500 m^{-2} where they consume up to 35% of the annual production of plant litter (Wood and Sands, 1977). They are obviously less numerous in deserts, although there is no quantitative data from the Sahara. However, at Feté Olé (16.5°N) in Senegal, Lepage (1974) recorded populations of 229 m^{-2} which consumed 10% of the annual litter production of 125 g m^{-2}.

Twenty-eight species of termites have been recorded from the Sahara desert, thirteen from the northern region and eighteen from the southern (Johnson and Wood, 1980). The Sahara is recognized as a distinct barrier between Palearctic and Ethiopian faunas and the northern and southern regions have only three species in common (Table 12.1). The fungus-growing termites (Macrotermitinae) evolved in Africa and although ten species are recorded from the southern region they are absent from the northern region.

The compilation of a species-list from such a vast region as the Sahara desert is faced with the problem of the presence within this arid area of local, semi-arid or humid areas associated with sea coasts, oases and rivers; the most important being the River Nile. All the Macrotermitinae are associated with these non-arid localities and there are only five species which can be considered as typical desert inhabitants (Table 12.1). Most of the literature on these species relates to studies in semi-arid areas and there have been very few investigations in the Sahara *sensu strictu*; in particular there is very little information from the Western Sahara.

12.4.1. Typical desert inhabitants

Anacanthotermes ochraceus. This species has a deep, diffuse, subterranean nest (Clement, 1954; Hosny and Said, 1980). It is restricted to areas where there is some clay or alluvium in the soil and in the sands of the Sahara is replaced by *Psammotermes*, the sand termite (Harris, 1970). It is often common in saline soils and is always present in oases (Bernard, 1954). It is widely distributed in North Africa (Harris, 1967) and the Nile valley (Zidan *et al.*, 1980) and has been recorded from Sudan (Harris, 1968) and the Fezzan area of Libya (Scortecci, 1936). It occurs up to 2000 m on the high Algerian plateau (Bernard, 1954). It feeds on dry grass and other vegetable debris and in oases and semi-desert conditions also consumes dung, the trash from grain crops and the trunks and leaves of date palms. After several years without rain its foraging has led to the destruction of desert pasture, e.g. south of Fezzan in 1944 and in the centre of Tassili n'Ajjer in 1949 (Bernard, 1954). Considerable damage is done to rural buildings constructed from mud/straw bricks by the removal of the straw and there are reports from Egypt (Kassab, Hassan, Shaarawi and Shahwan, 1960; Hafez, 1980) of the destruction of entire villages.

TABLE 12.1. Termite fauna of the Sahara (adapted from Johnson and Wood, 1980)

Species	Northern Sahara	Southern Sahara
Kalotermitidae		
Cryptotermes brevis (Walker) *	+	
Kalotermes flavicollis (Fabricius) *	+	
K. sinaicus Kemner*	+	
Hodotermitidae		
Anacanthotermes ochraceus (Burmeister)	+	+
Microhodotermes maroccanus (Sjöstedt)	+	
M. wasmanni (Sjöstedt)	+	
Rhinotermitidae		
Heterotermes aethiopicus (Sjöstedt) *		+
Psammotermes assuanensis Sjöstedt*	+	
P. fuscofemoralis (Sjöstedt)	+	+
P. hybostoma Desneux	+	+
Reticulitermes lucifugus (Rossi) *	+	
Termitidae—Termitinae		
Amitermes desertorum (Desneux) *	+	
A. messinae Fuller*		+
A. santschi Silvestri*	+	
Angulitermes nilensis Harris*		+
Eremotermes nanus Harris*		+
Microcerotermes eugnathus Silvestri*	+	
M. parvulus (Sjöstedt)*		+
—Macrotermitinae		
Macrotermes herus (Sjöstedt) *		+
M. subhyalinus (Rambur) *		+
Microtermes aluco (Sjöstedt) *		+
M. subhyalinus Silvestri*		+
M. thoracalis Sjöstedt*		+
M. traghardi (Sjöstedt) *		+
Odontotermes classicus (Sjöstedt) *		+
O. nilensis Emerson*		+
O. smeathmani (Fuller) *		+
O. sudanensis Sjöstedt*		+

* Associated with oases, rivers or coastal areas and therefore not strictly desert–dwelling species.

Microhodotermes maroccanus and *M. wasmanni.* Very little is known of these species, the former being recorded from Morocco and the latter from Tunisia, Libya and Egypt (Harris, 1970). They are known only from the desert fringes, have subterranean nests and feed on vegetable debris.

Psammotermes hybostoma. This species has a deep, subterranean nest. Known as the 'sand termite', it characteristically occurs on sandy soils though is not restricted to them. It is found throughout the desert and as far south as 14°N from Senegal to Sudan (Harris, 1968, 1970). In Algeria it is common east and south of the Atlas mountains below 1300 m (Bernard, 1948). In the Fezzan area of Libya it occurs in oases in arid steppes, around the base of sand dunes and even in areas devoid of vegetation where it survives on wind-blown debris, animal droppings and the sub-fossil relics of a humid Pleistocene flora (Harris, 1970). In semi-desert conditions it feeds on all kinds of vegetation, including the poisonous shrub *Calotropis procera* (Ait.) Aitf. and causes extensive damage to wooden structures (Hafez, 1980; Harris, 1970; Nour, 1980).

Psammotermes fuscofemoralis. This species is less well known than *P. hybostoma*, but it is similarly widely distributed and damaging to wooden structures. Harris (1968) considered these two species (and *P.*

assuanensis) to be conspecific whereas there is evidence (S. Bacchus, personal communication) that they are distinct.

These five species show various adaptations to the desert environment (Johnson and Wood, 1980). All have deep, subterranean nests and workers may go down to the water table at depths of 40 m. All are adapted to uncertain, seasonal productivity by foraging at the time of year when food is most abundant and by storing food in their nests. The problem of foraging on sparsely distributed food resources has been solved in various ways. The harvesting habit of foraging in the open on the soil surface, which has been adopted by *Microhodotermes* and *Anacanthotermes*, potentially exposes foraging termites to the hazards of ultraviolet light and desiccation and these species are normally restricted to foraging at night or during cooler parts of the day. The hazards of this type of foraging behaviour have been avoided by *Psammotermes* which construct protective soil sheeting over their food sources enabling them to forage during the day; *Anacanthotermes* sometimes adopt this behaviour.

12.4.2. Species inhabiting local, non-arid situations

Species such as *Psammotermes assuanensis* (possibly synonymous with *P. hybostoma*), *Eremotermes nanus*, *Amitermes desertorum*, *A. santschi* and *Microcerotermes eugnathus* characteristically occur only in the semi-arid areas fringing the desert although *M. eugnathus* has been recorded at Tamanrasset (Bernard, 1954). The three species of Kalotermitidae (dry wood termites) are limited to the coastal regions; *Cryptotermes brevis* is a 'tramp' species, occurring in both hemispheres where conditions are suitable and the two *Kalotermes*, together with *Reticulitermes lucifugus* are Mediterranean species. *Heterotermes aethiopicus* is limited to north-east Africa and Arabia while *Microcerotermes parvulus*, *Amitermes messinae* and all the Macrotermitinae have a wide distribution in the dry savannas south of the Sahara.

Several of these species have been recorded damaging crops, trees and structural timber on the desert fringes and their depradations can be expected to extend into areas of the Sahara where irrigation favours cultivation and settlement. This is particularly apparent along the Nile valley in Sudan where the fungus-growers, *Microtermes* and *Odontotermes*, are often abundant in riparian woodland or scrub and damage structural timber, seedlings of forestry trees, such as *Eucalyptus*, and crops such as sugar-cane, groundnuts and date-palms as far north as Dongola (Schmutterer, 1969).

12.4.3. References

BERNARD, F. (1948) Les insectes sociaux du Fezzan. *Misc. Sci. Fezzan Zool.* **5**, 87–200.

BERNARD, F. (1954) Role des insectes sociaux dans les terrains du Sahara. In *Biology of Deserts*, edited by J. L. CLOUDSLEY-THOMPSON, pp. 104–111. Institute of Biology, London.

CLEMENT, G. (1954) Contribution à l'étude de la biologie d'*Anacanthotermes ochraceous*. *Insectes Soc.* **1**, 194–198.

HAFEZ, M. (1980) Highlights of the termite problem in Egypt. *Sociobiology*, **5**, 147–153.

HARRIS, W. V. (1967) Termites of the genus *Anacanthotermes* in North Africa and the near East. *Proc. R. ent. Soc. Lond.* (B), **36**, 79–86.

HARRIS, W. V. (1968) *Termites of the Sudan.* Sudan Natural History Museum, Khartoum.

HARRIS, W. V. (1970) Termites of the Palearctic Region. In *Biology of Termites*, Vol. II, edited by K. KRISHNA and F. M. WEESNER, pp. 295–313. Academic Press, N.Y. and London.

HOSNY, M. M. and SAID, W. A. (1980) The nesting system and the chemical structure of the material coating the internal parts of the nests of *Anacanthotermes ochraceus* in Egypt. *Sociobiology*, **5**, 101–114.

JOHNSON, R. A. (1981) Termites: damage and control reviewed. *Span*, **24**, 108–110.

JOHNSON, R. A. and WOOD, T. G. (1981) Termites of the arid zones of Africa and the Arabian Peninsula. *Sociobiology*, **5**, 279–293.

KASSAB, A., SHAARAWI, A. M., HASSAN, M. I. and SHAHWAN, A. M. (1960) *The Termite Problem in Egypt, with Special Reference to Control.* Min. of Agric. Pub., Cairo, U.A.R.

LEPAGE, M. (1974) *Les Termites d'une savane sahelienne (Ferlo Septentrional, Senegal): peuplement, populations, consommation, rôle dans l'écosysteme.* Doctoral Thesis, University of Dijon, Dijon.

NOUR, H. O. ABD EL (1980) The natural durability of building wood and the use of wood preservatives in the Sudan. *Sociobiology*, **5**, 175–182.

SCHMUTTERER, H. (1969) *Pests of Crops in North-east and Central Africa*. G. Fischer, Stuttgart.

SCORTECCI, G. (1936) Note sui Termitidi del Fezzan. *Natura*, **27**, 1–12.

WOOD, T. G. (1977) Food and feeding habits of termites. In *Production Ecology of Ants and Termites*, edited by M. V. BRIAN, pp. 55–80. Cambridge Univ. Press.

WOOD, T. G. and SANDS, W. A. (1977) The role of termites in ecosystems. In *Production Ecology of Ants and Termites*, edited by M. V. BRIAN, pp. 245–292. Cambridge Univ. Press.

ZIDAN, Z. H., EL-HEMAESY, A. H. and SAID, W. A. (1980) The geographical distribution, swarming dates and wood preferences of the subterranean harvester termite, *Anacanthotermes ochraceus* (Burm.) in Egypt. *Sociobiology*, **5**, 295–305.

CHAPTER 13

Arachnids

J. L. CLOUDSLEY-THOMPSON

CONTENTS

13.1. INTRODUCTION

The Arachnida is a class of Arthropoda that includes scorpions (Scorpiones), whip-scorpions (Thelyphonida), wind-scorpions or camel-spiders (Solifugae), false-scorpions (Pseudoscorpiones), spiders (Araneae), harvestmen (Opiliones), mites and ticks (Acari)—as well as a few less-well-known orders. Arachnids are chelicerates, having the first pair of head appendages either pointed stylets or small pincers like miniature crab claws. Mandibles, as found in grasshoppers, beetles and other insects, are lacking as are antennae, while there are nearly always four* pairs of legs instead of the usual three to be found in insects. There is also a pair of pedipalps which form the claws of scorpions, pseudoscorpions and whip-scorpions, and are tactile sense organs in Solifugae and spiders (where, in the male, they are also

---

* The gall-mites (family Eriophyidae) have only two pairs of legs in the adult stage, while the larvae of other mites have three pairs of legs, the fourth pair appearing in the protonymph.

concerned with mating). The arachnid body is divided into two parts; the prosoma or cephalothorax (which includes the head and leg-bearing portions) and the opisthosoma or abdomen. In scorpions, the abdomen is divided into two portions: a broad pre-abdomen consisting of seven segments and a slender, tail-like, post-abdomen, at the end of which is a sting. The abdomen has lost its segmentation and evolved into a soft, extensible, sac in almost all spiders and mites while the number of abdominal segments varies greatly in those groups in which they are retained.

The legs are equipped with sensory hairs and are often used as feelers. In spiders, harvestmen and mites, the first or second pair of legs may be especially long while, in Solifugae and whip-scorpions, they are not only elongated and slender, but often lack claws. In addition to tactile hairs, arachnids possess tricho-bothria—long, delicate, movable, hair-like sense organs inserted in the centre of a circular membrane. In scorpions, these are innervated by the dendrite of a single nerve and move in one plane while, in spiders, they are innervated by the dendrites of three sense cells and can move in various directions. Slit sense organs are narrow crevices, often in small clusters, known as 'lyriform organs' because of their parallel arrangement. They are usually close to joints, and react to movement of the limbs, probably through stresses set up in the cuticle. In addition to this proprioceptive function, slit sense organs serve as vibration receptors. The eyes of arachnids are always simple ocelli: usually there are both median and lateral eyes, differing in structure and, sometimes, in function.

The arachnid fauna of the Sahara desert comprises mainly the orders Scorpiones or scorpions, and Solifugae—known variously as camel-spiders, false-spiders, or wind-scorpions. Spiders are far from numerous, while false-scorpions and harvestmen are generally scarce. Giant velvet mites (family Trombidiidae) appear for a week or two after rain in certain localities, while ticks are common parasites of birds and mammals, but free-living mites are uncommon. All these groups will be discussed in the following pages.

13.2. SCORPIONS

13.2.1. Biology

Scorpions are unfortunate animals, frequently maligned and almost universally detested because they are able to inflict powerful stings that sometimes cause human deaths. They are not naturally aggressive, however, and seldom use their stings except in self-defence or to subdue prey. On the contrary, their behaviour is timid to the extent of cowardice, and human beings are never stung except as the result of accident. Scorpion venom probably evolved, in the first instance, as a means of subduing prey: only secondly has it acquired a defensive function.

Mythology

Scorpions have influenced the imagination of the peoples of the Orient and of the Mediterranean since earliest times. According to legend, when Mithras, the Persian god of light, killed the sacred bull from whose blood all life sprang, the evil spirit Ahriman sent a scorpion to destroy the source of life by attacking the testicles of the animal. Scorpions are prominent in monuments to Mithras, whose worship was prolonged until the 3rd century A.D., especially in North Africa. In ancient Egypt, too, scorpions were frequently represented in tombs and monuments. They are mentioned in the Ebers Papyrus, and in several passages of the Book of the Dead. The Talmud and Bible also refer to scorpions as repugnant and formidable animals. According to Greek Mythology Orion, son of Zeus and a great hunter, was killed by a scorpion which the goddess Artemis produced when he defied her. Aristotle wrote that the stings of scorpions were harmless in some lands but invariably fatal in others. Other legends are quoted by Balozet (1971).

Current misconceptions about scorpions are often equally fanciful. One is that, if surrounded by a ring of fire, a scorpion will commit suicide by stinging itself to death. This must be nonsense because no animal, other than Man, could possibly have the imagination to realize that, by self-destruction it might avoid unnecessary pain. Moreover, a suicidal instinct would inevitably become eliminated by natural selection during the passage of time since it could convey no biological advantage and might, in certain circumstances, be disadvantageous. In any case, scorpions are immune to their own venom. What actually happens is that a scorpion, lashing frantically with its sting, may inadvertently strike its own body and this has been misinterpreted as suicide (Cloudsley-Thompson, 1965). It is also untrue that scorpions do not drink or that black scorpions are more venomous than yellow ones. All really dangerous species belong to the family Buthidae, most of whose members are yellow or brown in colour.

Occasionally, in the Sahara, one meets someone who will pick up scorpions with his hand, but the creatures make no attempt to sting. Either the man will be wearing a piece of the dried root of a plant, often hibiscus, as a charm or he will previously have drunk a decoction of it. Consequently, he handles the scorpions without nervousness, and thereby does not irritate them, because he has complete confidence in his magic. In some parts of the Sudan, scorpions are placed in sesame oil where they die and disintegrate. The oil is then kept as a remedy to be applied to the site of a sting. In other parts of the country, the afflicted area is rubbed with the charred toenail of a baboon, of which a dried foot is worn as a charm in readiness for use in an emergency (Cloudsley-Thompson, 1965, 1980).

General behaviour

Scorpions resemble other Arachnida in having an impermeable integument and efficient powers of water retention, which are especially marked in desert species. As will be discussed in Section 13.2.4, desert scorpions are also more strictly nocturnal than forest species. The lives of scorpions are comparatively simple. Although sub-social in the sense that the young ride on their mother's back until after the first moult, scorpions live solitary lives and show no other social instincts: they usually avoid one another or, when confined within a small space, may fight to the death. The days are spent in burrows or sheltered retreats, from which the animals emerge at night to wander around in search of prey. *Buthus occitanus* and *Androctonus australis* are often to be found in shallow scrapes under rocks, which they dig with their claws and legs, but *Leiurus quinquestriatus* (Fig. 13.1) and *Parabuthus hunteri* usually dig burrows a metre or so in

FIG. 13.1. *Leiurus quinquestriatus* with young (photo).

length and slanting to a depth of 10–15 cm. *Scorpio maurus*, whose enlarged pedipalps are probably specially adapted for this purpose, digs holes up to 75 cm in depth (Cloudsley-Thompson, 1958, 1965).

Food and feeding

Scorpions feed mainly on insects, spiders and even other scorpions. Probably the ability to survive for long periods without food is of especial service to scorpions living under the hazardous conditions of desert regions. No doubt Saharan species obtain sufficient moisture for their needs from the body fluids of their prey but they drink readily if water becomes available and especially when starved. Scorpions often take prey that unwisely shelters in their retreats or burrows. When hungry, however, they emerge at dusk and wander slowly across the desert, their claws open and extended, the tail raised and pointing forwards. The sense organs mainly used for the detection of the prey consist of fine sensory hairs or trichobothria on the pedipalps whose pincers grab it quickly. Should they miss their target at the first attempt, they may try again, but scorpions do not usually chase their prey. When this has been captured, it is held firmly in the claws and stung if it struggles. The chelicerae then tear the food apart, one alternately holding on while the other pinches and pulls.

As the prey is picked to pieces, the juices and soft tissues are drawn into the mouth by the pumping action of the pharynx. There they are digested by a fluid containing amylases, proteases and lipases, and then sucked into the alimentary canal. Undigested portions are held back by the setae of the preoral cavity, balled together with the empty exoskeleton, and discarded. Intracellular digestion continues in the midgut caecae of the digestive gland. The intersegmental membranes of the abdomen allow for extension so that considerable amounts of food may be taken up at one time (Kaestner, 1968).

Enemies

The greatest threats to a scorpion's existence are probably food shortage, drought, or human activities. At the same time, despite their poisonous stings, scorpions suffer from the attentions of a number of predatory enemies, including various large centipedes, spiders and Solifugae, as well as lizards, snakes, birds and other scorpions. African baboons have been observed catching large scorpions, tearing off the tails and greedily devouring the remainder of the bodies.

On the ventral side of the abdomen, immediately behind the genital opercula, is a pair of comb-like organs known as 'pectines' (Fig. 13.7) whose function has long remained problematical. They have been regarded as external respiratory organs and external genitalia, and it has even been claimed that the lamellae of the pectines of the male and female scorpion become interlocked and serve to hold the two sexes together during mating, or their function is to clean the extremities of the legs and tail, or to absorb moisture. It is now known, however, that they are used during the process of reproduction (see below). Pectines are found in both sexes and all developmental stages of scorpions, however, which suggests that they may have additional functions as well. As a result of experiments carried out on *Euscorpius germanus* from Italy, and the Saharan species *Buthus occitanus* and *Androctonus australis*, in which it was found that the response to a vibrating tuning fork decreased markedly after the pectines had been amputated or painted over, it was concluded that a function of these organs lies in the perception of ground vibrations. Probably they are used more as a warning of danger than in the detection of prey (Cloudsley-Thompson, 1955). This hypothesis has not yet been confirmed (Carthy, 1968).

The predatory enemies of scorpions have been reviewed by Polis *et al.* (1981). In order of importance they are: scorpions of the same or other species, vertebrates, and other large invertebrates such as camel-spiders, tarantulas (Mygalomorphae) and scolopendras (Scolopendromorpha). Many vertebrate predators specialize on scorpions; others prey on them only rarely. The former probably include owls, bats, mongooses and insectivores. Several mammals are reported to detach the sting of their scorpion prey before attempting to eat them. Some lizards, too, appear to be important predators of scorpions and a few species probably specialize on them. Most vertebrate predators are nocturnally active, like their prey: it has been argued that the primary adaptive function of nocturnalism in scorpions is the avoidance of large predatory enemies (Cloudsley-Thompson, 1960, 1965, 1968). Other defensive strategies of scorpions include burrowing behaviour and the use of enlarged claws as shields in burrows, stridulation as a warning or threat to potential enemies, and squirting or spraying venom at the eyes of an aggressor (Polis *et al.*, 1981).

Mating and reproduction. The mating habits of scorpions have long been a mystery, additionally confused by faulty observation (Cloudsley-Thompson, 1976). Indeed, it was not until 1955 that indirect sperm transfer—by means of spermatophores deposited upon the substrate—was first observed. Initial contact between the sexes appears to be almost accidental. The male scorpion crawls towards a female of the same species and seizes the nearest part of her body with his claws. Sometimes several males seize the same female, repulsing one another with their stings. Immediately he has touched the female, the male scorpion shifts the grip of his pedipalps until he is clasping those of his partner. From time to time he vibrates his pedipalps and body, an act that has been termed 'juddering' and which may stimulate the female. The two scorpions now move back and forward in a '*promenade-à-deux*', during which the pectines of the male are spread outward and moved over the substrate. In this way, the scorpion is able to detect when he is above a surface suitable for the deposition of a spermatophore (Carthy, 1968). Occasionally the male moves forward and 'kisses' the female with his chelicerae atop hers. As soon as the spermatophore has been deposited, the male moves backwards and pulls his mate towards it. She feels for the spermatophore with her pectines and orients her body so that its paired sperm containers are situated exactly in front of her genital opening. The cover of the spermatophore is then opened with a sudden movement and the sperms are injected into her genital atrium. The male immediately releases his partner and subsequently she consumes the empty spermatophore (Kaestner, 1968).

13.2.2. Venoms

The poison of scorpions is secreted by a pair of glands, each with a separate duct and leading to the base of the sting whence a common duct conveys it to a small, elliptical opening on the outer side, just below the point. Scorpions are economical in the use of their poison, merely ejecting sufficient to subdue the prey. Many small insects and spiders are devoured without being stung at all. Stings delivered in self-defence, however, usually contain a maximum dose of poison (Cloudsley-Thompson, 1980).

The biochemistry of scorpion venoms has attracted much attention (Balozet, 1971; Minton, 1968; Sheals, 1973). The venoms of scorpions are of two kinds. The first causes mainly a local reaction and is comparatively harmless to Man. This type is secreted by the Saharan scorpion *Scorpio maurus* (Fig. 13.2). The second produces not only local, but also general systemic reactions, especially on the nervous system. Its neurotoxic effects resemble those of some snake venoms: furthermore it has a haemolytic action, destroying red blood corpuscles. Some of the symptoms evoked by the stings of buthid scorpions resemble those of strychnine poisoning.

The symptoms caused by scorpion venom of the less virulent kind consist of sudden, sharp pain followed by numbness and local swelling. They pass away within an hour or two. The sting of scorpions whose poison is neurotoxic, on the other hand, produces intense local pain at the site of the sting, often without swelling or discoloration beyond a small area of gooseflesh. A feeling of tightness then develops in the throat, so that the victim tries to clear imaginary phlegm, the tongue develops a feeling of thickness

Fig. 13.2. *Scorpio maurus* showing large claws for burrowing (photo).

and speech becomes difficult. The patient becomes restless with involuntary twitching of the muscles. At this stage small children will not be still: some even attempt to climb up the wall or the sides of their cots. Sneezing spasms are accompanied by continuous flow of fluid from the nose and mouth which may form a copious froth. The rate of heart beat may increase considerably; convulsions occur, the arms are flailed about, and the extremities become quite blue before death occurs. This complex pattern of reactions may last from 45 minutes to 10–12 hours. If the victim recovers, the effects of the venom persist longest at the site of the sting which may remain hypersensitive for several days so that the slightest bump will send painful or tingling sensations throughout the surrounding area (Balozet, 1971; Cloudsley-Thompson, 1965, 1980).

Among buthid scorpions with neurotoxic venoms, *Androctonus australis*, the fat-tailed scorpion of the Atlas mountains and northern Sahara (Fig. 13.3), and *Leiurus quinquestriatus* from the Sudan and the south-eastern edge of the desert (Fig. 13.4) are often cited as being among the world's most dangerous species. Drop for drop, their venom is almost as toxic as that of a cobra and the sting of *A. australis* has been known to kill a man in 4 hours and a dog in 7 minutes (Millot and Vachon, 1949). Over half the children admitted to Omdurman Hospital suffering from the stings of *L. quinquestriatus* fail to recover, despite treatment. No doubt only the more serious cases are brought in.

In a review of scorpion poisoning in North Africa, Balozet (1964) reported that no less than 20,164 cases in southern Algeria over a period of 17 years had been sufficiently serious to warrant medical treatment. Of these, 386 died—a mortality rate of 1.9%. *A. australis* was considered to be responsible for 80% of the incidents and 95% of the fatalities although in this region *A. aeneas*, *A. amoreuxi*, *Buthus occitanus* and *Buthacus arenicola* are also considered to be dangerous.

The venoms of *A. australis*, *Buthus occitanus* and *L. quinquestriatus* have been studied in detail (reviewed by Balozet, 1971). Their toxic principles are soluble in water and partly dialysable. They are basic proteins of low molecular weight and have amongst the highest activity known. Comparatively little work has been carried out on their mode of action, but they must stimulate any nerve or muscle fibre that they reach, and they probably have a general effect on excitable membranes. Research has been directed towards producing antidotes, using the serum of horses that have been injected over periods of time with gradually increasing doses of poison.

Fig. 13.3. *Androctonus australis*, fat-tailed scorpion of the northern Sahara (photo).

Fig. 13.4. *Leiurus quinquestriatus* in defensive attitude (photo).

13.2.3. Saharan scorpions

Of the six families of scorpions known to science, only two—Scorpionidae and Buthidae—occur in the Sahara. A third family, Chactidae, is represented along the Mediterranean coast of Africa by three species of *Euscorpius*. These are small, black, scorpions with thin tails and very large claws. Rare in Africa, *Euscorpius* spp. are better known from southern Europe where they are quite common. The genus does not extend into the desert proper. *E. flavicaudis* (Fig. 13.5) is of special interest to British zoologists because it was introduced into south-eastern England during the 19th century, and is now firmly established in one or two restricted localities (Wanless, 1977). It can be distinguished from other species of *Euscorpius* by the yellow bulb of its telson.

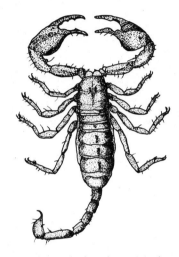

Fig. 13.5. *Euscorpius flavicaudis* (length 4 cm) (redrawn after Vachon).

Scorpionidae. The family Scorpionidae is represented in most of the Sahara by a single species, *Scorpio maurus* (Figs. 13.2, 13.6) of which numerous sub-species have been described (Vachon, 1952). This well-known species is the classical 'scorpio' of antiquity of which Pliny (*Nat. Hist.* XL. xxviii) wrote: 'they are a horrible plague, poisonous like snakes, except that they inflict a worse torture by despatching the victim with a lingering death lasting three days, their wound being always fatal to girls and almost absolutely so to women, but to men only in the morning. . . . Their tail is always engaged in striking and does not stop practising at any moment, lest it should ever miss an opportunity . . .' (transl. H. Rackham, 1940). In fact, however, no member of the Scorpionidae is dangerous to Man.

S. maurus can immediately be recognized by its massive claws. It may reach a length of 7.5 cm when adult, and its colour varies from pale yellow to dark brown. The front of the carapace is somewhat bilobed, while the tail is thick and shorter than the length of the body. Characters diagnostic of all Scorpionidae include the markedly pentagonal shape of the sternum (Fig. 13.7(a)) and the fact that only one spur is found on the last tarsal segment of the fourth pair of legs—and this is on the outside (Fig. 13.8(a)). There is no spur beneath the sting. The distribution of *S. maurus* extends from the Atlantic across the entire northern Sahara, and continues into Asia as far as Afghanistan.

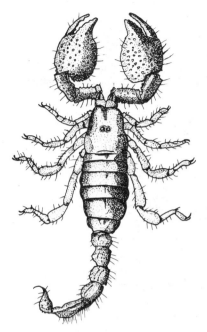

FIG. 13.6. *Scorpio maurus* (length 7.5 cm) (redrawn after Vachon).

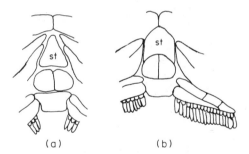

FIG. 13.7. Ventral view of the sternum (st) in (a) Scorpionidae, (b) Buthidae.

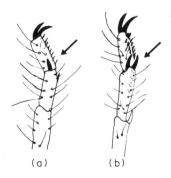

FIG. 13.8. Terminal tarsal segments of fourth legs in (a) Scorpionidae, (b) Buthidae.

Buthidae. Most Saharan scorpions belong to the family Buthidae. They can be recognized by the fact that the sternum is usually triangular in shape, narrowing towards the front (Fig. 13.7(b)), and there are one or two spurs, inside and outside at the base of the last tarsal segment of the fourth pair of legs (Fig. 13.8(b)). The venoms of buthid scorpions cause intense pain and are sometimes dangerous to Man. Vachon (1952) recognizes thirteen Saharan genera, twenty-nine species and numerous sub-species. Most of the genera and species are to be found throughout northern Africa and can be identified by reference to Vachon's comprehensive monograph. In the following paragraphs I shall mention some of the more common and easily recognizable species of Saharan buthids.

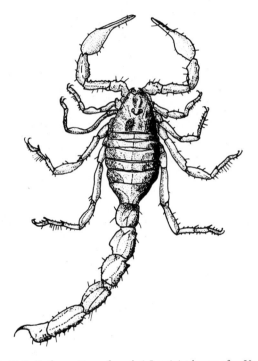

Fig. 13.9. *Buthus occitanus* (length 6.5 cm) (redrawn after Vachon).

Buthus occitanus (Fig. 13.9) is by far the most common species along the northern edge of the Sahara from Morocco to Egypt. It also occurs, but less frequently, in the south from Senegal to Ethiopia, and is also found in parts of south-west Europe. The venom of the European populations is much less toxic than that of the same species from the Sahara and the Middle East. The adult body length varies from 4.5 to 7.0 cm, and the colour is pale yellow. The claws are slender with rounded bulbs. The posterior median crest of the cephalothorax, with the bow-shaped median lateral crests, make a clearly defined lyre-shaped figure. Three other species of *Buthus* have been collected in Morocco, but all of them are rare: *B. atlantis*, 6.0–8.0 cm in length, *B. moroccanus*, 5.5 cm long from the region of Rabat, and *B. barbouri*, known only by a single specimen from Agadir.

Second only to *B. occitanus* in frequency of occurrence in the northern parts of the Sahara is the notorious *Androctonus australis* (Figs. 13.3, 13.10). This species has not been found in Morocco, but its distribution extends from Algeria, Tunisia, Libya and Egypt, eastwards into India and Baluchistan. With a body length of 9.0 cm or more, dark-brown colour and massive tail, *A. australis* is unmistakeable. Less well-known members of the genus include *A. amoreuxi*, a huge yellow species, 12.0 cm long and very widely

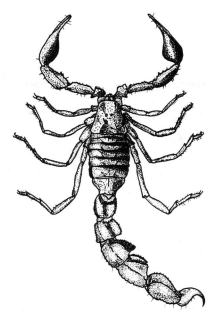

Fig. 13.10. *Androctonus australis* (length 9 cm) (redrawn after Vachon).

distributed throughout the Sahara, and five black species: *A. crassicauda*, about 8.5 cm in length with an exceptionally massive tail, *A. hoggarensis* from the region around the Hoggar mountains, *A. aeneas, A. sargenti* and *A. mauritanicus*, all of which reach lengths around 7.0 cm. The last three species are found mainly in the north-west of Africa, but *A. crassicauda* extends into Asia—I have found it to be common south of Kuwait.

Another black north-west Saharan buthid is *Buthotus franzwerneri* which, again, measures 7.0–8.0 cm in length and has claws with exceptionally long fingers. *B. hottentota* is a yellowish congeneric species found all along the West African coast, while *B. minax* is a smaller, dark species, never exceeding 7.0 cm in length and easily recognized by the granulated appearance of the cephalothorax. In the northern Sudan it appears to be confined to crevices in the bark of '*sunt*' trees (*Acacia nilotica*) or similar places but, further south and in Eritrea, it can be found under rocks and fallen trees.

Leiurus quinquestriatus is the most common species of the south-eastern regions of the Sahara but extends to the west as far as the Hoggar mountains. It is medium to large in size, reaching a maximum length of 9.5 cm, and can be recognized by the long, very slender hand of the pedipalp (Figs. 13.1, 13.4, 13.11) and five keels on the first abdominal tergite. *Parabuthus hunteri*, common in the Red Sea hills and coastal plain, is even larger, reaching a length of 10.0 cm. It has a thick tail, of which the two apical segments and poison vesicle are nearly black, while the remainder of the body is a rich yellow colour.

Compsobuthus werneri is a small central or southern Saharan species, reaching no more than 3 cm in total length. Its distribution stretches from Timbuktu to Nubia, southern Egypt and the Middle East. I collected specimens from Wadi Halfa before the town was submerged by Lake Nubia, after construction of the High Dam at Aswan. *Microbuthus* and *Butheoloides* spp. are even smaller—not longer than 2–5 cm at most when adult—while scorpions of the genera *Mesobuthus* and *Odontobuthus* never exceed 4 cm. They are all rare or of local provenance. Other Saharan buthid genera include *Anoplobuthus, Buthacus, Buthiscus, Cicileus, Lissothus* and *Odontobuthus. Orthochirus innesi* is the little bicoloured scorpion of the oases (Fig. 13.12).

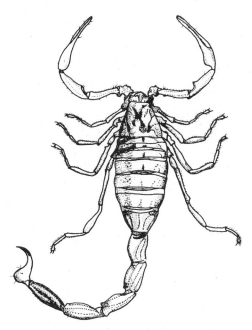

FIG. 13.11. *Leiurus quinquestriatus* (length 8 cm) (redrawn after Vachon).

13.2.4. Adaptations to desert life

As Hadley (1974) points out in an important review of their adaptive biology, desert scorpions rely on a combination of behavioural, morphological and physiological adaptations in adjusting to the harsh conditions found in these habitats. They can, however, escape the high temperatures or extreme drying powers of the surface air by burrowing or sheltering during the day and restricting surface activities to the night.

Burrowing. Burrowing behaviour is to some extent related to taxonomic position. Species belonging to the families Chactidae and Scorpionidae inhabit deep burrows which they either dig for themselves or which have been excavated by lizards or small rodents. Most members of the family Buthidae do not burrow but, as already mentioned, occupy scrapes beneath rocks, dried vegetation or surface litter. Species that burrow usually dig their holes near vegetation where the ground is softened by penetrating roots and the soil texture is favourable for packing. Other factors such as reduction in soil temperature, resulting from shading or increased soil moisture, may also be important in site selection.

Nocturnal surface activity. Day-active scorpions are restricted to regions of tropical and sub-tropical forest, while desert scorpions are exclusively nocturnal in habit. As mentioned above, this is probably a response to predation: in addition, however, the high temperature and low humidity of the desert surface during the day are thereby avoided. Endogenous rhythms of locomotion and oxygen consumption have been demonstrated in *Androctonus australis*, *Babycurus buthneri* (=*B. centrurimorphus*), *Buthotus minax*, *B. hottentota*, *B. occitanus*, *Leiurus quinquestriatus* and *Scorpio maurus* (for references, see Cloudsley-Thompson, 1978; Constantinou, 1980; Hadley, 1974).

Temperature tolerance. The upper lethal temperatures of desert scorpions are generally several degrees higher than those of other desert arthropods; that of *Leiurus quinquestriatus* being 47°C for a 24-hour exposure at a relative humidity below 10% (Cloudsley-Thompson, 1962).

Resistance to desiccation. Small animals have very large surface areas in proportion to their mass. Consequently, the conservation of water is extremely important to them (see discussion in Cloudsley-Thompson, 1968). Scorpions, like other arachnids and insects, avoid desiccation by secreting, in the outermost layer or epicuticle of their integuments, a thin waterproof sheet of crystalline wax. In many arthropods, this wax layer has a critical temperature, at which it becomes porous, in the region of 35–40°C. In desert scorpions, however, the critical temperature is some 20–25°C higher, while below this the integument is extremely impervious. Water-loss rates approaching 0.01% of their body weight per hour in dry air are the lowest reported for desert animals. Evidence suggests the restrictive mechanisms in the cuticle may supplement the physical barrier of the wax layer in controlling cuticular transpiration. Scorpions have extremely low metabolic rates which not only result in a reduced respiratory component of the total water loss, but also extend the time that scorpions can remain inactive during particularly stressful periods. Water loss is further minimized by the excretion of nitrogenous wastes in the form of insoluble guanine, and the production of extremely dry faecal pellets (for references, see Hadley, 1975).

Water uptake. Replenishment of lost water is provided primarily by the body fluids of captured prey, although drinking by some species may serve as a supplementary source of moisture when free water is available. Scorpions are able to withstand considerable dehydration (30–40% of their body weight) during adverse conditions and apparently tolerate the increased salinity osmotic pressure and ionic concentrations of their blood until body fluids can be replenished. In general, therefore, the physiological adaptations of scorpions to desert conditions consist of mechanisms for conserving rather than for obtaining water. No species of scorpion is able to absorb moisture vapour from unsaturated air—as can some insects and ticks—nor can scorpions take up water from a moist substrate. The majority of desert adaptations exhibited by scorpions are not unique to these arachnids, but are exceptionally well developed and efficiently utilized (Hadley, 1974).

13.3. FALSE-SCORPIONS

Pseudoscorpions and Opiliones or harvestmen are rare in deserts everywhere. They have not been recorded in the Namib, and are scarce in the North American desert regions. I have occasionally found both on the southern fringes of the Sahara, but only false-scorpions in the more arid regions. False-scorpions are therefore discussed below, but no further mention will be made of harvestmen.

13.3.1. Biology

Pseudoscorpions are small terrestrial arachnids, about 1–7 mm in length and superficially resembling true scorpions deprived of their post-abdomen and sting. Despite their small size and the fact that they are seldom found accidentally, they were known to Aristotle who wrote of *Chelifer cancroides*, 'this animal, living in books, bears chelae too . . . and is similar to those living in clothing (bed-bugs) and to scorpions but without a tail and very small' (*De Animalibus Historiae*, IV, 7). *C. cancroides* can, today, still be found in old libraries throughout the world, and is known colloquially as the 'book-scorpion'.

Like scorpions, false-scorpions are predators feeding on other small arthropods which they grasp with their claws. In many species these are supplied with poison glands whose venom subdues the prey. The victim is then conveyed to the chelicerae or jaws. Some pseudoscorpions masticate their food with the

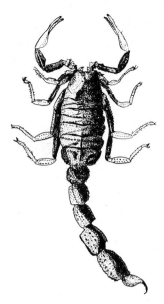

FIG. 13.12. *Orthochirus innesi* (length 3 cm) (redrawn after Vachon).

chelicerae but, in all the more advanced families—which include the desert species—a small hole is torn in the body wall of the prey. Through this is injected digestive fluid, which is later sucked back and ingested. During a meal, the prey is held only by the chelicerae. Pseudoscorpions normally walk forward rather slowly when hunting, but they have a surprising ability to dart backwards when disturbed, and to turn round extremely quickly—if touched at the posterior of the body—to present their chelae to the aggressor. They can move as easily on glass as on a rough surface, climb along a hair by means of the claws alone, and feign death when threatened by a larger enemy.

Grooming is important to false-scorpions. From time to time they stop walking, and clean the fingers of their claws by drawing them through the jaws. This usually takes place after a meal. Silken chambers are constructed for moulting, brooding and hibernation or aestivation. The silk is secreted by glands that open onto the spinnerets at the tip of the movable finger of the chelicerae. Some pseudoscorpions are able to suppress predatory aggression by members of their own species by spectacular displays or the

FIG. 13.13. *Dactylochelifer latreillei* (photo).

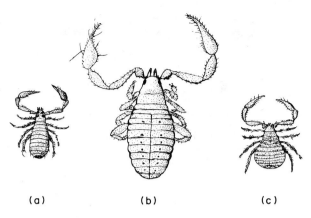

FIG. 13.14. (a) *Olpium* sp., (b) *Dacylochelifer* sp., (c) *Cheiridium* sp.

production of pheromones. *Dactylochelifer* spp. (Fig. 13.13), for instance, shake or beckon with their pedipalps, and even grasp one or both of the hands of another animal that approaches them. Cannibalism does not normally occur. Nor are false-scorpions parasitic, although some species cling to the legs of insects and are thus transported for relatively long distances, a mode of travel known as 'phoresy'.

In most species of false-scorpions the sexes are easily distinguished by the fact that, in the female, the genital area is only lightly sclerotized so that the area behind the coxae of the legs appears bright whereas, in males, it is dark like the rest of the body. Insemination results from indirect spermatophore transfer, with or without a complex mating dance. In *Dactylochelifer* spp., for instance, the male grasps the female's claws in his, vibrates his whole body and displays his ram's-horn organs. These are erectile tubes attached to the posterior wall of the genital atrium which are extended by a rise in blood pressure. They are not glandular, but are probably covered with an odorous substance from the genital chamber for the female orients herself to them. The male deposits a spermatophore and helps his mate to take up the sperm from it by opening her genital atrium with the specially modified tarsi of his forelegs. In *Cheiridium* spp., on the other hand, there is no mating. The males produce spermatophores in isolation. When the female finds one, she touches it with the tips of her pedipalpal fingers, steps over it on extended legs and takes up the sperm (Cloudsley-Thompson, 1976; Weygoldt, 1969). Pseudoscorpions carry their eggs and nourish the developing embryos in their bodies. In most species, the female builds a silken chamber and retreats into it at this time, remaining there without feeding until the protonymphs hatch.

13.3.2. Saharan false-scorpions

Of sixteen known families of Pseudoscorpiones only three appear to be represented in the Sahara. These are the Olpiidae, Cheliferidae and Cheiridiidae. Olpiidae (suborder Displosphyronida) are distinguished by the fact that there are four eyes and the tarsi are divided into a metatarsus and telotarsus, so that each leg has six articles excluding the coxa. In the suborder Monosphyronida to which the other two families belong, there are usually two or no eyes and the tarsi are not divided so that each leg has five segments excluding the coxa. In addition, the dorsum of the prosoma is tapered or rounded.

In the extremely arid region of the Sahara around Wadi Halfa, four species of false-scorpions have been recorded by Beier (1962), viz. *Olpium tenue*, *Dactylochelifer nubicus*, *Rhacochelifer nubicus* and *Cheiridium nubicum*. (*D. nubicus* was found phoretic on a noctuid moth, *R. nubicus* on the trunk of a palm tree and *C. nubicum* under vegetable litter.) Congeneric Saharan species include *Olpium savignyi* from Lower Egypt, *Dactylochelifer moroccanus*, and the Libyan *Rhacochelifer similis* (Fig. 13.14).

13.3.3. Adaptations to desert life

No detailed experimental studies have been carried out on the ecology of pseudoscorpions, on habitat selection, niche preferences or the requirements of the various species. Most forms seem to be rather restricted in their requirements. Different species of the same genus often occur in very similar habitats and are separated geographically, but there may be some overlapping at the periphery of their ranges. Some species, such as *Withius subruber, Cheiridium museorum* and *Toxochernes panzeri*, are often associated with stored food products in warehouses, where they inhabit extremely dry environments: they also occur in birds' nests in hollow trees. In view of their small size, it is obvious that their powers of water retention must be extremely efficient (Cloudsley-Thompson, 1958). It is safe to assume that the same must be true of Saharan species, but no experimental work has yet been carried out on them.

In view of their low numbers and scattered distribution, false-scorpions probably have little effect upon populations of detritus-feeding animals such as Collembola. At the same time, since they are not much preyed on by other carnivores, they may be regarded as tertiary consumers, occupying the terminal stage of the food chain. The ecological importance of this rôle is probably greatest in ecosystems with a simple community structure such as occur in deserts (Wallwork, 1970).

13.4. SOLIFUGAE (=Solpugida)

Predominantly inhabitants of arid tropical regions, Arachnida of the order Solifugae (sometimes known as 'false-spiders', 'wind-scorpions' or 'camel-spiders') are even more characteristic of desert regions than are scorpions. A relatively primitive group of Arachnida, the Solifugae or Solpugida is represented by numerous Saharan species but, because these animals are burrowers, usually secretive, and principally nocturnal in habit, solpugid material is not abundant in collections and the taxonomy of the order is little known. In addition, biological observations are fragmentary or incomplete.

13.4.1. Biology

The adaptational biology of Solifugae has been reviewed by Cloudsley-Thompson (1977). These impressive animals have been cited as world-wide indicators of deserts, since most species are adapted to desert conditions and occur in hot, arid environments. They avoid oases and other fertile places, seeming to prefer utterly neglected regions where the soil is broken and bare. At night, they sometimes enter the tents of travellers to catch flies and other insects, circulating so rapidly that it almost makes one dizzy to watch them.

According to Aelianus (*De Natura Animalium*) an area of 'Ethiopia' was deserted by its inhabitants on account of the appearance of an incredible number of scorpions or 'Phalangia' but Pliny (*Historia Naturalis*), in quoting the same story, replaced 'Phalangia' by 'Solfuga'. There has been some controversy as to whether there is any truth in the evil reputation of Solifugae for inflicting poisonous bites, but it is now generally assumed that they are not venomous since several authors have searched in vain for poison glands, such as those in the jaws of spiders, and a number of people have actually allowed themselves to be bitten without experiencing any ill effects. On the few occasions that poisoning has occurred, it has most probably been due to infection of the wound.

Like most other desert arachnids, Solifugae inhabit deep burrows from which they emerge only at night. Those of *Galeodes granti* (Fig. 13.15), for instance, usually follow a convoluted course at a depth of 10–20 cm, and often extend for several metres into the soil. During digging, sand is rapidly scraped backwards by the second pair of legs whilst occasionally the animal lowers itself into the hollow thus formed, bulldozing with the lower parts of its chelicerae (Fig. 13.15). In cooler, damp weather at the time

Fig. 13.15. *Galeodes granti* burrowing (photo).

Fig. 13.16. *Galeodes granti* attacking a scorpion, *Leiurus quinquestriatus* (photo).

of the rains, *G. granti* may be found near the entrance to its burrow but, in hot, dry conditions it retreats to the innermost depth, first closing the entrance with a plug of dead leaves. Although nocturnal, Saharan solifugids have high lethal temperatures and extremely low rates of water loss by transpiration. *G. granti* can withstand 50°C for 24 hours at a relative humidity below 10% (Cloudsley-Thompson, 1962b).

Feeding. Solifugae are exclusively predatory and carnivorous, having an extraordinary voracity and feeding until their abdomens are so distended that they can scarcely move (Cloudsley-Thompson, 1968) (Fig. 13.16). Food-searching behaviour involves random running, and congregating in areas where prey is likely to occur—such as the neighbourhood of lights at night or the nests of prey. The food may be located by orientation to tactile and visual stimuli, and by vibrations of the substrate (Cloudsley-Thompson, 1961a). Most species chase or ambush their prey, and some have been observed to kill and eat large spiders, scorpions, insects, lizards, mice and even small birds. Cannibalism is almost inevitable if more than one individual is confined in a restricted space, females usually overpowering the weaker males (Cloudsley-Thompson, 1958).

When living prey has been caught, it is held crossways in both jaws and masticated by alternate movements of the scissor-like chelicerae. At the same time, it is ground between the inner surfaces of the two chelicerae which move alternately backwards and forwards. Although drinking sometimes occurs when free water is available, it seems probable that Solifugae normally obtain sufficient moisture for their needs from the body fluids of their prey (Cloudsley-Thompson, 1962b). The anus is terminal, and liquid excretary matter can be ejected forcibly from it.

The reactions of *G. granti* have been classified as sleep, alertness, low-intensity threat (the chelicerae move soundlessly and the animal rocks or sways on its legs), and high-intensity threat (usually accompanied by stridulation), leading to attack or flight (often accompanied by displacement sand-digging) (Cloudsley-Thompson, 1961a). Threat is strikingly exaggerated in species with long legs which permit accentuated rocking movements. Threat behaviour and feeding responses are obviously related, and the victor of a fight usually devours the loser.

Although adult solifugids are so aggressive, the first and second instars sometimes exhibit a curious form of social behaviour when they gather into loose clusters in which the movement of a single individual causes a slight shift or progression of the entire group. Continued mild stimuli may result in the cluster moving away from the source of irritation, but stronger stimulation causes it to disperse altogether. First instar nymphs do not feed—as in other arachnid classes, they are sustained by the yolk remaining in their alimentary canals—but second instar nymphs may show communal feeding behaviour (Cloudsley-Thompson, 1977). Little is known of the natural enemies of Solifugae, although these probably include reptiles, birds, and small mammals. Encounters with scorpions usually result in destruction of the latter, unless the solifugid is much smaller than its adversary. Solifugae are able to raise their abdomens almost vertically. This no doubt protects that vulnerable portion of the anatomy. At the same time, such behaviour may represent a form of mimicry, since it results in a scorpion-like appearance, particularly in the short-legged Rhagodidae (Cloudsley-Thompson, 1980) (Fig. 13.17).

Reproduction. The most significant difference between mating in Solifugae, and in scorpions or false-scorpions, consists in the immediate transfer of the spermatophore which is not fixed to the ground. After a brief courtship, in which the male strokes, or merely touches the female with its pedipalps, the female becomes lethargic, lifts her abdomen and allows the male to grasp her with his jaws. A spermatophore is produced, and the male inserts it with his chelicerae into the genital opening of the female. This form of indirect copulation, using a spermatophore, probably represents an intermediate stage between indirect sperm transfer via the substrate, and indirect free sperm transfer, as found in spiders (Cloudsley-Thompson, 1976).

Fig. 13.17. *Rhagodessa melanocephala* (photo).

The flagella are curious organs of unknown function situated on the dorso-distal region of the chelicerae in male Solifugae. Junqua (1966) investigated their possible importance in courtship and mating, but found that severance of the shaft at the base had no observable effect on such behaviour. More recently, Lamoral (1974) presented evidence that the flagellum may operate for temporary storage and emission of an exocrine secretion, possibly a pheromone, which may play a rôle in brief displays of territoriality among males during the mating season. The sensory system of the malleoli or racquet-organs—stalked mallet-shaped structures on the ventral sides of the proximal segments of the last pair of legs—suggests that the sense organs in the sensory grooves may be chemoreceptors (Brownell and Farley, 1974). If so, the malleoli might respond to the exocrine secretions of the flagellum. It has also been suggested that they may serve to detect vibrations of the substrate, and this could account for the greater sensitivity of males, in which sex they are significantly larger (Cloudsley-Thompson, 1961a). The pedipalpal organ was at one time believed to be a receptor for airborne odours but it is now known to be a sucker used in the capture of prey and for climbing rocks or trees.

The number of eggs produced is related to the size of the mother, as in other arachnids. The eggs usually hatch within 3 or 4 weeks. Newly hatched post-embryos are translucent, white, and almost immobile: they moult in about a week to non-feeding, first-instar nymphs which resemble their parents in shape and appearance. These complete development in a further week, and moult again to burrowing, feeding,

FIG. 13.18. Second instar *Galeodes granti* (photo).

second–instar nymphs (Fig. 13.18). The total number of instars is about nine. The mother protects her eggs and first-instar nymphs in a haphazard sort of way, but dies within 5 or 6 weeks of their birth (Cloudsley-Thompson, 1977). Despite their large size, Solifugae appear to live for only a year in the Sahara. The young are produced in summer at the time of the annual rains when there is plenty of food available. It is not known where the juvenile instars pass the winter, for they are seldom to be found. It seems not unlikely that they may secrete themselves in the nests of termites where they can obtain abundant food even though the desert outside is almost lifeless.

13.4.2. Saharan Solifugae

Of ten known families of Solifugae, at least five occur in the Sahara. These are as follows: (a) Rhagodidae (Fig. 13.19)—mainly black, short-legged species in which the anus is ventral. (In all other families it is dorsal.) The remaining families include yellow, long-legged species: (b) Galeodidae (Fig. 13.20)—in these the tarsal claws are hairy. (c) Karschiidae—in these there are two tarsal claws on leg 1, but they are not hairy; tarsi 2 to 4 each have only one segment. (d) Solpugidae—tarsal claws on leg 1 are lacking; tarsi of legs 2 and 3 have four articles, that of leg 4 has at least six. (e) Daesiidae—tarsal claws are lacking on leg 1; the tarsi of legs 2 and 3 have one or two segments; the cheliceral flagellum of the male is awl-shaped and can be turned through 180°.

FIG. 13.19. *Rhagodes* sp. (Rhagodidae) (length 6 cm).

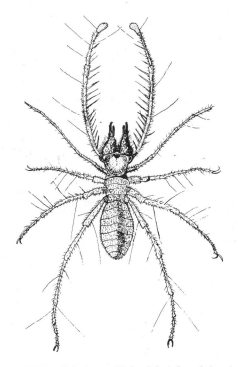

FIG. 13.20. *Galeodes* sp. (Galeodidae) (length 8 cm).

Identification of these bizarre arachnids requires specialized knowledge. In many cases, species can only be distinguished with certainty by the dentition of the chelicerae and the shape of the flagellum of the male (Fig. 13.21). Recourse has to be made to Roewer (1932–1934; 1941) in the absence of any modern revision of the order.

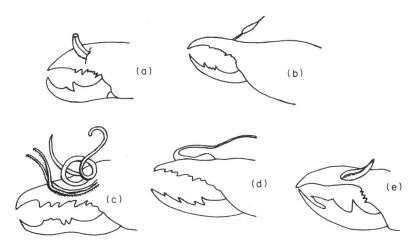

FIG. 13.21. Chelicera of male showing flagellum: (a) Rhagodidae, (b) Galeodidae, (c) Karschiidae, (d) Solpugidae, (e) Daesiidae.

13.5. SPIDERS

Although tarantulas and trapdoor spiders (Mygalomorphae) are the dominant spiders of the deserts of America and Australia, they are not well represented in the Sahara. Here the most numerous Araneae include mainly ground hunting spiders (Gnaphosidae), giant crab spiders (Sparassidae), crab spiders (Thomisidae), bark spiders (Hersiliidae) (Fig. 13.22), jumping spiders (Salticidae) (Fig. 13.23), wolf-spiders (Lycosidae), sheet-web spiders (Agelenidae), comb-footed spiders (Theridiidae), and orb-weavers (Tetragnathidae and Argiopidae) (Fig. 13.24).

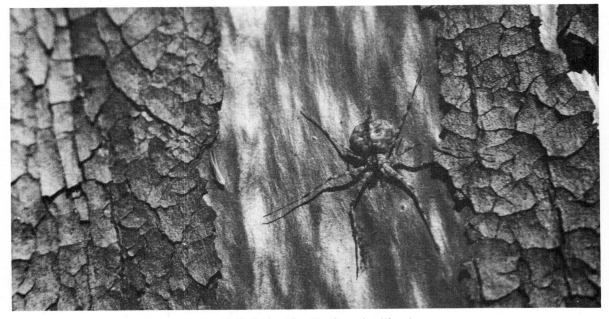

FIG. 13.22. Bark spider, *Hersilia caudata* (photo).

Fig. 13.23. Jumping spider, *Cosmophasis nigrocyanea* female: a mimic of mutillid wasps (photo).

13.5.1. Biology

Spiders are predatory arachnids feeding mainly on insects and other arthropods. Some attack large insects, others capture smaller kinds; some specialize in crawling insects, others on those that fly. Some hunting spiders seek their prey by day, trusting to their good sight, while others, including Gnaphosidae, are active at night and depend mainly on the sense of touch. Jumping spiders (Salticidae) have the keenest sight of all and stalk their prey from afar. Equally at home on perpendicular surfaces or the undersides of stones or branches, they are able to maintain their position by means of an adhesive tuft of hairs, the 'scopula', on each of the tarsi. As it moves, a jumping spider trails behind it a fine thread of silk which is attached at frequent intervals like a climbing rope so that, in the event of a slip, the spider does not fall to the ground. The prey is approached stealthily until the jumping spider is close enough for a sudden spring, siezing its prey with the front pair of legs. In contrast, wolf spiders (Lycosidae) capture their quarry by sheer strength and speed. The typical crab spiders (Thomisidae) wait motionless, often in vegetation, for passing insects which are seized by the powerful outstretched legs. Having buried their jaws in the head or thorax of the prey, they draw their limbs backwards out of danger of the victim's bite or sting. Many species show a remarkable degree of resemblance to the colour of their background.

Tetragnathidae, Theridiidae and Argiopidae often spin their aerial webs in vegetation or among rocks, but sheet-web spiders (Agelenidae) are mostly found down cracks in the soil. Their funnel-shaped

FIG. 13.24. Orb-web spinner *Argiope lobata*, an inhabitant of desert dunes (photo).

cobwebs consist of a triangular sheet with its apex rolled into a tube in which the spider waits for its prey. The threads of the web are not adhesive, but trip passing insects which fall on the sheet. Before they have time to recover, the owner of the web darts from its tube and gathers them in. The webs of Theridiidae and Argiopidae contain sticky threads which entangle the prey (Cloudsley-Thompson, 1958).

Spiders are nearly always on the defensive and ready to kill and eat most animals of suitable size that come within range. Mating is therefore a hazardous undertaking, fraught with real danger to the male who is usually smaller and weaker than his intended mate. The function of courtship is, therefore, probably not merely to stimulate the female but also to establish the male's identity so that he is not mistaken for prey. Consequently, whichever of the senses is the one on which the spider chiefly relies for the capture of prey is the sense most employed in courtship. Male jumping spiders and wolf-spiders make use of visual signals, short-sighted and nocturnal species of contact stimuli, web spinners make distinctive tweaks and vibrations of the thread of the female's snare, and so on. When mature, the male spider weaves a small pad of silk on which a drop of sperm is deposited and this is sucked up by the specially modified pedipalps which, in due course, are inserted into the vagina (epigyne) of the female. Each species has a palp with its own distinctive shape, a diagnostic character essential for accurate identification of male spiders. Females are identified by the shape of the epigyne, a chitinous process covering the genital opening (Cloudsley-Thompson, 1976).

Spiders lay their eggs in cocoons which they construct of silk, and often mount guard over them until young have hatched. Wolf-spiders (Lycosidae) carry their globular cocoons attached to their spinnerets and, after hatching, the young climb onto their mother's back where, like baby scorpions, they remain for several days. The number of moults necessary to attain maturity varies widely. Most of the smaller species live for only a year.

13.5.2. Saharan spiders

African spiders have been studied comparatively little, and anyone who collects spiders in the Sahara will find that up to one-quarter of the species he takes are unknown to science! Of ninety-seven species collected at Siwa Oasis in 1935, twenty-five were new (Denis, 1947).

The general appearance of the families mentioned in the preceding paragraphs are indicated in Figs. 13.25–13.29. The Gnaphosidae are nocturnal hunting spiders with eight eyes in two rows, and long and narrowish abdomens usually without markings. The spinnerets project clearly from the end of the abdomen, the anterior ones being distinctly separated. Tarsal claws, two. Most species are dull grey or black, living under stones or bark where they build silken retreats.

Sparassidae (Fig. 13.25) and Thomisidae (Fig. 13.26), although not closely related, are alike in holding their legs sideways like crabs. They have only two claws on each leg, and their eyes are in two rows. *Eusparassus walckenaeri* is a large species, widely dispersed throughout the Sahara. Salticidae (Fig. 13.27) also possess two claws and can be recognized by their squat shape and large anterior median eyes. Lycosidae (Fig. 13.28) have three leg claws and more elongated bodies. Hersiliidae (Fig. 13.22) also have three tarsal claws and the posterior spinnerets have an unusually long distal segment. These flattened spiders sit pressed closely to bark or rocks in a head-down position. *Hersilia caudata*, which reaches about 1.5 cm in length, is common on the bark of acacia trees in the Sahel savanna region on the southern fringe of the desert. Members of the family Agelenidae spin typical cobwebs radiating from holes in the ground in which their owners live. There are eight eyes in three rows, not greatly unequal in size, and three-toothed tarsal claws. The posterior spinnerets are clearly larger than the anterior ones, and stick out

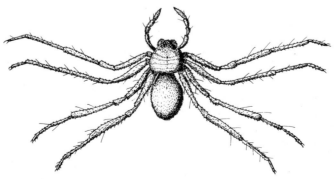

FIG. 13.25. *Eusparassus* sp. (Sparassidae) (length 2 cm).

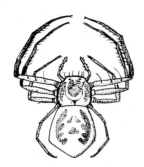

FIG. 13.26. *Xysticus* sp. (Thomisidae) (length 1 cm).

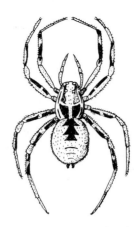

FIG. 13.28. *Lycosa* sp. (Lycosidae) (length 1 cm).

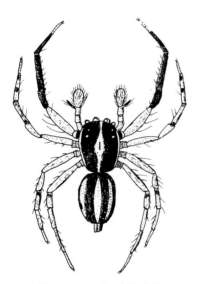

FIG. 13.27. *Plexippus* sp. (Salticidae) (length 5 mm).

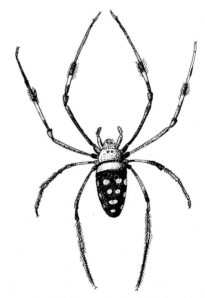

FIG. 13.29. *Nephila* sp. (Argiopidae) (length 1.5 cm).

behind the abdomen. Spiders of the family Tetragnathidae can be recognized by the elongated bodies and long legs. The eight eyes are in two rows and not greatly unequal in size. The chelicerae are unusually large and have strong teeth. The webs are nearly always spun on grass. Theridiidae possess a comb of serrated bristles on the ventral sides of the fourth legs which are used to throw threads of silk over prey caught in the web. They are usually considerably smaller than orb-weavers (Argiopidae) (Figs. 13.24, 13.29).

13.5.3. Adaptations to desert life

In proportion to web-building species, hunting spiders are more numerous in deserts than in other terrestrial biomes. The predominance of Salticidae and Thomisidae over web-spinners in the Sahara, for instance, is probably due to the sparseness of the vegetation which provides few sites for webs and leaves Argiopidae exposed to predators and to the strong winds which are so destructive to their webs.

Many desert spiders inhabit burrows, and share the same morphological adaptations although they belong to different families. They are usually large, with relatively big bodies and thick-set legs. The surface-to-volume ratio is less than in smaller animals, so that they have a comparatively smaller area through which evaporative water loss can occur. A second ecological grouping includes small, nomadic hunting spiders belonging mainly to the families Gnaphosidae, Salticidae, and the Thomisidae, whose ways of life are somewhat different from those of congeneric species in tropical and temperate regions.

Like scorpions and Solifugae, many spiders escape the high temperatures and saturation deficiency of the desert surface by burrowing or seeking some form of cover during the day, and restricting their activities to the hours of darkness. Some desert Sparassidae dig deep tubes in the sand and cement the walls of the burrow with a criss-cross of webbing. Others are among the dominant species on the trunks of acacia trees in the Sudan, along with Hersiliidae (Fig. 13.22). Most spiders that do not burrow take shelter under loose litter and rocks, down cracks in the ground or beneath the bark of trees. Many desert spiders are able to compensate metabolically for high temperature although this has not yet been investigated among Saharan species.

Desert spiders, vulnerable to predatory vertebrates, are invariably cryptically coloured. Many of those that emerge from their burrows to feed or mate show disruptive coloration while those that inhabit sandy deserts without much contrast in background colour are usually white or pale without markings. *Argiope lobata*, for instance, is not only the colour of sand, but its irregular shape tends to render it inconspicuous— despite its large size (Fig. 13.24).

Several desert species of Salticidae are especially notable in that they mimic ants, with which they are associated, and thus escape the attentions of hungry spider predators and probably avoid ant predators by running away when alarmed. *Myrmarachne* spp. are among the outstanding mimics of ants on the fringes of the Sahara (Wanless, 1978). *Cosmophasis nigrocyanea* is common around Khartoum. The males resemble ants, while the females look like wingless mutillid wasps. Both sexes mimic ants in their behaviour (Cloudsley-Thompson, 1968) (Fig. 13.23). The resemblance is so exact that it is difficult to recognize the animals as spiders. In some ant mimics, the appearance of a narrow waist is created by a strip of dark colour on a white background. In *Myrmarachne*, it is achieved by wedge-shaped bands of white bars against a black background. Furthermore, the spider walks in the jerky manner of an ant, and holds up its front legs so that they look like antennae. Ant mimicry is found among several families of spiders in all terrestrial biomes, but is particularly striking in desert regions where ants, and their mimics, are especially conspicuous.

Desert wolf-spiders, Zodariidae and Sparassidae, tend to have life histories of at least 2 years, unlike many of their relatives in less harsh environments, are able to fast for long periods and seal themselves up in their retreats. Spiders that do not burrow are usually annual species. Although their existence is more precarious than that of burrowers, they have the advantage of being able to recolonize by aerial dispersal of juveniles from other areas, localities that have been devastated by drought or fire.

13.6. MITES AND TICKS (ACARI)

Mites are generally of minor ecological significance in the ecology of deserts, although ticks (Ixodoidea) may be important vectors of disease among domesticated animals. Mites and ticks comprise the sub-class Acari. In these animals, the distinction between the prosoma and opisthosoma is obscure, but there is usually a well-developed head region, known as the 'gnathosoma' or 'capitulum'. There are nearly always four pairs of legs in the adult and nymphal stages, while the larvae have only three.

13.6.1. Free-living and parasitic mites other than ticks

Most mites require a moist environment. For this reason, damp soil and débris support denser populations than dry sand. During wet weather, many species move upwards to the surface of the soil, but burrow deeply as this dries out.

The soil fauna of hot deserts is a very specialized one (Wallwork, 1976), but has been little studied even in North America, and not at all in the Sahara. A small, pale mite, *Erythraeus nigritiensis*, occurs in the northern Sudan but its life-cycle has not been investigated. Many small mites are phoretic on insects and other arthropods, some are predacious, others feed on plants or detritus. Some are parasites of vertebrates—Entonyssidae inhabit the respiratory passages of snakes, Rhinonyssidae the nasal passages of birds while Haemogamasidae are parasitic on small mammals. Many of them transmit various diseases to their hosts.

Giant velvet mites, *Dinothrombium tinctorium* (Fig. 13.30), appear on the sandy surfaces of the Sahara after rain and probably feed on insects. The larvae are parasitic on grasshoppers. They are day-active but become crepuscular as the soil dries out. A high rate of water loss by transpiration suggests that they are not particularly well adapted to drought. The adult mites are positively phototactic and negatively geotactic in dry sand—moving towards the light and upwards—but dig burrows where it is damp. Their scarlet coloration has an aposematic or warning function, and is associated with repugnatorial dermal glands (Cloudsley-Thompson, 1962a).

FIG. 13.30. *Dinothrombium tinctorium* (photo).

13.6.2. Ticks

In contrast to mites, the ixodid ticks of the Sahara have been studied extensively (Arthur, 1965; Hoogstral, 1956). They are partly of Ethiopian, partly of Eurasian origin. Those whose hosts are bats or birds naturally have wider geographical distribution than the ticks of mammals such as rodents, foxes, hedgehogs, badgers and ungulates whose hosts are less mobile. Some species parasitize many different hosts, including reptiles, others are much more restricted. Ticks of the genus *Hyalomma* (Fig. 13.31), parasites of camels, sheep and goats, are locally common on the southern fringe of the Sahara and Sahel savanna. They are unusually resistant to water loss and have been found on sand dunes in the full heat of the summer sun. Ixodid ticks transmit several diseases to Man and domestic animals including turalemia, and can even cause paralysis by their bites. Argasid or soft ticks are also of considerable economic and medical importance. *Ornithodoros moubata* is the notorious vector of the spirochaetes of relapsing fever (*Borrelia* spp.), while *Argas persicus* is not only a troublesome domestic pest in parts of the Sahara, but also infests poultry all over the world, causing much harm by its bites. Argasid ticks are extremely resistant to water loss. *O. savignyi*, for instance, is to be found in dry sand where camels rest. Other well-known Saharan ticks are dog ticks (*Rhipicephalus* spp.) and cattle ticks (*Boophilus* spp.), both transmitters of disease.

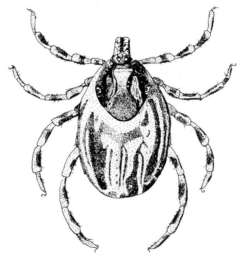

Fig. 13.31. *Hyalomma* sp. (Ixodidae) (length 5 mm).

13.7. SUMMARY

The ways in which arachnids and other invertebrates contribute to desert ecosystems has been summarized by Crawford (1981) in an impressive analysis of an aspect of desert biology that has scarcely been touched upon in the present chapter. Despite a recent surge of knowledge regarding the ecology of arid regions, we are still far from any comprehensive understanding of the rôles of desert invertebrates in general, and of arachnids in particular. At present our knowledge of the arachnid fauna of the Sahara is, with few exceptions, extremely limited and it is to be hoped that the defects apparent in this chapter will encourage further work in this most interesting field of study.

I would like to thank my friends Keith Hyatt and Fred Wanless of the British Museum (Natural History) for their constructive comments on this manuscript.

13.8. REFERENCES

ARTHUR, D. R. (1965) *Ticks of the Genus* Ixodes *in Africa*. Athlone Press, London.

BALOZET, L. (1964) Le scorpionisme en Afrique du Nord. *Bull. Soc. Path. exot.* **57**, 33–38.

BALOZET, L. (1971) Scorpionism in the Old World. In *Venomous Animals and their Venoms*, Vol. 3. *Venomous Invertebrates*, edited by W. BÜCHERL and E. E. BUCKLEY, pp. 349–371. Academic Press, New York.

BEIER, M. (1962) Ergebnisse der Zoologischen Nubien-Expedition 1962. Teil III. Pseudoscorpionidea. *Ann. Naturhistor. Mus. Wien*, **65**, 297–303.

BROWNELL, P. H. and FARLEY, R. D. (1974) The organization of the malleolar sensory system in the solpugid, *Chanbria* sp. *Tissue & Cell*, **6**, 471–485.

CARTHY, J. D. (1968) The pectines of scorpions. *Symp. Zool. Soc. Lond.* No. 23, pp. 251–261.

CLOUDSLEY-THOMPSON, J. L. (1955) On the function of the pectines of scorpions. *Ann. Mag. Nat. Hist.* (12), **8**, 556–560.

CLOUDSLEY-THOMPSON, J. L. (1958) *Spiders, Scorpions, Centipedes and Mites*. Pergamon Press, Oxford (2nd ed., 1968).

CLOUDSLEY-THOMPSON, J. L. (1960) Adaptive functions of circadian rhythms. *Cold Spring Harb. Symp. Quant. Biol.* **25**, 345–355.

CLOUDSLEY-THOMPSON, J. L. (1961a) Some aspects of the physiology and behaviour of *Galeodes arabs*. *Entomolgia exp. appl.* **4**, 257–263.

CLOUDSLEY-THOMPSON, J. L. (1961b) Observations on the natural history of the 'camel-spider' *Galeodes arabs* C L Koch (Solifugae: Galeodidae) in the Sudan. *Entomologist's mon. Mag.* **97**, 145–152.

CLOUDSLEY-THOMPSON, J. L. (1962a) Some aspects of the physiology and behaviour of *Dinothrombium* (Acari). *Entomolgia exp. appl.* **5**, 69–73.

CLOUDSLEY-THOMPSON, J. L. (1962b) Lethal temperatures of some desert arthropods and the mechanism of heat death. *Entomolgia exp. appl.* **5**, 270–280.

CLOUDSLEY-THOMPSON, J. L. (1965) The scorpion. *Science J.* **1** (5), 35–41.

CLOUDSLEY-THOMPSON, J. L. (1968) The Merkhiyat jebels: a desert community. In *Desert Biology*, edited by G. W. BROWN, Jr. Vol. 1, 1–20. Academic Press, New York.

CLOUDSLEY-THOMPSON, J. L. (1975) Adaptations of Arthropoda to arid environments. *Ann. Rev. Ent.* **20**, 261–283.

CLOUDSLEY-THOMPSON, J. L. (1976) *Evolutionary Trends in the Mating of Arthropoda*. Meadowfield Press, Sheldon, Co. Durham.

CLOUDSLEY-THOMPSON, J. L. (1977) Adaptational biology of Solifugae (Solpugida). *Bull. Br. Arachnol. Soc.* **4** (2), 61–71.

CLOUDSLEY-THOMPSON, J. L. (1978) Biological clocks in Arachnida. *Bull. Br. Arachnol. Soc.* **4** (4), 184–191.

CLOUDSLEY-THOMPSON, J. L. (1980) *Tooth and Claw. Defensive Strategies in the Animal World*. J. M. Dent, London.

CONSTANTINOU, C. (1980) Entrainment of the circadian rhythm of activity in desert and forest-inhabiting scorpions. *J. Arid Environments*, **3**, 133–139.

CRAWFORD, C. S. (1981) *Biology of Desert Invertebrates*. Springer-Verlag, New York.

DENIS, J. (1947) Results of the Armstrong College Expedition to Siwa Oasis (Libyan Desert), 1935. Spiders (Araneae). *Bull. Soc. Fouad 1ᵉʳ Entom.* **31**, 17–103.

HADLEY, N. F. (1974) Adaptational biology of desert scorpions. *J. Arachnol.* **2**, 11–23.

HOOGSTRAL, H. (1956) *African Ixodoidea. 1. Ticks of the Sudan*. Washington, Res. Rep. NM 005, 050, 29, 07, U.S. Naval Med. Res. Unit Cairo, No. 3. 1–1101.

JUNQUA, C. (1966) Recherches biologique et histophysiologiques sur un solifuge saharien *Othoes saharae* Panouse. *Mem. Mus. Natn. Hist. nat. Paris* (A), **43**, 1–124.

KAESTNER, A. (transl. H. W. LEVI and L. R. LEVI) (1968) *Invertebrate Zoology*, Vol. 2. *Arthropod Relatives, Chelicerata, Myriapoda*. Wiley, New York.

LAMORAL, B. H. (1974) The structure and possible function of the flagellum in four species of male solifuges of the family Solpugida. *Proc. 6th int. Congr. Arachnol.*, pp. 136–141.

MILLOT, J. and VACHON, M. (1949) Ordre des Scorpions. In *Traité de Zoologie, Anatomie, Systématique, Biologie*, edited by P.-P. GRASSÉ, Tome 6, pp. 386–436. Masson, Paris.

MINTON, S. A., Jr. (1968) Venoms of desert animals. In *Desert Biology*, edited by G. W. BROWN, Jr., Vol. 1, pp. 487–516. Academic Press, New York.

POLIS, G. A., SISSOM, W. D. and McCORMICK, S. J. (1981) Predators of scorpions: field data and a review. *J. Arid. Environm.* **4**, 309–326.

ROEWER, C. F. (1932–1934) Solifugae, Palpigradi. In *Dr H. G. Bronn's Klassen und Ordnungen des Tierreichs*. **5** (4), 4, 1–723. Leipzig.

ROEWER, C. F. (1941) Solifugen 1934–1940. *Veröff. Deutsch. Kolen.-u. Uebersee-Mus. Bremen.* **3**, 97–192.

SHEALS, J. G. (1973) Arachnida (scorpions, spiders, ticks, etc.). In *Insects and other Arthropods of Medical Importance*, edited by K. G. V. SMITH, pp. 417–472. British Museum (Natural History), London.

VACHON, M. (1952) *Études sur les Scorpions*. Institut Pasteur d'Algérie, Alger.

WALLWORK, J. A. (1970) *Ecology of Soil Animals*. McGraw-Hill, London.

WALLWORK, J. A. (1976) *The Distribution and Diversity of Soil Fauna*. Academic Press, London.

WANLESS, F. R. (1977) On the occurrence of the scorpion *Euscorpius flavicaudis* (De Geer) at Sheerness Port, Isle of Sheppey, Kent. *Bull. Br. arachnol. Soc.* **4**, 74–76.

WANLESS, F. R. (1978) A revision of the spider genera *Belippo* and *Myrmarachne* (Araneae: Salticidae) in the Ethiopian region. *Bull. Br. Mus. Nat. Hist. (Zool.)*, **33** (1), 1–139.

WEYGOLDT, P. (1969) *The Biology of Pseudoscorpions*. Harvard University Press, Cambridge, Mass.

CHAPTER 14

Amphibians and Reptiles

M. R. K. LAMBERT

c/o British Herpetological Society, Zoological Society of London, Regent's Park, London

CONTENTS

14.1. INTRODUCTION

Amphibians and reptiles have a generally poor ability to withstand changes in habitat and climate, and they have limited powers of dispersal. Being particularly dependent on environmental conditions for survival, they are therefore indicator species of habitat and climate. Few have actually evolved adaptations peculiar to life in the Sahara Desert in the way species, especially Amphibia, have in the ancient Australian deserts. Reptiles are, however, generally well adapted to resist arid conditions and are probably the most numerous of the vertebrates in the Sahara Desert. Over 100 herpetofaunal species and subspecies have continued to inhabit this hot, dry environment since the Pleistocene.

The Sahara Desert covers a large area of the African Continent. It is fringed to the north and north-west by climatically Mediterranean North Africa, to the south by the Sahel of Africa, to the east by the Red Sea with the riverine habitat of the Nile Valley transecting the Nubian Desert and Upper Egypt, and to the west by the Atlantic Ocean with the cold Canaries Current influencing the climate of the coast. Between the Mediterranean zone in the north, with winter rains, and the tropical zone in the south, with summer rains, there is a large area where the often non-existent rains do not have a seasonal rhythm. Between these zones and the Sahara proper, there is a 'sahel'—a 'beach' of the desert 'sea'—where the mean annual rainfall is 100–400 mm in the north and 100–800 mm in the warmer, more Equatorial south. The definite east–west climatic belt providing savanna conditions across tropical Africa and known as the Sahel strongly influences the species surviving in the true desert immediately to the north.

The distribution of the herpetofauna in the Sahara Desert of today is very recent in geological terms and has established itself in only about the last 2000 years. Until 10,000 years ago, excellent grazing for wild animals used to extend over what is now sterile desert in the northern Sahara, but a marked downturn in rainfall about 4000 years ago destroyed these habitats. Extensive areas, however, were still cultivated or provided good pasture until about 2000 years ago. Indeed, the region provided for the granaries of the Roman Empire until its decline at the start of the Christian Era and remains of such cities as Sabratha, Leptis Magna, Apollonia and Cyrene are evidence of this in Libya today. Plainly, many species of herpetofauna less well adapted to desert conditions must have become extinct in the central Sahara and the present distributions of species are invariably relicts of formerly greater ranges.

Influenced by habitat and vegetation (itself influenced by various factors), climate and zoogeographical affinity, temperate, Palaearctic species have remained extended from the north, and a few tropical, Ethiopian species from the south. Some species and subspecies are endemic, confined to the Sahara, and adapted to withstand hot, dry conditions and desertification. Some species occur either in the north or the south, while the riverine habitat of the Nile Valley has allowed the northward extension of some tropical species. Saharan oases and high mountain ranges (e.g. the Hoggar, the Tibesti) have allowed some temperate species to survive in the south beyond their normal ranges.

The Sahara Desert straddles the Palaearctic and Ethiopian zones, but most species are in fact Palaearctic. Braestrup (1947) has considered herpetofaunal exchange across the Sahara based on geological evidence and recent locality records, for depending on the current position of the 100 mm isohyets to the north and south of the Sahara, desertification has inevitably favoured better adapted desert species progressively driving back Mediterranean and tropical forms leaving only relatively few to penetrate the desert area to any extent with the greater part of their ranges outside the Sahara.

Many species of reptiles in particular depend in their Saharan habitats on the presence of vegetation to provide refuge and food and shelter for their insectivorous and other prey. Quezel (1965) has also used the 100-mm isohyet to define the Sahara vegetationally.

14.2. LITERATURE

Publications on the herpetofauna of the Sahara were only fragmentary during the nineteenth century and based on expedition or even word-of-mouth reports. Early expeditions were made by the Baron Aucapitaine, Duveyrier (1864) and Erwin von Bary in 1878. Foureau (1905) reported the presence of crocodiles (*Crocodylus niloticus*) in the Tassili N'Ajjer plateau (southern Algeria). Subsequently, the Saharan herpetofauna has been specifically documented upon mainly by French, German and Italian investigators from infrequent expeditions to isolated areas like the Hoggar (southern Algeria) and Tibesti ranges (northern Tchad). Egyptian and Sudanese species were investigated mainly by British workers, and species of the Western Sahara (Rio de Oro or Spanish Sahara until 1976) by Spanish workers.

Quite recently (1970s), systematic ecological, and also physiological, investigations were carried out by French workers on north-west Saharan species while based at the Centre de Recherches sur les Zones Arides in the Béni-Abbès region (Algeria).

The references to publications mainly on herpetofaunal systematics in the different countries making up the Sahara Desert are given in Table 14.1. Although the list is comprehensive, it does not claim to be complete, but forms the basis for the descriptive biology and from which a provisional check list of the Saharan herpetofauna has been drawn up (Table 14.2). The subdivision of the Sahara Desert into West, Central and East zones is based on the work of Schiffers (1971). The Saharan species and subspecies have been the subject of some taxonomic confusion in the past and are still due for further clarification.

TABLE 14.1. Reference to Saharan herpetofauna: taxonomy, distribution and general ecology

Country/region	Subject matter	Author(s)
Mauritania	Reptile collection lists	Pellegrin (1910), Chabanaud (1924), Angel (1938)
	Snakes	Villiers (1950)
	General, including herpetofauna	Dekeyser and Villiers (1956)
Western Sahara	Reptile collection list	Günther (1903)
	General, including herpetofauna	Valverde (1957, 1965)
	Herpetofaunal contribution	Salvador and Peris (1975)
NW Africa	NW African lizards	Guibé (1950a)
Morocco	Amphibia	Pasteur and Bons (1959)
	SW Moroccan lizards	Bons (1959)
	Reptile check list	Pasteur and Bons (1960)
	Reptile key	Bons and Girot (1962)
	Herpetofaunal biogeography	Bons (1967)
	Herpetofaunal check list	Bons (1972)
	Biogeography of herpetofaunal populations	Bons (1973)
Algeria	Herpetofaunal check list	Olivier (1894)
	Herpetofauna of the Hoggar	Pellegrin (1930, 1934)
	Reptiles of the Hoggar	Werner (1937)
	NW Saharan reptile ecology	Gauthier (1967a)
	NW Algerian reptiles	Gauthier (1967b)
	NW Saharan reptile ecology	Grenot and Vernet (1972a, b), Vernet and Grenot (1972)
Tunisia	Reptile check list	Olivier (1896)
	Herpetofaunal check list	Mayet (1903)
	Herpetofaunal contribution	Mertens (1929)
	Herpetofaunal check lists	Mosauer (1934), Blanc (1935)
	Snakes	Chpakowsky and Chneour (1953)
Libya	Cyrenaican herpetofauna	Zavattari (1930)
	Reptiles of Kufra	Vicinguerra (1931)
	Fezzan herpetofauna	Scortecci (1934a)
	Philochortus sp. nov. described	Scortecci (1934b)
	Tripolitanian herpetofauna	Scortecci (1936)
	Snakes	Kramer and Schnurrenberger (1963)
Egypt	Herpetofaunal list with descriptions	Anderson (1898)
	Herpetofaunal annotated check list	Flower (1933)
	Lizard and snake key	Marx (1956)
	Herpetofaunal check list	Marx (1968)
Sudan	Snakes	Corkill (1935)
	Nile Valley herpetofauna	Eiselt (1962)
Tchad	Herpetofauna of the Tibesti	Pellegrin (1936)
	Herpetofaunal collection	Wake and Kluge (1961)
Niger	Amphibia of the Aïr	Guibé (1950b)
	Lizards of the Aïr	Angel (1950)
	Snakes and chelonians of the Aïr	Villiers (1950)
Central Sahara	W Saharan herpetofaunal collection	Hartert (1913)
(S Algeria,	Herpetofaunal collection	de Witte (1930)
N Mali, N Niger,	Lower vertebrate collection	Pellegrin (1931)
E Mauritania)	Herpetofaunal collection	Andersson (1935), Scortecci (1937)
	Herpetofaunal check list	Angel and Lhote (1938)
	Herpetofaunal contribution	Angel (1944)
	TransSaharan faunal exchange	Braestrup (1947)
	The 'Empty Quarter', general, including reptiles	Monod (1958)
	Central W African reptiles	Papenfuss (1969)
	Reggan ecology, including reptiles	Grenot and Niaussat (1967)
Sahara general	General description of herpetofauna	Gruber (1971)
	Frogs in S Sahara	Böhme (1978)
	W African snakes	Villiers (1975)
	African bufonids	Hulselmans (1977)
	W African herpetofaunal collection	Joger (1981)

TABLE 14.2. A preliminary check list of the Saharan herpetofauna, showing zoogeographical affinity, distribution and Saharan habitat

Species and subspecies	Zoogeographical affinity	Overall range						Saharan range			Saharan habitat						
		Europe	North Africa	SW Asia	Nile Valley	Sahel	Tropical Africa	West	Central	East	Erg (dune)	Reg (rocks)	Jebel (hills, range)	Hamada (steppe)	Wadi (dry river-bed)	Oasis: palmery, houses	Water, oued (river-bed), daya (seasonal pool), guelta (deep, still pool)
AMPHIBIA																	
Ranidae																	
Rana ridibunda perezi Seoane, 'green frog'	P	+	+	+	?			+	+	+							+
Rana (Euphlyctis) occipitalis (Günther), western bull frog	E		+		+	+	+	+	+								+
Ptychadena floweri (Boulenger), Flower's ridged frog	E				+	+	+	+	+								+
Bufonidae																	
Bufo viridis viridis Laurenti, green toad	P	+	+	+	+			+	+	+					+	+	+
Bufo brongersmai Hoogmoed, *Brongersma's toad	P		+					+								+	+
Bufo mauritanicus Schlegel, Mauritanian toad	P		+			+		+	+						+	+	+
Bufo dodsoni Boulenger, Dodson's toad	E					+	+		+						+	+	+
Bufo xeros Tandy, Tandy, Keith & Duff-Mackay, *savanna toad	E					+		?	+							+	+
Bufo regularis Reuss, African or leopard toad	E				+	+	+	+	+	+						+	+
REPTILIA																	
Crocodilidae																	
Crocodylus niloticus Laurenti, Nile crocodile	E				+	+	+	?	+								+
Trionychidae																	
Trionyx triunguis (Forskål), Nile soft-shelled turtle	E			+	+	+			+								+
Emydidae																	
Mauremys caspica leprosa (Schweigger), Spanish or striped-neck terrapin	P	+	+	(+)	?			+	+								+
Gekkonidae																	
Geckonia chazaliae Mocquard, helmeted gecko	Sah.		+					+			+						
Quedenfeldtia trachyblepharus (Boettger), *thorny-eyelidded gecko	?P		+					+					+	+			
Saurodactylus mauritanicus brosseti Bons & Pasteur, **Brosset's lizard-toed gecko	P		+					+						+	?	?	
Stenodactylus petrii Anderson, Petrie's gecko	Sah.		+	+	+	+		+	+	+	+					+	+
Stenodactylus sthenodactylus (Lichtenstein), elegant gecko	Sah.		+	+	+	+		+	+	+	+			+			
Gymnodactylus scaber (Heyden), rough-skinned, rough-scaled or keeled rock gecko	P			+	+				+				+	+			
Ptyodactylus hasselquistii (Donndorf), fan-footed gecko	P		+	+	+		+	+	+	+			+	+			
Pristurus flavipunctatus Rüppell, *yellow-spotted gecko	?E			+	+				+				+	+		+	
Hemidactylus turcicus turcicus (L.), Turkish gecko	P	+	+	+	+				+							+	
Hemidactylus flaviviridis Rüppell, Cocteau's or yellow-bellied house gecko	P			+	+				+							?	+
Tarentola neglecta Strauch, neglected gecko	P		+						+		+					+	
Tarentola mauritanica deserti Boulenger, Moorish desert gecko	P	(+)	+					+								+	
Tarentola ephippiata-annularis complex, 'desert geckos'	E				+	+	+	+	+	+						+	+
Tarentola parvicarinata Joger, *small-keeled gecko	E					+	+	+					+	+			
Tropiocolotes steudneri (Peters), Steudner's gecko	P			+	+	+			+	+						+	+
Tropiocolotes tripolitanus tripolitanus Peters, Tripoli gecko	P		+						+	+					?	+	
Tropiocolotes tripolitanus algericus Loveridge, *Algerian gecko	P		+					+	+					+	+	+	
Tropiocolotes tripolitanus occidentalis Parker, singing gecko	P							+						+	+		

TABLE 14.2—*continued*

Species and subspecies	Zoogeographical affinity	Europe	North Africa	SW Asia	Nile Valley	Sahel	Tropical Africa	West	Central	East	Erg (dune)	Reg (rocks)	Jebel (hills, range)	Hamada (steppe)	Wadi (dry river-bed)	Oasis: palmery, houses	Water, *oued* (river-bed), *daya* (seasonal pool), *guelta* (deep, still pool)
REPTILIA (*continued*)																	
Scincidae																	
Scincus scincus (L.), skink	P		+		+	+		+	+		+						
Scincus albifasciatus Boulenger, skink	P		+		+	+		+	+		+						
Scincopus fasciatus Peters, banded skink	P				+			+	+	+				+			
Chalcides ocellatus subtypicus Werner, eyed skink	P	(+)	+	+	+	+		+	+	+					+	+	+
Chalcides polylepis polylepis Boulenger, *multiscaled skink	P		+					+						+			
Eumeces schneideri algeriensis Peters, Algerian skink	P		+					?								+	?
Sphenops delislei (Lataste), de Lisle's skink	E					+	+	+	+							+	+
Sphenops boulengeri (Anderson), Audouin's skink	P		+	+		+		+	+	+	+						+
Sphenops sphenopsiformis (Duméril), Senegal skink	Sah.		+			+	+	+			+						+
Mabuya quinquetaeniata (Lichtenstein), five-lined skink	E				+		+		+								+
Lacertidae																	
Lacerta lepida pater (Lataste), eyed lizard	P	+	+					+	+							+	+
Latastia longicaudata longicaudata Reuss, long-tailed lizard	E			+	+		+	+	+	+						+	
Acanthodactylus boskianus asper (Audouin), Bosc's sand-racer or fringe-toed lizard	Sah.		+	+	+	+		+	+	+					+	+	+
Acanthodactylus scutellatus (Audouin), mottled or Nidua sand-racer or fringe-toed lizard	P		+	+	+	+		+	+	+	+	+			+	+	+
Acanthodactylus aureus Günther, *Western Sahara sand-racer or fringe-toed lizard	P							+			+						
Acanthodactylus longipes (Gray), *long-footed sand-racer or fringe-toed lizard	Sah.		+					+			+						
Acanthodactylus pardalis pardalis (Lichtenstein), leopard sand-racer or fringe-toed lizard	P		+	+				+	+					+			
Acanthodactylus busacki Salvador (?syn. *A. pardalis bedriagae* Lataste)	?P		?					+						+			
Acanthodactylus maculatus Gray, *mottled sand-racer or fringe-toed lizard	P		+					+	+					+			
Acanthodactylus spinicaudus Doumergue, *spiny-tailed sand-racer or fringe-toed lizard	Sah.		?					+	?								+
Eremias mucronata (Blanford), Anseba lizard	Sah.				+				+		+						
Mesalina guttulata guttulata (Lichtenstein), small-spotted lizard	P		+	+	+	+		+	+	+				+			
Mesalina rubropunctata (Lichtenstein), red-spotted lizard	P		+	+	+	+		+	+	+	+			+			
Mesalina olivieri olivieri (Audouin), *Olivier's small-spotted lizard	P		+	+				+	+					+		+	
Mesalina olivieri simoni (Boettger), *Simon's small-spotted lizard	P		+					+								+	
Mesalina pasteuri (Bons), *Sahara lizard	Sah.							+	+		+						+
Ophisops occidentalis Boulenger, western snake-eyed lizard	.P		+					+	+	?						+	
Ophisops elegans Menetries, snake-eyed lizard	P								+	+						+	
Ophisops elbaensis Schmidt & Marx, Mount Elba snake-eyed lizard	?E								+						+	?	
Philochortus zolii Scortecci (?syn. *P. intermedius* Boulenger or *P. lhotei* Angel)	E			+		+	+	+	?							+	
Chameleonidae																	
Chameleo chameleon chameleon (L.) (*Chameleo chameleon saharicus* Mertens), common chameleon	P	+	+	+	+			+	+							+	

TABLE 14.2—*continued*

Species and subspecies	Zoogeographical affinity	Overall range						Saharan range			Saharan habitat							
		Europe	North Africa	SW Asia	Nile Valley	Sahel	Tropical Africa	West	Central	East	Erg (dune)	Reg (rocks)	Jebel (hills, range)	Hamada (steppe)	Wadi (dry river-bed)	Oasis: palmery, houses	Water, oued (river-bed), daya (seasonal pool)	guelta (deep, still pool)
REPTILIA (*continued*)																		
Chameleonidae (*continued*)																		
Chameleo africanus Laurenti, basilisk chameleon	E		+		+	+		+	+	+						+		
Agamidae																		
Agama agama L., common agama	E					+	+	+	+	+		+	+			+		
Agama bibroni Duméril, Bibron's agama	Sah.		+			+		+	+				+					
Agama mutabilis Merrem, changeable agama	Sah.		+	+	+	+		+	+	+				+				
Agama flavimaculata flavimaculata (Rüppell), Savigny's agama	?P		+	+						+	+							
Agama flavimaculata tournevillei Lataste, dune agama	Sah.							+	+		+							
Agama sinaita von Heyden, Sinai agama	P		+	+	?					+	?	+						
Uromastyx acanthinurus Bell, spiny-tailed lizard, dob or Bell's dab-lizard	Sah.		+	+		+		+	+				+	+				
Uromastyx geyri Müller, Geyr's dab-lizard	Sah.					+		+					+					
Uromastyx aegyptius aegyptius (Forskål), Egyptian dab-lizard	P		+	+						+			+					
Uromastyx ocellatus Lichtenstein, eyed dab-lizard	P		+			+				+			+	+				
Varanidae																		
Varanus griseus griseus (Daudin), grey or desert monitor	P		+	+				+	+	+	+					+	+	
Varanus niloticus (L.), Nile monitor	E				+	+	+			+							+	
Leptotyphlopidae																		
Leptotyphlops macrorhynchus (Jan), worm snake	P		+	+		+		+	+							+		
Leptotyphlops cairi (Duméril & Bibron), Cairo earth snake	E				+	+				+						+	+	
Boidae																		
Eryx muelleri subniger Angel, sand boa	E					+		+			?			+				
Eryx colubrinus colubrinus (L.), Theban sand boa	E				+	+	+	+	+		?			+				
Colubridae																		
Natrix maura (L.), viperine snake	P	+	+					+	+							+	+	
Boaedon fuliginosum (Boie), common African house snake	E				+		+	+								+	+	
Lycophidion capense (A. Smith), Cape wolf-snake	E				+	+	+		+							+	+	
Coluber florulentus complex, 'flowered snake'	P		+		+	+		+	+	+			+	+	+	+		
Coluber rhodorhachis rhodorhachis (Jan), Jan's desert racer	P		+	+	+	+			+	+			+	+			+	
Sphalerosophis diadema cliffordi (Schlegel), Clifford's snake	P		+	+	+	+		+	+	+			+	+			+	
Sphalerosophis dolichospilus Werner, *long-marked snake	P		+					+					+					
Lytorhynchus diadema (Duméril & Bibron), diademed sand snake	P		+	+	+			+	+	+	+						+	
Telescopus dhara obtusus (Reuss), Egyptian cat-snake	E			+	+	+			+	+						+	+	?
Malpolon moilensis (Reuss), Moila snake	P		+	+	+	+		+	+	+				+			+	
Malpolon monspessulanus monspessulanus (Hermann), Montpelier snake	P	+	+	+	+			+								+	?	+
Macroprotodon cucullatus cucullatus (Geoffroy St.-Hilaire), Mediterranean hooded or false smooth snake	P	+	+	+	+	?		+	+	+						+	?	+
Macroprotodon cucullatus brevis (Günther), false smooth snake	P		+					+								+	?	+
Psammophis schokari (Forskål), Schokari sand snake	P		+	+	+	+		+	+	+			+	+		+	+	
Psammophis sibilans sibilans (L.), African beauty snake	E			+	+	+		+	+	+			+	+		+	+	
Dasypeltis scabra (L.), rough-keeled or egg-eating snake	E			+	+	+	+	+		+						+		

TABLE 14.2—*continued*

Species and subspecies	Zoogeographical affinity	Overall range						Saharan range			Saharan habitat								
		Europe	North Africa	SW Asia	Nile Valley	Sahel	Tropical Africa	West	Central	East	Erg (dune)	Reg (rocks)	Jebel (hills, range)	Hamada (steppe)	Wadi (dry river-bed)	Oasis: palmery, houses	Water, oued (river-bed)	daya (seasonal pool)	guelta (deep, still pool)
REPTILIA (*continued*)																			
Elapidae																			
Naja haje haje (L.), Egyptian or black cobra	E		+	+	+	+		+		+						+			+
Naja nigricollis nigricollis Reinhardt, black-necked cobra	E				+	+			+							+	+		
Naja mossambica pallida Boulenger, *pale cobra	E				+	+	+		+							+			
Walterinnesia aegyptia Lataste, Innes's snake	P		+							+						+			
Viperidae																			
Echis carinatus pyramidum (Geoffroy St.-Hilaire), carpet viper	P		+	(+)	+	+		+	+							+			
Echis leucogaster Roman, carpet viper	?E				+	+	+	+	+	+						+			
Echis coloratus Günther, Burton's carpet viper	P		+							+						+	+	+	
Cerastes cerastes (L.), greater horned, sand or carpet viper	P		+	+	+	+		+	+	+		+	+	+		+			
Cerastes vipera (L.), lesser horned, sand or carpet viper	P		+	+	+			+	+	+	+	+	+	+	+	+			
Vipera lebetina mauritanica (Gray), blunt-nosed viper (? Cleopatra's asp)	P	(+)	+	(+)	?			+		?						+	+		

Abbreviations: P—Palaearctic, E—Ethiopian, Sah.—Saharan. (+)—distribution of other subspecies not mentioned.
* Proposed English names, if none.

Desertification in recent years, moreover, may also have outdated the information given in some of the earlier papers, especially with regard to the position of the southern boundary of the Sahara Desert proper against the Sahel.

14.3. AMPHIBIA

Amphibians depend on water for their early stages of development. Not surprisingly, only a few occur in the Sahara Desert since few localities have standing water for many months of the year. Besides oases and montane areas, the Nile River valley cuts into the desert and provides riverine habitat. No Urodela have been recorded within Saharan confines. There are three frogs and six toads.

Besides northern Algerian, Moroccan and Western Saharan oases, *Rana ridibunda perezi* (one of the 'green frog' complex with an extensive West Palaearctic range, although the North African form may be separate) has apparently been recorded near Ghat (south-western Libya) and in the Hoggar and Tassili N'Ajjer. In Djanet oasis (southern Algeria), the frog frequents springs and irrigation channels in areas of cultivation, apparently having a preference for running water (Fig. 14.1). *Rana (Euphlyctis) occipitalis*, in contrast, inhabits deep pools of still water (*gueltas*) in the Adrar (Mauritania). It has also been recorded at Bilma (Niger) and in the Ennedi (eastern Tchad), and otherwise occupies wet habitats in tropical Africa. It also occurs in Morocco and Algeria, probably as an introduction. The third frog, *Ptychadena floweri*, is also tropical eastern African, but has been recorded in the Tassili N'Ajjer and is not otherwise found further north except in the Nile Valley, where it is common, and found as far north as the Nile Delta in Lower Egypt.

FIG. 14.1. *Rana ridibunda perezi*, an adult (about 125 mm long) and two subadults in an oasis stream. Tinerhir, southern Morocco, 19.iv.1969.

The toad, *Bufo v. viridis*, has, like *R. r. perezi*, a vast West Palaearctic range. It occurs in oases in southern Morocco and is often seen crawling among palm bases and near water holes at night (e.g. Rissani oasis, pers. obs., 29.vii.1966). Unexpectedly, the toad also inhabits the Hoggar, abounding where there are human dwellings and areas of water, and has been found up to 2500 m. The Hoggar was connected with North Africa by the Oued Igharghar after the Pleistocene, but it is curious it does not occur in the Tassili N'Ajjer, where *B. mauritanicus* occurs instead, in gueltas. *B. mauritanicus* has mainly been recorded in Morocco (Fig. 14.2) and northern Algeria, but extends south into Western Sahara and otherwise recorded in the Adrar. Pasteur and Bons (1959) have described the biology of both species of toads and *R. r. perezi* in Morocco. *B. brongersmai* was found in 1971 by Hoogmoed (1972) in south-western Morocco.

Bufo dodsoni has been recorded at Jebel Elba (south-eastern Egypt) and also in present Somalia, and *B. xeros* in the Hoggar and possibly also Ghat, as well as further south in the Sahel and eastern Africa.

Bufo regularis is a common and widespread species in tropical and southern Africa extending its range to Saharan confines. It is recorded in the Adrar, and also the Aïr (Niger) by Guibé (1950b). It has only extended further north into Lower Egypt by following the Nile. Its ecology and breeding biology in tropical Africa have been described by Winston (1955), Webb (1958) and Chapman and Chapman (1958).

A physiological comment on Saharan Bufo *sympatry:* In northern Sudan, *B. regularis* only breeds in irrigation ditches and seasonal pools (July–September) in the desert where these are just beyond the cultivated area fringing the Nile (Cloudsley-Thompson and Chadwick, 1964). It has not become adapted to conserve water by reduced evaporation or increased rate of water uptake and in these respects compares with temperate *Bufo bufo* L. The subspecies, *B. bufo spinosus* Daudin and *B. mauritanicus*, are sympatric in western North Africa at a southerly point of the former's range in the High Atlas (Morocco) where they appear to hydridize (e.g. Imlil, pers. obs., 9.vii.1966; Tandy and Keith, 1972), but *B. mauritanicus* (also sympatric with *B. brongersmai*), like *B. v. viridis*, continues southwards. In localities in Mali (River Niger valley, the Adrar des Ifohras) in the Sahel, it is sympatric with both *B. v. viridis* and *B. regularis*, and with the latter also in the Aïr. *B. mauritanicus* has a higher evaporation rate and is able to maintain a greater temperature differential with the environment by evaporative cooling than *B. regularis* (Table 14.3).

FIG. 14.2. *Bufo mauritanicus*, a pair in amplexus in a water tank. Near Ait Melloul, Souss Valley, south-western Morocco, 14.iv.1969. Note sexual dimorphism. Adults achieve about 155 mm in length.

TABLE 14.3. Water loss (% body weight per hour ± standard error) and mean oral temperatures (N = 5) of *Bufo mauritanicus* in dry air at various ambient temperatures compared with similar figures for *Bufo regularis* (Cloudsley-Thompson, 1967). After Cloudsley-Thompson (1974)

Ambient temperature (°C)	*Bufo mauritanicus*		*Bufo regularis*	
	Water loss (%)	Body temperature (°C)	Water loss (%)	Body temperature (°C)
21	2.4 ± 0.44	20.2	—	—
30	4.3 ± 1.05	25.8	0.5	28
35	5.9 ± 1.48	26.4	2.4	31
40	10.2 ± 1.63	29.3	3.8	34
45	11.2 ± 1.56	31.2	5.8	37

Although it cannot survive for long in a hot, dry atmosphere, this ability to maintain a lower body temperature for short periods in dry air may be useful in hotter habitats, perhaps explaining it's and *B. v. viridis*'s presence in the Sahara. The widespread latter species seems to be still more resistant to hot, dry environments and it would be of interest to compare experimentally the water relations of each, and perhaps those of *B. xeros* and *B. dodsoni*, to explain their distributions.

14.4. REPTILIA

Varyingly adapted to arid conditions, reptiles are probably the best represented of all vertebrate groups in the Sahara Desert. Although nearly 100 species and subspecies occur there (Table 14.2), terrestrial chelonians are excluded. *Geochelone sulcata* (Miller), the grooved tortoise, is a savanna species confined to the Sahel, and *Testudo kleinmanni* Lortet, Leith's or the Egyptian tortoise, to the Mediterranean littoral of

Cyrenaica (Libya) and Egypt (mean annual rainfall 100–200 mm). *Mauremys caspica leprosa*, common in the Iberian Peninsula and north-western Africa, only occurs within Saharan confines in the running water of southern Moroccan oases (e.g. Meski, pers, obs., 20.iv.1969). Absent from the central Sahara, it has been reported to the south in the Adrar des Ifohras, the Aïr and in Mauritania, possibly crossing the Sahara during the Pleistocene, but equally likely (Papenfuss, 1969) to have been introduced by Tuareg camel caravans from North Africa; Agadez, the only specific locality for the Aïr, being a major southern caravan terminus. *Chameleo c. chameleon*, *Lacerta lepida pater* and *Natrix maura* are other Mediterranean species only occurring in oases in southern Morocco and the southern Oranais (Algeria), together with Saharan subspecies of Mediterranean forms: *Chalcides ocellatus subtypicus* and *Tarentola mauritanica deserti*. A specimen of *Lacerta l. pater* has been found recently even as far south in the Algerian Sahara as Amguid (Joger, 1981), possibly the result of an introduction by a tourist during a transSaharan trek.

14.4.1. Tropical Nile Valley reptiles

Thirteen reptiles are not Saharan desert forms, all being tropical Ethiopian, and only occur within the 100-mm isohyet because their ranges have been extended northwards by the riverine habitat of the Nile Valley.

Crocodylus niloticus	*Lycophidion capensis*
Trionyx triunguis	*Telescopus dhara obtusus*
Mabuya quinquetaeniata	*Dasypeltis scabra*
Eryx c. colubrinus	*Naja haje haje*
Chameleo africanus	*Naja nigricollis nigricollis*
Varanus niloticus	*Naja mossambica pallida*
Leptotyphlops cairi	

None is truly a desert-loving species. *Naja h. haje* in North Africa has been recorded in Saharan confines in the Lower Dra valley of southern Morocco and *Dasypeltis scabra* in the northern Western Sahara (possibly as an introduction), but both snakes range extensively in Africa south of the Sahara.

Crocodylus niloticus has become extinct in the central Sahara only since the beginning of this century. The savanna region south of the Sahara extended north during the Pleistocene and the western part of the River Niger was connected with the Tassili N'Ajjer by the Oued Tafasasset. Bones of the crocodile have been found in Neolithic layers south of the Tibesti and in Mauritania (Standinger, 1929) suggesting it was once common. Crocodiles are depicted in the ancient frescoes of the Tassili N'Ajjer. The Tuaregs call the crocodile *arochaf* or *agânba* (Tamachek) and was therefore known to them. Duveyrier (1864) confirmed Aucapitaine's earlier report (in Monod, 1931) that the crocodile occurred towards the end of the last century in the central Sahara by recording its inhabiting the Oued Ihmirou (Tassili N'Ajjer). Erwin von Bary found its tracks in 1880, and Foureau (1905) mentions indigene's information that it occurred in the small lakes of the Tassili N'Ajjer. Finally, in 1910, a Captain Niéger killed a young crocodile in a pocket of water of the O. Ihmirou and in 1925, a Lieutenant Beauval caught a 2-m specimen in the Oued Ahir, a tributary of the Ihmirou. Lavauden (1926) foresaw its disappearance, however, and André Lhote during long periods surveying for it (1928–37) in the Ihmirou and tributaries found no trace, local people confirming that to their knowledge it existed there no longer. The crocodile was found most recently in pools of the Tagant (Mauritania), as reported by R. Chudeau in 1937, but now it must be considered extinct in the Sahara.

14.4.2. Widely distributed Saharan reptiles

About thirty species or subspecies of reptiles occur in a greater proportion of the Sahara and can be considered to be characteristic. Often abundant, none is threatened by desertification.

Stenodactylus petrii	*Agama mutabilis*
Stenodactylus sthenodactylus	*Uromastyx acanthinurus*
Ptyodactylus hasselquistii	*Varanus griseus griseus*
Tarentola ephippiata-annularis complex	*Leptotyphlops macrorhynchus*
Tropoiocolotes tripolitanus	*Coluber florulentus* complex
Scincus scincus	*Sphalerosophis diadema cliffordi*
Scincus albifasciatus	*Lytorhynchus diadema*
Scincopus fasciatus	*Malpolon moilensis*
Chalcides ocellatus subtypicus	*Macroprotodon cucullatus*
Sphenops boulengeri	*Psammophis schokari*
Acanthodactylus boskianus asper	*Psammophis sibilans sibilans*
Acanthodactylus scutellatus	*Echis carinatus pyramidum*
Mesalina guttulata guttulata	*Echis leucogaster*
Mesalina rubropunctata	*Cerastes cerastes*
Agama bibroni	*Cerastes vipera*

Some of the species are so well adapted to Saharan conditions that it is worth considering their biology in more detail.

Stenodactylus petrii and *S. sthenodactylus* are nocturnal geckos which adopt a curious gait when traversing bare sand, unlike *Ptyodactylus hasselquistii*, which may be seen in rock crevices during the day, although not being strictly nocturnal nor crepuscular, and sometimes entering houses. *Tarentola ephippiata* O'Shaughnessy and *T. annularis* (Geoffroy St,-Hilaire) have been confused and described as a complex by Grandison (1961). They are almost completely sympatric, but the former is absent from the Nile Valley. There has also possibly been some past confusion with *Tarentola mauritanica deserti* and *T. parvicarinata* in the literature.

The taxonomy of *Scincus* has recently been revised by Arnold and Leviton (1977) and allotted in the Sahara to *S. scincus* and *S. albifasciatus*. The genus burrows fish-like extremely rapidly in loose sand and is hunted remorselessly as a food item by the Tuaregs, who call it *tejolémagne* (Tamachek). *Chalcides ocellatus subtypicus* (Fig. 14.3) is a more surface-dwelling species, inhabiting oueds in southern Morocco and common in Algerian oases, seeking refuge beneath slabs of rock and amongst vegetation during the warmest hours of the day. Like *Sphenops boulengeri*, which is a burrowing species, it is temperature sensitive; nocturnal during the summer, diurnal in spring and autumn.

Small lizards like *Acanthodactylus boskianus asper* are very active and widespread in Saharan desert conditions, being well adapted and ubiquitous. *A. boskianus asper* is one of the commonest lizards in southern Morocco, feeding like *Mesalina g. guttulata*, depending on size on habitat, on a range of insect orders (Robson and Lambert, 1980). It has a bright colour pattern; reddish grey with four lines dorsally, spangles of blue and deep pink with brownish spots in the female, orange with black spots in the male, white ventrally and the end of the tail pink. In the young, the whole tail is pink, becoming orange at its extremity. Further south, it is known by the Tuaregs as *timkelkelt* (Tamachek) or *cermomilla* (Arabic). The genus *Acanthodactylus* is still under revision at present and *A. scutellatus* comprises several former species and subspecies (E. N. Arnold, pers. comm.). It is one of the commonest species in the north-western

FIG. 14.3. *Chalcides ocellatus subtypicus*, an adult (snouth–vent length about 115 mm). 12 km south of Aouinet-Torkoz, Lower Dra Valley, south-western Morocco, 20.iii.1969.

Sahara, extending south to the Hoggar and Tibesti, and like *A. boskianus asper*, forms 30–40 cm deep burrows in the sand amongst the roots of 'drinn' (*Stipagrostis pungens*), in contrast to *Mesalina g. guttulata*, which has a well-defined rocky habitat. *Mesalina rubropunctata* is an important desert lizard being one of the most characteristic of the Sahara and the only one in parts of the Tidikelt/Tanezrouft region, although with a rather lower density than other lizards elsewhere. Characteristically, the dorsal surface bears scattered large red spots, and the species depends strongly on sunshine while continuously active from 10.00 to 18.00 hours during spring and autumn.

The Agamidae are well represented in the Sahara. Being a cliff and *jebel* dweller, the diurnal *Agama bibroni* jumps and climbs over rocks with amazing agility. *A. mutabilis* (Fig. 14.4) is still more desert loving, inhabiting the *hamada*, and being extremely resistant to insolation, high temperatures and lack of food, becomes quite thin by the end of summer. Voracious when food (beetles, ants and termites) is once again available after the rains, it also actively chases adult locusts, and regains its normal weight after its summer and dry season fast. Strictly diurnal, it often selects a high vantage-point on a rock about a metre above the surrounding terrain. It is common in the Hoggar up to 1500 m, occurring in the *reg*, on steppes and grassland, its colour harmonizing well. Inoffensive like the Sahelian *A. agama*, it is known as *amechtartar* (Tamachek) to the Tuaregs, who regard both species as venomous and kill them whenever they are seen.

Uromastyx acanthinurus (Fig. 14.5) is a large, widespread and abundant Saharan agamid, known as *daub* (Arabic), which shows considerable colour polymorphism from individual male to male (Grenot, 1974). It varies dorsally from yellow and green to reddish-orange with a pattern of spots, lines, ocellations or stripes, the various morphs probably developing on account of population immobility and localization (e.g. Grenot and Vernet, 1973). The females are usually a dirty yellow colour or like the young greyish. Males are especially brightly coloured during the season of courtship. Large specimens are scarce near human habitation in north-western Sahara because they are actively hunted as food. The lizard has an unusual diet in that vegetable matter is quite as important as insects and more important at certain times of the year. While a preference is shown in the north-western Sahara for *Rumex vesicarius* and various crucifers, *Bubonium graveolens, Anvillea radiata* and various grasses and umbellifers can form a food base. Solitary desert locusts, *Schistocerca gregaria*, also provide an appreciable proportion of the diet and a large number consumed in a day when actively hunted. Further details on the unusual feeding regime are given by Dubuis *et al.* (1971).

Fig. 14.4. *Agama mutabilis*, a subadult resting in late afternoon on a *maader* (dry flood pan). Lower Dra Valley, south-western Morocco, 20.iii.1969.

The ecophysiology of *U. acanthinurus* has been the subject of a detailed study by Grenot (1976) in the north-western Sahara (Béni-Abbes region). The species is diurnal and its activity, like other species, is related to sunshine and temperature, the thermoregulatory activity and physiology having been investigated by Grenot (1967) and Grenot and Loirat (1973). Several papers have subsequently been written on the water and electrolyte balance as well (e.g. Lemire, Grenot *et al.*, 1979, 1980, 1981; Lemire, Vernet *et al.*, 1980). The lizard seeks refuge in large burrows whose entrances are often protected by a large stone. The single entrance may lead to two or three passages descending about a metre and often reaching 2 m in length, with several sharp bends, the first after 0.5 m. Spring mating in May, six to fifteen white, oval eggs of 30 × 20 mm, with parchment-like shells are laid by females in June and July. Hatchlings appear in October and start to look for their own territory and make a burrow.

The ecology and thermoregulatory activity of the large desert monitor, *Varanus g. griseus*, has been the subject of a special study by Vernet (1977). Reaching 100–120 cm in length, it frequents sandy areas in north-western Sahara, occupying oueds, but most frequently dunes where firm areas between ridges offer sure cover and sufficiently numerous prey. It feeds on lizards mostly, but snake remains have been found amongst the stomach contents as well as those of birds. Small rodents like gerbils and jerboas are also eaten when their burrows are raided. In the Hoggar further south in the Sahara, the monitor inhabits granitic outcrops, and is known as *ourane* (Chaamba Arabic) and *arata* (Tamachek) to the Tuaregs. Exclusively diurnal, it often makes great sorties of 4–5 km around its burrow, seeking shelter in a clump of vegetation during the hottest hours of the day in summer. The Tuaregs hunt the animals for their skin to make-up small bags, while both Tuaregs and Arabs often carry the head as a kind of talisman against the bite of vipers.

Leptotyphlops macrorhynchus is one of the twelve characteristic Saharan snakes and is worm-like, 205 mm long, pale pink in colour, with the little ecological information available suggesting it to be subterranean and aquatic. *Coluber florulentus* has been variously described under different names. It is a widely distributed species, variable and requires taxonomic revision. *Sphalerosophis diadema cliffordi* is in contrast to *L. macrorhynchus* a 1.6-m-long colubrid and like *Lytorhynchus diadema* includes lizards in its diet, and preys on rodents, especially *Meriones lybicus*, in whose burrows refuge is also sought. Of the opisthoglyphs, *Malpolon moilensis* occupies a wide range of habitats, although absent from the central Sahara. Its diet and activity are like *S. d. cliffordi*, not feeding during cool, damp weather, and nocturnal from the end

FIG. 14.5. *Uromastyx acanthinurus*, adult specimen. Lengths range from 280 to 380 mm. Near Ouarzazate, southern Morocco, 19.iv.1969. (a) Running in the *hamada*, (b) ventral view.

of April to mid–October. It is diurnal during winter in the north-west Sahara, becoming crepuscular as the temperature increases. In captivity it can be ophiophagous and, although timid, can be aggressive, inflating its neck cobra–like and biting vigorously. *Macroprotodon c. cucullatus* in contrast, although also an opisthoglyph, soon becomes quiet when handled, reaching 65 cm in length and often found under stones in rocky, scrubby areas during the middle of the day. *Psammophis schokari* (Fig. 14.6) is possibly the most widespread of Saharan species and is a long, thin snake, smaller than the more tropical *P. s. sibilans*, and inhabits arid areas with scattered vegetation, hunting diurnally for small lizards among desert bushes. *P. s. sibilans* extends to South Africa and in Egypt spends most of its time in trees in the gardens of houses, where, although of striking appearance, it remains unmolested by humans. It is known to the Tuaregs as *achel* (Tamachek), which covers several colubrid species and is synonomous with 'snake'.

The viperids include *Echis carinatus pyramidum* in the northern Sahara and *E. leucogaster* in the south, but

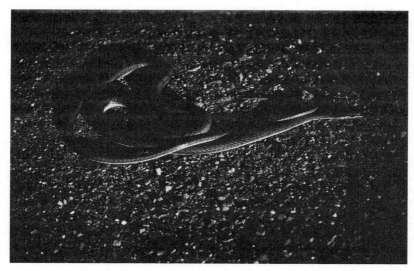

FIG. 14.6. *Psammophis schokari*, a snake of rocky scrubland reaching about 1½ m in length. West of Goulimime, south-western Morocco, 22.iii.1969.

both are rare or local over a wide area. The former is tawny orange in colour with thirty rather irregular cross-bars of pale-yellow scales tending to be edged black on the neck and body, fading on the tail. The top of the head has an indistinct, dark-brown V-shaped marking, Crepuscular or nocturnal, it feeds on small mammals and large insects, striking extremely fast and launching its body into the attack. The venom is very active, inevitably causing death in untreated human cases of snake-bite. Another widely distributed, but often abundant viperid, is *Cerastes cerastes*, inhabiting the *hamada* in north-western Sahara and finding its way into other habitats during nocturnal movements of several kilometres when in search of prey. It has a particular liking for *oued* habitats with clumps of 'drinn' and occurs up to 1000 m in the Hoggar. It hunts for lizards (e.g. *Acanthodactylus pardalis*, *Agama mutabilis*) in a characteristic way during the day by burying all but the top of the head in sand and lying in wait for a passing prey. It suddenly strikes and ingests the animal after it has died from the venom. It reaches 750 mm in length and is sluggish in movement. The snake is much feared by the Tuaregs and deaths from the bite are common. In parts of the Tassili N'Ajjer, the Tuaregs look for the snakes using a metal hook on the end of a stick to remove animals hidden under rocks. Like the smaller *Cerastes vipera*, which only reaches 490 mm, its vernacular names include *tachelt* or *tachilet* (Tamachek), *lefaa* (Arabic) and *kububus* (Hausa). *C. vipera* is more agile and paler in colour, generally a sandy yellow or pale brick-red marked with spots, black in the middle of the body becoming brown towards the tail. The terminal scales of the tail are black. The snake typically inhabits dunes, and like *C. cerastes* buries itself in the sand except for the top of the head by adopting an S-shape and making a series of lateral movements starting from the tail. It feeds exclusively on lizards, skinks and small agamids, but unlike *C. cerastes*, clings on to its prey with a bull-dog bite, not relaxing its grip until the victim has died from the venom, since prey like *Scincus* (not eaten by *C. cerastes*) try to escape by burrowing in the sand. Crepuscular during spring and autumn in north-western Sahara, it is strictly nocturnal during summer, burying itself during the day in the lightly shaded sand at the entrances of rodent burrows.

14.4.3. Localized or endemic Saharan reptiles

About fourteen species and subspecies have ranges which are restricted or confined to the Sahara Desert. Their ranges vary in size. In some areas, the species can be the most abundant of reptiles.

Geckonia chazaliae
Tarentola neglecta
Tarentola mauritanica deserti
Tropiocolotes tripolitanus algericus
Tropiocolotes tripolitanus occidentalis
Sphenops sphenopsiformis
Acanthodactylus scutellatus

Acanthodactylus aureus
Acanthodactylus longipes
Acanthodactylus spinicaudus
Eremias pasteuri
Ophisops elbaensis
Philochortus ?zolii/intermedius/lhotei
Agama flavimaculata tournevillei

Geckonia chazaliae is one of the few endemic species of the western Sahara. It is confined to within 5 km of the Atlantic coast in the zone of high humidity and lower temperatures influenced by the cold Canaries current. It inhabits coastal dune areas and is nocturnal, traversing open ground with a slow gait similar to *Stenodactylus* species. Another Saharan gecko subspecies, *Tropiocolotes tripolitanus occidentalis* (widespread as a species), *Acanthodactylus aureus* and the skink, *Sphenops sphenopsiformis*, also only occur along the Atlantic coastal fringe; the last also enters the Souss Valley of south-western Morocco.

Tarentola neglecta is a dune species frequenting clumps of *Ephedra alata* and *Calligonum azel* in southern Algeria and north-western Fezzan (Libya), while *T. mauritanica deserti* occurs in Algerian high plateaux, the Hoggar and Tibesti. In some localities in the Algerian Sahara, it frequents the palm groves of oases and enters houses in search of noctuid moths on walls and ceilings. *Tropiocolotes tripolitanus algericus* is typically Saharan and occurs in southern Morocco, mainly in the Algerian Sahara and in the Hoggar (*T. t. tripolitanus*, the nominate subspecies, occurs in Egypt west to Tunisia and in the Aïr south of the Sahara). It feeds on termites almost exclusively in the *hamada* and can be found under stones during the day.

The genus *Acanthodactylus* has become differentiated into a large number of forms in northern Africa. *A. scutellatus*, as mentioned earlier, has a wide Saharan distribution. *A. longipes* is a dune species and, like *A. scutellatus*, takes refuge in sandy burrows amongst the roots of 'drinn' in the Béni-Abbès region. The nominate subspecies occurs in the northern Sahara of Tripolitania and in Algeria, while *Acanthodactylus longipes panousei* Bons & Girot, recorded in the Erg Chebbi, southern Morocco, is possibly an isolated subspecies. *A. spinicaudus* only occurs in the southern Oranais and *Mesalina pasteuri* in the southern Oranais also, as well as the Hoggar, southern Moroccan Hamadas and south of the Sahara, having been recorded at Bilma and in the Aïr. *Ophisops elbaensis* is only known from the type locality of Jebel Elba.

Philochortus zolii has been described from a specimen from Ghat. *Philochortus* is a highly fragmented genus with isolated Saharan forms. *P. zolii* is possibly synonomous with *P. lhotei* from In Abezou (Niger) between the Hoggar and Agadez, or more probably *P. intermedius*, also occurring in eastern Africa.

Agama flavimaculata tournevillei is an active agamid of north-western Sahara recorded in the Great Eastern and Western Dunes (the nominate subspecies occurs in north-eastern Egypt, Israel and western Arabia). Illustrated by Vernet and Grenot (1972), this distinctive species is pale ventrally with a yellow sandy colour dorsally, brown cross-bands on the eyes, two longitudinal bands from the head to neck and two others each side of the head. The back has a series of regular, brown square marks across, separated by less conspicuous pale longitudinal lines. The bands disappear with higher temperatures as a thermoregulatory response and the animal becomes a uniform pale sandy yellow. Exclusively diurnal, it is a dune dweller, its activity varying according to temperature and season. It climbs actively, feeding on any insects found, and during the hottest hours of the day in summer seeks the shade of a plant or more often climbs to the top of a bush. It descends to the ground to catch prey, but returns to the bushes afterwards. During winter, it undergoes complete hibernation. Clutches of five to seven eggs are laid varying in size and weight from 16 × 8 to 19 × 10 mm and 0.88 to 1.21 g, respectively, and deposited in a 20-cm-long burrow at the base of plants.

14.4.4. Reptiles between the Nile Valley and the Red Sea (Arabian Desert)

About thirty-five species and subspecies of Saharan reptiles occur east of the Nile Valley. Some are not found west of the Nile.*

Stenodactylus sthenodactylus	Agama agama
*Gymnodactylus scaber	Agama mutabilis
Ptyodactylus hasselquistii	*Agama f. flavimaculata
*Pristurus flavipunctatus	*Agama sinaita
Hemidactylus t. turcicus	*Uromastyx ocellatus
Tarentola ephippiata-annularis complex	Varanus g. griseus
Tropiocolotes steudneri	Leptotyphlops cairi
Chalcides ocellatus subtypicus	Eryx c. colubrinus
Sphenops boulengeri	Coluber r. rhodorachis
Latastia l. longicaudata	Sphalerosophis diadema cliffordi
Acanthodactylus boskianus asper	Lytorhynchus diadema
Acanthodactylus scutellatus	Telescopus dhara obtusus
*Eremias mucronata	Psammophis schokari
Mesalina g. guttulata	Echis carinatus pyramidum
Mesalina rubropunctata	*Echis coloratus
*Ophisops elbaensis	Cerastes cerastes
Ophisops elegans	Cerastes vipera

Gymnodactylus scaber includes the Arabian Desert within its range from north-western India to Egypt and south to Sudan and possibly Eritrea. *Pristurus flavipunctatus*, like *Ophisops elbaensis*, is recorded at Jebel Elba, but continues southwards to Somalia. *Eremias mucronata* is distributed along the west coast of the Red Sea from Upper Egypt and the Sudan to Somalia, and is abundant on the great maritime plain. It also ascends the hills to about 1250 m. *Agama f. flavimaculata* (synonomous with *Agama savignyi* Duméril & Bibron) includes the northern Arabian Desert within its range, while *A. sinaita* is East Saharan, otherwise occurring in the Near East, and does in fact just cross the Nile in the Nubian Desert and has been collected near El Faiyum. The head is a bright blue and the body becoming khaki. The lizard is found up to 1700 m and individuals often perch on rocks to bask and await prey. Refuge is sought in crevices. The Shagia Arabs call the lizard *marg-ai-ain* and know it to be harmless. *Uromastyx ocellatus* is another agamid occurring east of the Nile in northern Sudan and Upper Egypt. It inhabits rugged country, apparently living in pairs in holes among the rocks, emerging in the early morning to feed on green leaves, especially an Acacia species. Retiring during the heat of the day, agamids are hunted as food by children and poorer people, and are called *dun-dun-e* by the Shagia Arabs.

Finally, *Echis coloratus* is a viperid not uncommon in the desert area east of Cairo and south to the latitude of Aswan. The type specimen from Arabia was collected by Sir Richard Burton in the nineteenth century and presented to the British Museum (Natural History), London. The scales on the snout and vertex of the head are smoother than in *E. carinatus*

14.4.5. Reptiles of restricted distribution extending into the Sahara Desert and threatened by desertification

The very survival of about four species and subspecies of reptiles, whose distributions extend into the fringe of the Sahara Desert, could be threatened by further climatic change and desert encroachment both to the north and south. All otherwise have very restricted ranges.

Quedenfeldtia trachyblepharus	*Uromastyx geyri*
Saurodactylus mauritanicus brosseti	*Sphalerosophis dolichospilus*

Quedenfeldtia trachyblepharus is an endemic Moroccan species occurring up to 3600 m in the western High Atlas and also present in the Jebel Ouarkziz south of the lower Dra Valley, south-west Morocco. Surviving essentially in montane conditions, it would inevitably become extinct with further warming of the climate already probably now having a relict distribution at high altitudes. *Saurodactylus mauritanicus brosseti* is less montane than the preceding species, but also a Moroccan endemic and would probably become extinct in extreme north-western Africa with further drying of the climate. Both are small, fragile geckos.

Uromastix geyri is unusual in possibly being threatened by desertification in the southern Sahara. It superficially resembles *U. acanthinurus*, a far more abundant agamid, with which it has probably been confused in the past. *A. geyri* seems to have a restricted distribution with scattered locality records which include the Hoggar, Tassili N'Ajjer, and further south the Aïr and Adrar des Ifohras, suggesting it to be primarily montane.

Sphalerosophis dolichospilus is a North African species occurring on the southern slopes of the Anti-, High and Saharan Atlas, and would probably be threatened by a warmer, drier climate.

Another subspecies in south-western Morocco (not Saharan) is *Chalcides mionecton trifasciatus* Chabanaud, which could also become extinct with further warming of the climate. The other subspecies of this endemic Moroccan form is *C. m. mionecton* (Boettger), which like several other endemic species (Bons, 1967) is confined to the Atlantic coastal region with a cooler, damper climate than elsewhere in North Africa. The range of some other Mediterranean species in North Africa adapted to temperate conditions are also likely to be reduced by progressive desertification and evidence exists of this already having taken place, especially in northern Libya where the Sirte Desert abuts the Mediterranean Sea. The present distributions of several species can only be explained in terms of being relicts of formerly greater ranges after the Pleistocene until about 4000 years ago. The Mediterranean spur-thighed tortoise, *Testudo g. graeca* L., for example, is now confined to the north-west of Morocco, north of the Saharan Atlas in Algeria and Tunisia, and then the hill country of northern Cyrenaica, when about 2000 years ago it probably formed a belt across northern Africa.

14.4.6. Reptiles as indicators of Saharan conditions

The distributions of some Saharan reptiles reflect Saharan bioclimatic conditions. The biogeographical limits of the Sahara Desert in relation to the position of the 100-mm isohyet are defined by them in north-western Africa (Fig. 14.7). Such species as the viperids, *Cerastes cerastes* and *C. vipera*, and gecko, *Ptyodactylus hasselquistii*, together with the lizards, *Mesalina pasteuri* and *M. rubropunctata*, the colubrid, *Malpolon moilensis*, and widespread gecko, *Tropiocolotes tripolitanus*, are restricted to south of the High and Anti-Atlas in southern Morocco and the Saharan Atlas in the south Oranais, Algeria. Their survival depends on Saharan habitat and climate.

14.4.7. Adaptations of Saharan reptiles

The adaptations of reptiles to hot, dry conditions have been described by Cloudsley-Thompson (1971) and the most detailed study in the Sahara on *Uromastyx acanthinurus* by Grenot (1976).

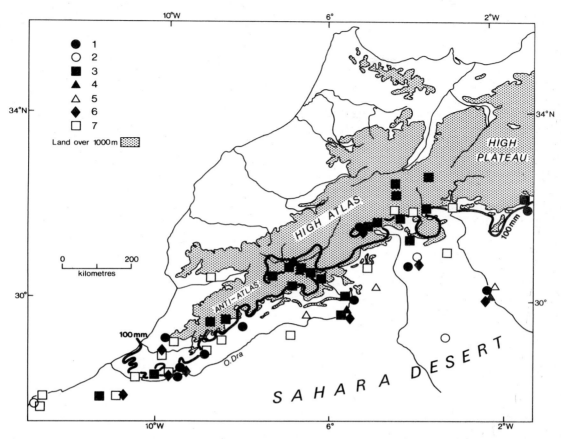

FIG. 14.7. The distribution of some Saharan reptiles, in relation to the 100-mm isohyet, as indicator species in southern Morocco and the southern Oranais, Algeria. 1. *Cerastes cerastes*, 2. *Cerastes vipera*, 3. *Ptyodactylus hasselquistii*, 4. *Mesalina pasteuri*, 5. *Mesalina rubropunctata*, 6. *Malpolon moilensis* and 7. *Tropiocolotes tripolitanus*. (Derived from Bons, 1957.)

Thermoregulation. The thermoregulatory activity of Saharan species is similar to reptiles of other of the world's deserts (e.g. Cloudsley-Thompson, 1972). Depending on species, homeostatic thermal activity varies both diurnally and seasonally. The hottest hours of the day with air temperatures of 35–44°C in full sunshine are usually avoided during summer and species range from being completely nocturnal or crepuscular, to diurnal except between 11.00 and 16.00 hours. In the northern Sahara, with temperatures dropping to 10°C or below during the cold winter months, species may undergo hibernation from the end of October to March, or become diurnal between about 11.00 and 16.00 hours instead of nocturnal. *Cerastes cerastes* is an example of the latter. During spring and autumn, intermediary activity is undertaken with nocturnal species becoming crepuscular and crepuscular species diurnal or only active between about 11.00 and 16.00 hours.

In the southern Sahara, the temperatures are more tropical and less seasonally limiting to activity. Seasonal rhythms depend more on rainfall than temperature. With northward and southward movements of the Intertropical Front, rains are received (if at all) between June and September, and insect food in response to the growth of green vegetation becomes more readily available. Depending on individual thermal requirements, reptiles in the southern Sahara tend either to be nocturnal or diurnal, avoiding the hottest hours of the day towards the end of the hot, dry season. Seasonal fluctuations in activity are relatively small. Rains occur during the warm period of the year and end with the cool season, which soon warms up again in late December or January.

Water and electrolyte relations. Salt and water balance in desert lizards have been considered by Templeton (1972), especially with regard to Australian and American desert species. Following the special study on *Uromastyx acanthinurus* in north-western Sahara by Grenot (1976), Lemire *et al.* (1982) concluded that the changes in the body condition of the herbivorous agamid, although more pronounced than in the American species, *Sauromalus obesus* (the chuckwalla) (Nagy, 1972), are closely related to climatic factors and influenced by habitat vegetation. The body mass decreases significantly during the dry season corresponding with the kind of plant material ingested. With green vegetation again available following rain, a total body water content of 75.4% of body mass is achieved by the accumulation of extracellular fluid amounting to 33% body mass. While a normal level of hydration is maintained during the dry season, the decrease in total body water is the result of extracellular fluid being reduced. Blood and plasma volume remain almost unchanged. Body solids, especially in the form of fat, increase during spring. The balance of electrolyte salts (sodium and potassium) in the plasma also varies seasonally. The sodium concentration increases significantly during the dry season in relation to the decrease in extracellular fluid, and returns to normal after the rains when green vegetation low in sodium is eaten again. Potassium in the plasma does not vary seasonally and is apparently regulated in the body by cloacal excretion and the exudation of salt by the nasal salt gland, possessed by both *U. acanthinurus* and *S. obesus.*

Little work on the water and salt relations of Saharan reptiles has been undertaken on other species.

14.5. CONCLUSION

The habitats of the Sahara Desert within the 100-mm isohyet vary from a great river and shifting sand dunes to high mountain ranges. While Amphibia only occur within the proximity of water, reptiles can survive in a wide range of habitats. Some Ethiopian species of tropical origin and totalling about thirteen only occur within Saharan confines in the riverine habitat of the Nile Valley and are absent from the hot, dry habitats of the desert. The survival of the species in drier habitats depends on their physiological and behavioural adaptations to hot, dry conditions. While a few species are able to withstand desiccation, others have adopted special thermoregulatory behaviour. Some species are nocturnal, others crepuscular and a substantial proportion fully diurnal, only seeking shade during the very hottest hours of the day even in mid-summer. Many species have seasonal activity rhythms depending more on temperature variation in the northern Sahara and rainfall in the south.

The recent Saharan habitats: *oued, guelta, daya,* oasis, *wadi, hamada, jebel, reg* and *erg* are the choice reptile species have to survive in. Mountain ranges have a cooler climate attracting increased rainfall at high altitudes and provide pockets in which temperate species can survive in addition to some water-dependent tropical forms at lower altitudes. Many poorly adapted species must plainly have become extinct in the central Sahara since the Pleistocene and perhaps the most classic case of this in recent times (even during the early part of this century) is the Nile crocodile, *Crocodylus niloticus.* Desertification is a process continuing to this day, exacerbated in the south of the Sahara by the destruction of woodland by nomadic peoples and overgrazing by their animal herds, and it is surprising that so many reptiles have adaptations to survive the increasingly warm, dry conditions of the Sahara and be so relatively numerous, and in parts even the most abundant of species.

Far from being threatened, the status of Saharan reptiles, possibly through the reduction of predators, can only be enhanced within the limitations of rainfall by further desertification. Only species not wholly adapted to Saharan conditions at the periphery of the desert are likely to be threatened. South of the Sahara, the problem is less acute for Sahelian species adapted to the conditions of the savannah would probably simply extend their ranges southwards. Species of wet tropical conditions could, however, be affected by further drying of the climate, especially in West Africa. Mediterranean species in North Africa could be further threatened for the Sahara Desert abutts the Mediterranean coast in some areas and the present relict distributions and contracted ranges are evidence of progressive desertification and northward encroachment of the Sahara since the Pleistocene. Where north-western Africa abutts the Atlantic Ocean, the climate is more oceanic, and species have become endemic in Morocco and further south in Western Sahara and Mauritania confined to the extreme Atlantic coastal fringe. These species probably ranged across north-western Africa more extensively during the Pleistocene.

So, far from being a threatened habitat requiring conservation effort and management, the Sahara Desert with its population of well-adapted reptile species is actually expanding and threatens the range of habitats and species assemblages to the north and possibly the south.

14.6. ACKNOWLEDGEMENTS

I would like to thank Drs C. Grenot and R. Vernet, Laboratoire de Zoologie, École Normale Supérieure, Paris, France (late of the Centre de Recherches sur les Zones Arides (CNRS), Béni-Abbès, Algeria), the latter also of the Société herpétologique de France, for providing an extensive range of reprints of their recent publications. I am also grateful to staff in the Amphibian and Reptile Sections, Department of Zoology, British Museum (Natural History), London, namely Dr E. N. Arnold and Mr C. J. McCarthy (reptiles), and Mr B. T. Clarke (amphibians) for help in clarifying the taxonomic nomenclature.

14.7. REFERENCES

ANDERSON, J. (1898) *Zoology of Egypt. 1. Reptilia and batrachia*. Bernard Quaritch, London.

ANDERSSON, L. G. (1935) Reptiles and batrachians from the central Sahara. *Göteborgs K. Vetensk.-o. VitterSamh. Handl. ser. B*, **4**, 3–19.

ANGEL, F. (1938) Liste des reptiles de Mauritanie recuellis par la mission d'études de la biologie des acridiens en 1936 et 1937. Description d'une sous-espéce nouvelle d'*Eryx muelleri*. *Bull. Mus. natn. Hist. nat., Paris, sér. 2*, **10**, 485–487.

ANGEL, F. (1944) Contribution a l'étude de la faune herpetologique du Sahara central. *Bull. Mus. natn. Hist. nat., Paris, sér. 2*, **16**, 418–419.

ANGEL, F. (1950) Lézards in Contribution a l'étude de l'Aïr (Mission L. Chopard et A. Villiers), edited by T. MONOD, pp. 331–336. *Mém. Inst. fr. Afr. noire*, **10**, 1–562.

ANGEL, F. and LHOTE, H. (1938) Reptiles et amphibiens du Sahara central et du Soudan. *Bull. Com. Étud. hist. sciént. Afr. occid. fr.* **21**, 345–384.

ARNOLD, E. N. and LEVITON, A. E. (1977) A revision of the lizard genus *Scincus* (Reptilia: Scincidae). *Bull. Br. Mus. nat. Hist. (Zool.)*, **31**, 187–248.

BLANC, M. (1935) Reptiles et batraciens in *Faune tunisienne*. Tunis, dactylograph.

BÖHME, W. (1978) Die Identität von *Rana esculenta bilmaensis* Angel, 1936, aus der südlichen Sahara. *Revue suisse Zool.* **85**, 641–644.

BONS, J. (1959) Les lacertiliens du Sud-Ouest marocain. *Trav. Inst. sciént. chérif., sér. zool.* **18**, 1–130.

BONS, J. (1967) *Recherches sur la biogeographie et la biologie des amphibiens et des reptiles du Maroc*. Thèse Doct. Sc. Montpellier, no. enreg. CNRSAO 2345.

BONS, J. (1972) Herpetologie marocaine. 1. Liste commentée des amphibiens et reptiles du Maroc. *Bull. Soc. Sci. nat. phys. Maroc*, **52**, 107–126.

BONS, J. (1973) Herpetologie marocaine. 2. Origines, évolution et particularités du peuplement herpetologique du Maroc. *Bull. Soc. Sci. nat. phys. Maroc*, **53**, 63–110.

BONS, J. and GIROT, B. (1962) Clé illustrée des reptiles du Maroc. *Trav. Inst. sciént. chérif. sér. zool.* **26**, 1–62.

BRAESTRUP, F. W. (1947) Remarks on faunal exchange through the Sahara. *Vidensk. Meddr dansk naturh. Foren.* **110**, 1–15.

CHABANAUD, P. (1924) Reptiles recueillis par Th. Monod en Mauritanie et aux îles du Cap-Vert. *Bull. Mus. natn. Hist. nat., Paris* **24**, 54–56.

CHAPMAN, B. M. and CHAPMAN, R. F. (1958) A field study of a population of leopard toads (*Bufo regularis regularis*). *J. Anim. Ecol.* **27**, 265–286.

CHPAKOWSKI, N. and CHNEOUR, A. (1953) Les serpents de Tunisie. *Bull. Soc. Sci. nat. Tunis* **6**, 125–145.

CLOUDSLEY-THOMPSON, J. L. (1967) Diurnal rhythm, temperature and water relations of the African toad, *Bufo regularis*. *J. Zool. Lond.* **151**, 43–54.

CLOUDSLEY-THOMPSON, J. L. (1971) *The Temperature and Water Relations of Reptiles*. Merrow, Watford.

CLOUDSLEY-THOMPSON, J. L. (1972) Temperature regulation in desert reptiles. In *Comparative Physiology of Desert Animals*, edited by G. M. O. MALOIY, pp. 39–59. Symposia of the Zoological Society of London No. 13. Academic Press, London.

CLOUDSLEY-THOMPSON, J. L. (1974) Water relations of the African toad, *Bufo mauritanicus*. *Br. J. Herpet.* **5**, 425–426.

CLOUDSLEY-THOMPSON, J. L. and CHADWICK, M. J. (1964) *Life in Deserts*. Foulis, London.

CORKILL, N. L. (1935) *Notes on Sudan Snakes. A Guide to the Species Represented in the Collection in the Natural History Museum*. Sudan Gov. Mus. (Nat. Hist.), Pub. 3. Sudan Government, Khartoum.

DEKEYSER, P. L. and VILLIERS, A. (1956) Contributions a l'étude du peuplement de la Mauritanie. Notations écologiques et biogéographiques sur la faune de l'Adrar. *Mém. Inst. fr. Afr. noire*, **44**, 1–222.

DUBUIS, A., FAUREL, L., GRENOT, C. and VERNET, R. (1971) Sur la régime alimentaire du lézard saharien *Uromastix acanthinurus* Bell. *C. r. hebd. Séanc. Acad. Sci., Paris*, **273**, 500–503.

DUVEYRIER, H. (1864) *Les Touaregs du Nord*. Paris.

EISELT, J. (1962) Ergebnisse der zoologischen Nubien-Expedition 1962. 2. Amphibien und Reptilien. *Annln naturh. Mus. Wien*, **65**, 281–296.

FLOWER, S. S. (1933) Notes on the recent reptiles and amphibians of Egypt, with a list of the species recorded from that kingdom. *Proc. zool. Soc. Lond.* **3**, 734–851.

FOUREAU, F. (1905) *Documents scientifiques de la mission saharienne, mission Foureau-Lamy*, Vol. 2. Masson, Paris.

GAUTHIER, R. (1967a) Écologie et éthologie des reptiles du Sahara nord-occidental (région de Béni-Abbès). *Mus. Roy. Afr. centr., Tervuren, sér. 8⁰, Sci. zool.* **155**, 1–83.

GAUTHIER, R. (1967b) La faune herpétologique du Sahara N.O. Algérien. Additions et mises à jour. *Bull. Mus. natn. Hist. Nat., Paris*, **5**, 821–828.

GRANDISON, A. G. C. (1961) Preliminary notes on the taxonomy of *Tarentola annularis* and *T. ephippiata* (Sauria: Gekkonidae). *Zool. Meded., Leiden* **38**, 1–13.

GRENOT, C. (1967) Observations physio-écologiques sur la régulation thermiques chez le lézard saharien *Uromastix acanthinurus* Bell. *Bull. Soc. zool. Fr.* **92**, 51–66.

GRENOT, C. (1974) Polymorphisme chromatique du lézard agamidé *Uromastix acanthinurus* Bell dans les populations du Sahara nord-occidental. *Bull. Soc. zool. Fr.* **99**, 153–164.

GRENOT, C. (1976) *Écophysiologie du lézard saharien* Uromastix acanthinurus *Bell, 1825 (Agamidae herbivore)*. Publ. Lab. Zool., E.N.S. 7, Paris.

GRENOT, C. and LOIRAT, F. (1973) L'activité et le comportement thermorégulateur du lézard saharien *Uromastix acanthinurus* Bell. *Terre Vie*, **27**, 435–455.

GRENOT, C. and NIAUSSAT, P. (1967) Aperçu écologique sur une région hyperdésertique du Sahara central (Reggan). *Sci. Nat., Paris*, **81**, 12–26.

GRENOT, C. and VERNET, R. (1972a) Place des reptiles dans l'écosystème du désert pierreux, au Sahara occidental. *Naturalistes orléan., sér. 3*, **5**, 25–47.

GRENOT, C. and VERNET, R. (1972b) Les reptiles dans l'écosysteme du Sahara occidental. *C. r. Somm. Séanc. Soc. Biogéogr.* **433**, 96–112.

GRENOT, C. and VERNET, R. (1973) Sur une population d'*Uromastix acanthinurus* Bell isolée au milieu du Grand Erg Occidental (Sahara algérien). *C. r. hebd. Séanc. Acad. Sci., Paris*, **276D**, 1349–1352.

GRUBER, U. F. (1971) Reptilien. Amphibien. Das Niltal in Die Fauna der Sahara 3, edited by G. NIETHAMMER, pp. 533–542. In *Die Sahara und ihre Randgebiete. Darstellung eines Naturgrosstraumes 1. Physiogeographie 12*, edited by H. SCHIFFERS, pp. 499–603. Ifo-Institut für Wirtschaftsforschung Afrika-Studien 60. Weltforum Verlag, München.

GUIBÉ, J. (1950a) Les lézards de l'Afrique du Nord (Algérie, Tunisie, Maroc). *Terre Vie*, **1**, 16–38.

GUIBÉ, J. (1950b) Batraciens in Contribution a l'étude de l'Aïr (Mission L. Chopard et A. Villiers), edited by T. MONOD, pp. 329–331. *Mém. Inst. fr. Afr. noire*, **10**, 1–562.

GÜNTHER, A. (1903) Reptiles from Rio de Oro, western Sahara. *Novit. zool.* **10**, 298–299.

HARTERT, E. (1913) Expedition to the Central Western Sahara. 5. Reptiles and batrachians. *Novit. zool.* **20**, 76–84.

HOOGMOED, M. S. (1972) On a new species of toad from southern Morocco. *Zool. Meded., Leiden*, **47**, 49–64.

HULSELMANS, J. L. J. (1977) Further notes on African Bufonidae, with descriptions of new species and subspecies (Amphibia, Bufonidae). *Rev. Zool. Bot. afr., Bruxelles*, **91**, 512–524.

JOGER, U. (1981) Zur Herpetofaunistik WestAfrikas. *Bonn. zool. Beitr.* **32**, 297–340.

KRAMER, E. and SCHNURRENBERGER, H. (1963) Systematik, Verbreitung und Ökologie der Libyschen Schlangen. *Revue suisse Zool.* **70**, 453–568.

LAVAUDEN, L. (1926) *Les Vertebrés du Sahara*. Imprimerie Guénard, Tunis.

LEMIRE, M., GRENOT, C. and VERNET, R. (1979) La balance hydrique d'*Uromastix acanthinurus* Bell (Sauria, Agamidae) au Sahara dans des conditions semi-naturelles. *C. r. hebd. Séanc. Acad. Sci., Paris*, **288D**, 359–362.

LEMIRE, M., GRENOT, C. and VERNET, R. (1980) Balance hydrique du Lézard agamidé *Uromastix acanthinurus* au Sahara Nord-Occidentale. *Bull. Soc. zool. Fr.* **105**, 261–264.

LEMIRE, M., GRENOT, C. and VERNET, R. (1982) Water and electrolyte balance of the free-living Saharan lizard *Uromastix acanthinurus* (Agamidae). *J. comp. Physiol.* **146**, 81–93.

LEMIRE, M., VERNET, R. and GRENOT, C. (1980) Electrolyte excretion by the nasal gland of an herbivorous Saharan lizard, *Uromastix acanthinurus* (Agamidae). Effects of isolated salt-loads. *J. Arid Environ.* **3**, 325–330.

MARX, H. (1956) *Keys to the Lizards and Snakes of Egypt.* Research report. U.S. Naval Medical Research Unit No. 3, Cairo.

MARX, H. (1968) *Check List of the Reptiles and Amphibians of Egypt.* Special publ. U.S. Naval Medical Research Unit No. 3, Cairo.

MAYET, V. (1903) *Catalogue raisonné des reptiles et batraciens de la Tunisie.* Explor. scient. Tunis.

MERTENS, R. (1929) Beiträge zur Herpetologie Tunisiens. *Senckenbergiana* **11**, 41–75.

MONOD, T. (1931) Reptiles in L'Adrar Ahnet. *Contributions a l'étude physique d'un district Saharien,* edited by T. MONOD, p. 29. *Revue Géogr. Phys. Géol. dyn.* **4**.

MONOD, T. (1958) Majâbat Al-Koubra. Contributions a l'étude de l''Empty Quarter' Ouest-Saharien. *Mém. Inst. fr. Afr. noire,* **52**, 1–406.

MOSAUER, W. (1934) The reptiles and amphibians of Tunisia. *Univ. Calif. Publs biol. Sci.* **1**, 49–64.

NAGY, K. A. (1972) Water and electrolyte budgets of a free living desert lizard, *Sauromalus obesus. J. comp. Physiol.* **79**, 26–62.

OLIVIER, E. (1894) Herpétologie algérienne ou catalogue raisonné des reptiles et des batraciens observés jusqu'a ce jour en Algérie. *Mém. Soc. zool. Fr.* **7**, 98–131.

OLIVIER, E. (1896) Materiaux pour la faune de Tunisie. 1. Catalogue des reptiles. *Revue. Scient. Bourbon. Cent. Fr.* **9**, 117–128.

PAPENFUSS, T. J. (1969) Preliminary analysis of the reptiles of arid central West Africa. *Wasmann J. zool.* **27**, 249–325.

PASTEUR, G. and BONS, J. (1959) Les batraciens du Maroc. *Trav. Inst. sciént. chérif., sér. zool.* **17**, 1–240.

PASTEUR, G. and BONS, J. (1960) Catalogue des reptiles actuels du Maroc. Révision des formes d'Afrique, d'Europe et d'Asie. *Trav. Inst. sciént. chérif., sér. zool.* **21**, 1–132.

PELLEGRIN, J. (1910) Mission en Mauritanie occidentale. 3 Partie zoologique, Reptiles. *Act. Soc. linn. Bordeaux* **64**, 21–25.

PELLEGRIN, J. (1930) *Reptiles, batraciens et poissons de la région du Hoggar.* Association français d'Avancement des Sciences, Congrès d'Alger.

PELLEGRIN, J. (1931) Reptiles, batraciens et poissons du Sahara central recueillis par le Prof. Seurat. *Bull. Mus. natn. Hist. nat., Paris, sér. 2,* **3**, 216–218.

PELLEGRIN, J. (1934) Reptiles, batraciens et poissons du Sahara central in Mission du Hoggar, *Mém. Soc. Hist. nat. Afr. N.* **3**, edited by L.-G. SEURAT, pp. 50–57.

PELLEGRIN, J. (1936) Mission au Tibesti. Étude preliminaire de la faune du Tibesti. Reptiles, batraciens. *Mém. Acad. Sci. Inst. Fr.* **62**, 50–52.

QUÉZEL, P. (1965) *La Végétation du Sahara. Du Tchad à la Mauritanie.* Gustav Fischer Verlag, Stuttgart.

ROBSON, G. M. and LAMBERT, M. R. K. (1980) Observations on the insect food of some semi-desert lizards in southern Morocco. *J. Arid. Environ.* **3**, 141–151.

SALVADOR, A. and PERIS, S. (1975) Contribution al estudio de la fauna herpetologica de Rio de Oro. *Bol. Estac. centr. ecol.* **4**, 49–60.

SCHIFFERS, H. (Ed.) (1971) *Die Sahara und ihre Randgebiete. Darstellung eines Naturgrossraumes. 1. Physiogeographie.* Ifo-Institut für Wirtschaftsforschung Afrika-Studien, 60. Weltforum Verlag, München.

SCORTECCI, G. (1934a) Cenni sui risultati di una campagna di ricerche zoologia nel Fezzan. *Riv. Sci. nat.* **25**.

SCORTECCI, G. (1934b) Descrizione preliminare di una nuova specie del genere *Philochortus. Atti Soc. ital. Sci. nat.* **73**, 305.

SCORTECCI, G. (1936) Gli anfibi della Tripolitana. *Atti Soc. ital. Sci. nat.* **75**, 129–226.

SCORTECCI, G. (1937) Relazione preliminare di un viaggio nel Fezzan sud oriente e sui Tassili. *Atti Soc. ital. Sci. nat.* **76**, 105–194.

STAUDINGER, P. (1929) Krokodile in der Inner Sahara und Mauritanien. *Sber. Ges. naturf. Freunde Berl.* **5–7**, 141–142.

TANDY, M. and KEITH, R. (1972) African *Bufo.* In *Evolution of the genus Bufo.,* edited by W. F. BLAIR, pp. 119–170. University of Texas Press, Austin.

TEMPLETON, J. R. (1972) Salt and water balance in desert lizards. In Comparative Physiology of Desert Animals, edited by G. M. O. MALOIY, pp. 61–77. Symposia of the Zoological Society of London No. 31. Academic Press, London.

VALVERDE, J. A. (1957) *Aves del Sahara espanol (estudio ecologico del desierto).* Instituto de Estudios Africanos, Madrid.

VALVERDE, J. A. (1965) Expedición zoologica en la Provincia del Sáhara. *Archos.*

VERNET, R. (1977) *Recherches sur l'écologie de Varanus griseus Daudin (Reptilia, Sauria, Varanidae) dans les écosystèmes sableaux du Sahara Nord-Occidental (Algérie).* Thèse Univ. P. et M. Curie, Dipl. Doct. 3ᵉᵐᵉ Cycle.

VERNET, R. and GRENOT, C. (1972) Place des reptiles dans l'écosystème de l'erg au Sahara nord-occidental. *Naturalistes orléan., sér. 3,* **5**, 48–61.

VILLIERS, A. (1950) Reptiles ophidiens et chéloniens. In *Contribution a l'étude de l'Aïr (Mission L. Chopard et A. Villiers),* edited by T. MONOD, pp. 337–344. *Mém. Inst. fr. Afr. noire,* **10**, 1–562.

VILLIERS, A. (1950) Contribution a l'étude du peuplement de la Mauritanie. Ophidiens. *Bull. Inst. fr. Afr. noire sér. A,* **12**, 984–998.

VILLIERS, A. (1975) Les serpents de l'Ouest africain. *Init. Ét. afr.* **2** (3ᵉ édition), 1–195. Univ. Dakar, Inst. fond. Afr. noire, Nouvelles Editions africaines, Dakar.

VINCIGUERRA, D. (1931) Spedizione scientifica all'oasi di Cufra (marzo-luglio 1931). Rettili, pp. 1–11. *Estr. anna. Mus. civ. St. nat. Genova,* **55**, 248–258.

WAKE, D. B. and KLUGE, A. G. (1961) The Machris expedition to Tchad, Africa. Amphibians and reptiles. *Contr. Sci.* **40**, 1–12.

WEBB, G. C. (1958) *The Common African Toad,* Bufo regularis. University Press, Ibadan (3rd edition).

WERNER, F. (1937) Uber Reptilien aus dem Hoggar-Gebirge (West Sahara). *Zool. Anz.* **118**, 31–35.

WINSTON, R. M. (1955) Identification and ecology of the toad *Bufo regularis. Copeia,* 1955, pp. 293–302.

WITTE, G.-F. de (1930) Mission saharienne Augieras-Draper, 1927–28. Reptiles et batraciens. *Bull. Mus. natn. Hist. nat., Paris, ser. 2,* **2**, 614–618.

ZAVATTARI, E. (1930) Erpetologia della Cirenaica. *Archo zool. ital. Torino,* **14**, 253–287.

CHAPTER 15

Breeding Birds

P. J. CASSELTON

Department of Botany, Birkbeck College, University of London

CONTENTS

15.1. INTRODUCTION

Most people would expect the Sahara, in which aridity is combined with considerable daily fluctuations in temperature, to be inhospitable to birdlife. Moreover, this desert separates two of the faunal regions of the world, the Western Palaearctic to the north and the Afrotropical to the south. This is more than a convenient convention for zoogeographers, at least as far as the distribution of birds is concerned. Only some thirty-three of the nearly 1500 species that constitute the Afrotropical avifauna also breed regularly in Mediterranean Africa (Moreau, 1966). The question that arises is whether any birds are characteristic of the desert environment which occupies an area in North Africa larger than the continent of Australia. Fortunately there is a clue: the Sahara forms part of the Great Palaearctic Desert that extends from the Atlantic Ocean through India to China. Thus any associated birds should show predominantly east–west distributions and Moreau (1966) identified a group of twenty-five species using this criterion (Table 15.1). These birds seem to form a genuine biological entity as attempts to relate them to either the Western Palaearctic or Afrotropical faunas have been unconvincing (Serventy, 1971). Most do not differ sufficiently throughout their individual ranges to merit division into subspecies, although the Desert Lark, with at least thirteen races described from Africa alone, is a notable exception. Three of the species in Table 15.1 are absent from the Western Sahara (Sooty Falcon, Sand Partridge and Hooded Wheatear) and two (Dunn's Lark and the Black-crowned Finch Lark) from the north of the desert. Of the remainder, several do not breed in the Central Sahara.

TABLE 15.1. The Sahara Desert Avifauna (after Moreau, 1966)

| Species | Occurrence | | Breeding (Etchécopar and Hue, 1964; Harrison, 1975) | | | |
	Central Sahara	Mountains	Season	Nest site	Nest	Eggs
Sooty Falcon (*Falco concolor*)	+	—	July–August	Ground	Unlined scrape	2–3 (4)
Sand Partridge (*Ammoperdix heyi*)	—	—	April–May	Ground	Shallow hollow	5–7
Houbara Bustard (*Chlamydotis undulata*)	?	—	March–April	Ground	Unlined hollow	2–3 (4)
Cream-coloured Courser (*Cursorius cursor*)	+	+	March–June	Ground	Unlined hollow	2 (3)
Crowned Sandgrouse (*Pterocles coronatus*)	+	+	May–July	Ground	Unlined hollow	2–3
Spotted Sandgrouse (*Pterocles senegallus*)	+	—	April–July	Ground	Unlined hollow	2–3
Hoopoe Lark (*Alaemon alaudipes*)	+	+	February–May	Ground (small bush)	Cup of twigs; lined	2–4
Bar-tailed Desert Lark (*Ammomanes cincturus*)	+	+	March–May	Ground (protected)	Cup; pebble edge	2–4 (5)
Desert Lark (*Ammomanes deserti*)	+	+	January–May	Ground (protected)	Cup; pebble edge	3–5
Dunn's Lark (*Ammomanes dunni*)	—	—	January–March	Ground (?)	—	—
Temminck's Horned Lark (*Eremophila bilopha*)	?	—	March–May	Ground (protected)	Cup in a hollow	2–4
Black-crowned Finch Lark (*Eremopterix nigriceps*)	—	+	April–June (September)	Ground (protected)	Cup; pebble edge	2 (3)
Thick-billed Lark (*Rhamphocorys clot-bey*)	—	—	March–May	Ground (protected)	Cup; supported	2–5 (6)
Brown-necked Raven (*Corvus ruficollis*)	+	+	January–April	Tree or cliff	Stick nest	1–5 (7)
House Bunting (*Emberiza striolata*)	+	+	March–April (September)	Hole or crevice	Cup of grass; lined	(2) 3 (4)
Trumpeter Finch (*Rhodopechys githaginea*)	+	+	March–June	Beneath stone (plant)	Cup of grass; lined	4–6
Pale Crag Martin (*Hirundo obsoleta*)	+	+	February–April	Cliff (bridge)	Mud half cup	2–3
Desert Sparrow (*Passer simplex*)	+	+	March–April (September)	Hole (tree)	Spherical; lined	2–5
Scrub Warbler (*Scotocerca inquieta*)	?	—	March–May	Low scrub	Spherical; lined	4–8 (9)
Desert Warbler (*Sylvia nana*)	+	+	January–May	Bush	Thick cup; lined	2–4
Fulvous Babbler (*Turdoides fulvus*)	+	+	November–July (September)	Scrub	Bulky cup; lined	3–6
Desert Wheatear (*Oenanthe deserti*)	?	—	April	Hole or burrow	Cup; lined	3–5 (6)
White-crowned Black Wheatear (*Oenanthe leucopyga*)	+	+	February–April	Hole or crevice	Cup; lined (rampart)	2–3 (5)
Morning Wheatear (*Oenanthe lugens*)	—	+	March–May	Hole or burrow	—	3–5 (6)
Hooded Wheatear (*Oenanthe monacha*)	—	—	May	—	—	3–4 (?)

The existence of a group of birds apparently adapted to life in the desert does not exhaust the possibilities for species breeding within the geographical limits of the Sahara. Firstly, a number of species have taken the opportunity to colonize the desert. Obvious examples are the Eagle and Little Owl. Secondly, the existence of ecological islands, particularly oases, enables additional species to breed successfully. There are also two areas which are markedly atypical. A bushy vegetation sustained largely by dew at the western end of the desert allows a southward projection of the western Palaearctic region to occur parallel with the Atlantic Ocean (Valverde, 1957). By contrast, the existence of the Nile Valley enables a number of typically sahel species to spread northwards. The Chestnut-bellied Sandgrouse (*Pterocles exustus*) and Nubian Nightjar (*Caprimulgus nubicus*) are in the latter category. Even if such birds are omitted, a considerable number remain. For example, Heim de Balsac (1936) listed some seventy species from the north-western corner of the Sahara alone. However, in addition to those found only in the Algerian oases close to the edge of the desert, he included species like the Red-rumped Wheatear (*Oenanthe moesta*) and the Blue-cheeked Bee-eater (*Merops superciliosus*) which are more characteristic of the subdesert, lying between the 100-mm and 200-mm isohyets, in which the undoubtedly desert birds also occur. By contrast, Table 15.2 lists only species which have been reported as breeding in the open desert, the more remote oases or the mountains of Ahaggar and Tibesti, but which are not included in Table 15.1. The relatively rich bird fauna of Aïr, a northward projection of the Afrotropical region owing its existence to higher rainfall (Bruneau de Miré, 1957), is excluded.

TABLE 15.2. Species Breeding Regularly within the Sahara

| Species | Occurrence | | Breeding | | |
	Desert/ Oases	Mountains	Nest site	Nest	Eggs
Long-legged Buzzard (*Buteo rufinus*)	—	+	Rock ledges, etc.	Twig nest; lined	2–3 (5)
Tawny Eagle (*Aquila rapax*)	—	+	Thorn tree, etc	Twig nest; lined	1–3
Egyptian Vulture (*Neophron percnopterus*)	+	+	Rock outcrop	Twig nest; lined	1–3
Lappet-faced Vulture (*Torgos tracheliotus*)	+	+	Thorn tree, etc.	Stick nest; lined	1
Lanner Falcon (*Falco biarmicus*)	+	+	Rock ledges, etc.	Unlined scrape	3–4 (5)
Barbary Falcon (*Falco pelegrinoides*)	+	+	Rock ledge	Small hollow	3–4
Kestrel (*Falco tinnunculus*)	—	+	Rock ledges, etc.	Small hollow	4–5 (9)
Barbary Partridge (*Alectoris barbara*)	—	+	Ground	Hollow; ±lined	8–16
Moorhen (*Gallinula chloropus*)	+	+	By water	Bulky platform	2–5 (11)
Stone Curlew (*Burhinus oedicnemus*)	+	—	Ground	Shallow scrape	1–3
Lichtenstein's Sandgrouse (*Pterocles lichtensteinii*)	—	+	Ground (protected)	Unlined hollow	2–3
Rock Dove (*Columba livia*)	+	+	Cave	±twig nest	(1) 2
Turtle Dove (*Streptopelia turtur*)	+	+	Tree	Twig platform	(1) 2
Palm Dove (*Streptopelia senegalensis*)	+	+	Bush (building)	Twig platform	2
African Collared Dove (*Streptopelia roseogrisea*)	—	+	Bush	Twig platform	2

SD-P

Table continued

TABLE 15.2—*continued*

| Species | Occurrence | | Breeding | | |
	Desert/Oases	Mountains	Nest site	Nest	Eggs
Barn Owl (*Tyto alba*)	—	+	Rock crevice	± Small hollow	3–7 (11)
Eagle Owl (*Bubo bubo*)	?	+	Rock ledge, cave, etc.	Unlined scrape	1–2 (6)
Little Owl (*Athene noctua*)	+	+	Hole or burrow	Unlined cavity	2–4 (8)
Egyptian Nightjar (*Caprimulgus aegyptius*)	+	—	Ground	Unlined scrape	2
Crested Lark (*Galerida cristata*)	?	+	Ground	Cup in a hollow	3–6
House Martin (*Delichon urbica*)	?	—	Cliff (building)	Mud half cup	2–6
Common Bulbul (*Pycnonotus barbatus*)	—	+	Bush or tree	Grass cup; lined	2–4
Great Grey Shrike (*Lanius excubitor*)	+	+	Thorn bush	Bulky cup; lined	5–7 (9)
Olivaceous Warbler (*Hippolais pallida*)	+	+	Bush	Twig cup; lined	2–4 (5)
Cricket Warbler (*Spiloptila clamens*)	+	—	Bush	Domed; lined	3
Rufous Bushchat (*Cercotrichas galactotes*)	+	+	Bush	Untidy cup; lined	2–5
Blackstart (*Cercomela melanura*)	—	+	Rock crevice, etc	Cup; pebble edge	3–5
Pygmy Sunbird (*Anthreptes platura*)	—	+	Bush	Oval suspended structure; lined	2–4
Spanish Sparrow (*Passer hispaniolensis*)	+	—	Tree	Domed	4–6 (8)
Golden Sparrow (*Auripasser luteus*)	—	+	Tree	Oval; lined	3–4
Fan-tailed Raven (*Corvus rhipidurus*)	—	+	Rock ledge, hole, etc.	Stick nest	3–4

15.2. THE DESERT AVIFAUNA

15.2.1. Description

It is perhaps not surprising that larks and wheatears, capable of living more or less constantly on the ground, account for eleven of the nineteen passerine species listed in Table 15.1. Similarly the majority of the remaining species are also fundamentally terrestrial. Other features of these birds illustrate two zoogeographical rules. Bergman's rule, implying that a desert bird will be smaller than a related form living in a cooler climate, is supported when Temminck's Horned Lark is contrasted with the distinctly larger Shore Lark occurring in the Atlas Mountains (*Eremophila alpestris atlas*). The Brown-necked Raven is also smaller than the Common Raven (*Corax corax*) of which it has been considered a subspecies. On the other hand, Gloger's rule, that birds of dry and cool habitats are lighter in colour than the same or similar species living in warm and humid habitats, has led to controversy. Nevertheless sandy coloration is characteristic of many of the desert birds, for example, the Cream-coloured Courser, the Desert Warbler and most of the larks. It results from negligible carotenoid or melanistic pigmentation of the feathers so that overall little contrast between different parts of the plumage is produced.

The obvious explanation for the paleness of typical desert birds is that of camouflage, so-called cryptic coloration. This implies an evolutionary response to the colour of the environment, rather than to its dryness, which might reduce predation by either bird or mammalian carnivores (Dorst, 1971). This interpretation is supported by the variation in colour of Desert Larks in the Middle East. Birds inhabiting areas of lava are considerably darker than those inhabiting lighter coloured substrata. Likewise Meinertzhagen (1934) described a lark from the Ahaggar which matched the rocks covered with desert varnish, although its mate was paler and resembled birds found in other parts of the Sahara. On dissection the dark bird proved to be the male. Comparable variations in colour are not shown by the sympatric Bar-tailed Desert Lark, presumably because it has different ecological requirements. The adaptive view of pale plumage was questioned by Buxton (1954) as follows: 'the desert-coloured owl chasing a desert-coloured bat over desert-coloured soil under a jaundiced moon'. However, the view that an owl would receive no advantage from cryptic coloration ignores vital activities like nesting or roosting and Serventy (1971) concluded that raptors could be expected to develop the characteristic desert coloration because it effaces them. This remains the best explanation of reduced pigmentation rather than some unknown physical or metabolic cause associated with the climate. Nevertheless pale plumage should also combat a bird's body temperature rising rapidly due to solar radiation which poses the question of the success of apparently non-cryptic birds such as the Brown-necked Raven, the White-crowned Black Wheatear and the male Black-crowned Finch Lark. In fact, as was recognized during the Second World War, black can obscure the outline of either military vehicles or animals in the desert. Further, the black and white plumage of male wheatears appears to be coupled with unpalatability which reduces predation. Thus the occurrence of dark birds, which are actually at an advantage at low temperatures, does not necessarily argue against the advantages to be gained from sandy coloration in the desert.

The desert avifauna can be considered from the point of view of the food-sources exploited by the different species. Since a strictly granivorous diet will necessitate a regular water-supply, insectivores or omnivores should be at an advantage. Nevertheless, at least five species in Table 15.1 are fundamentally seed-eaters (both Sandgrouse, House Bunting, Trumpeter Finch and Desert Sparrow) whereas the Sand Partridge, Houbara Bustard and some of the larks take invertebrates in addition to vegetable matter, including buds and fruits. The remainder are less dependent on plants, although the warblers and babblers occasionally eat berries to supplement their mainly invertebrate diet. The Sooty Falcon, Brown-necked Raven and Pale Crag Martin are respectively raptor, scavenger and aerial insectivore. Insects and their larvae, supplemented by other arthropods and lizards, are taken by the Cream-coloured Courser, Thick-billed Lark and the Wheatears. This latter group probably needs to drink rarely, if at all, whereas sandgrouse are limited to locations within 50 km of a source of water to which they make visits at characteristic times. For example, the Crowned and Spotted Sandgrouse usually drink in the morning whereas Lichtenstein's Sandgrouse drinks before sunrise or at dusk. It has been suggested that these regular gatherings offer the possibility of birds receiving information as to the direction of favourable feeding areas, groups of birds from such places departing immediately after drinking (Ward, 1972).

By definition all of the species in Table 15.2 occur in habitats outside the desert and since roughly half breed in Spain a strong Western Palaearctic component is indicated. Afrotropical species at the northern extreme of their ranges and species, like the Tawny Eagle and Long-legged Buzzard, which are found in the Middle East as well as the Maghreb, comprise the remainder. The appearance of raptors and vultures on the list reflects the ability of such birds to exploit a food-source in whatever environment it arises. Lanner Falcons and Barbary Falcons are the main predators of watering Sandgrouse. The Rock Dove is another highly adaptable species which can breed among rocks close to water, whereas the Moorhen is undoubtedly a relic from the period when the Sahara was generally more favourable to birdlife. Table 15.2 offers further examples to support both Bergman's and Gloger's rules. The Eagle Owl is smaller than the typical form found in northern forests and all the desert owls are usually light in colour. There is also general agreement that the Turtle Doves breeding in the oases are paler than those that cross the desert. It would support those who seek a metabolic explanation of reduced pigmentation, rather than a cryptic one, if the doves that undergo their early development in oases winter in an environment in which they are more conspicuous.

15.2.2. Ecology

The Sahara, apart from the oases and mountain blocks, consists of plains covered with either sand, stones (reg or serir) or larger fragments of rock (hamada). Since sandy deserts are more impoverished of birds than are stony ones (Dorst, 1971), it is significant that only about a quarter of the Sahara is covered with sand dunes (ergs); regs constitute about twice as much. There is also the possibility of additional bird species occurring where the dune hollows or wadis have favoured the growth of vegetation. For example, the Fulvous Babbler is usually found in association with bushes of *Acacia*, *Balanites* or *Zizyphus*. On the other hand, there are parts of the Fezzan where the clay soil does not produce vegetation even after heavy rainfall and further east the overall bareness of the Libyan Desert contrasts markedly with much of the Western Sahara (Moreau, 1934; Guichard, 1955). Consequently travellers in the desert often report proceeding for long periods without encountering any resident birds at all:

(a) 'During 4 March en route for Tibesti from Gatroun in the Fezzan, not a bird was seen during a day's driving of 222 kilometers.' (Guichard, 1955.)
(b) 'Towards midday a Painted Lady butterfly zigzagged across our path. The sight of something living after two days of empty reg was an event, a rumour of change in this apparently changeless landscape.' (Swift, 1975.)

Indeed soon afterwards Swift saw a White-crowned Black Wheatear amongst scattered bushes of *Cornulaca monacantha* at Timonissoao Well. This bird is the typical wheatear of the desert; the Desert Wheatear and Mourning Wheatear are largely restricted to the subdesert during the breeding season.

Since animals are expected to occupy distinct ecological niches within their environment, it would be interesting to analyse the desert avifauna with regard to habitat utilization. In practice, although both the nature of the vegetation and of the substratum are clearly important, this will not be easy. For example, even the Desert Lark may visit sandy deserts to feed (Jennings, 1981), although it is frequently recorded from stony deserts and occurs in particular on rocky outcrops. Similar complexities are found with sandgrouse. The Crowned Sandgrouse apparently favours stony deserts and mountains whereas the Spotted Sandgrouse is commonly associated with sandy deserts bearing scrub. Nevertheless before breeding in the spring the latter move from their feeding areas, covered with *Euphorbia guyoniana*, to adjacent hamadas (George, 1969). The general habitat preferences of some desert birds are outlined in Table 15.3. Species particularly characteristic of sand are the Desert Sparrow and the Hoopoe Lark although the latter is sometimes found on hamadas and also in crops. Particularly characteristic of hamadas is the Thick-billed Lark whereas Temminck's Horned Lark, which also avoids ergs, is found in either rocky or sandy deserts, which are strewn with rocks and carry some vegetation, notably *Anabasis articulata*. The Bar-tailed Desert Lark occurs on either sandy or gravelly desert, particularly when a low vegetational cover is present, whereas a number of species, including the wheatears, Sand Partridge, House Bunting and Trumpeter Finch, are found characteristically on rocky hillsides, among ruins or in wadis but rarely on level desert. The amount of vegetation in a given locality will vary from year to year depending on the occurrence of rain but in the north-western Sahara meagre vegetation usually extends for about 150 km beyond the 100-mm isohyet, at least in the wadis (Smith, 1968).

One advantage of the mountains to wildlife is the existence of pools of standing water (*gueltas*). These are visited for drinking and bathing by birds, including Trumpeter Finches, Desert Larks and sandgrouse. The gueltas also provide insects for the resident martins and may be visited briefly by migrating swallows.

TABLE 15.3. Broad Ecological Preferences of some Saharan Birds

Species	Topography				Vegetation		
	Wadis	Rocky areas	Stony desert	Sandy desert	Bushes	Herbs	± Absent
Desert Lark	+	+	+			+	+
White-crowned Black Wheatear	+	+				+	+
Mourning Wheatear	+	+					+
House Bunting	+	+		(+)		+	
Striated Scrub Warbler	+	+	+		+		
Great Grey Strike	+			+	+		
Desert Sparrow	+			+	(+)	+	
Fulvous Babbler	+			+	+		
Desert Warbler	+		(+)	+	+		
Sand Partridge		+				+	
Crowned Sandgrouse		+	+			+	
Lichtenstein's Sandgrouse		+	+		+		
Rock Dove		+					+
Eagle Owl		+					+
Pale Crag Martin		+					+
Trumpeter Finch		+	+			+	
Thick-billed Lark			+				+
Temminck's Horned Lark	+		+				+
Houbara Bustard			+	+	(+)	+	
Hoopoe Lark			+	+	(+)	+	(+)
Bar-tailed Desert Lark			+	+		+	
Black-crowned Finch Lark			+	+	+	+	
Spotted Sandgrouse			(+)	+	+	+	
Cream-coloured Courser			(+)	+	(+)	+	
Dunn's Lark				+	+	+	
Egyptian Nightjar				+		+	

Another feature of the mountains is the occurrence of very old trees including oleanders, olives and cypresses. Nevertheless, despite considerable vegetational cover, neither the Ahaggar nor Tibesti apparently support many resident passerines. With regard to the former, Meinertzhagen (1934) recorded Trumpeter Finches, House Buntings, Desert Larks and Fulvous Babblers with Pale Crag Martins and White-crowned Black Wheatears abundant in early March; Great-grey Shrikes and Desert Warblers occurred in small numbers in the better vegetated wadis. The Bar-tailed Desert Lark is the commonest species apart from the White-crowned Black Wheatear in the Tassili N'Azzer (Swift, 1975). The birdlife of the higher parts of Tibesti resembles that of the Ahaggar, although the Desert Lark may be less prominent. However, the palmeries and wadis are more rewarding and support a number of Afrotropical species (Guichard, 1955; Simon, 1965); Golden Sparrows and African Collared Doves are associated like Palm Doves with trees, whereas the Common Bulbuls and Blackstarts are largely found in wadis containing bushes (Turk, 1959). Lichtenstein's Sandgrouse and Pygmy Sunbirds apparently exploit particular plants, *Acacia seyal* and *Calotropis procera* respectively. The other breeding species include three summer visitors (Turtle Dove, Rufous Bushchat and Olivaceous Warbler) and a bustard (either *Neotis denhami* or *Ardeotis arabs*). Besides ravens the mountains support a number of raptors and vultures (Table 15.2). The Long-legged Buzzard and Barbary Falcon, like the Desert Warbler, reach their southern limit in the Ahaggar and are probably not found in Tibesti, whereas the occurrence of Tawny Eagles and Lanner Falcons in both areas is likely.

15.2.3. Breeding

Most birds resident in the Northern Sahara breed in the spring, corresponding more or less with similar species in the Western Palaearctic. Nevertheless November to March is generally the coolest and wettest period of the year in this part of the desert and rainfall, besides promoting the growth of vegetation, may stimulate the commencement of avian breeding activity, as it clearly does in the extremely arid environment found in parts of Australia. Certainly marked differences in sexual activity among birds of the same species occupying adjacent areas of the Sahara can be observed even in the more accessible parts, such as southern Morocco (Smith, 1965). It is also clear that lack of rain leads to fewer eggs and single broods. For example, the larks had clutches about half the normal size in the very dry year of 1936 (Guirtchitch, 1937). In really unfavourable years the gonads of many species atrophy and the birds remain in feeding flocks rather than pairing; in 1969 only a small proportion of the Spotted Sandgrouse population bred (George, 1969). In fact, although it is obvious that breeding only occurs sporadically in the Sahara and should reflect the distribution of rainfall in any particular year, not much information about the actual numbers involved is available. Birds recorded by travellers often represent only circumstantial evidence of breeding and frequently relate to single observations. Hence, for example, a suggestion that the Desert Sparrow is commoner in the Fezzan than further west (Summers-Smith and Vernon, 1972). However, it appears that a number of relevant species, including the Houbara Bustard, Temminck's Horned Lark and the Striated Scrub Warbler, mostly nest in the subdesert so that their occurrence beyond the 100-mm isohyet is limited (Table 15.1). Others, notably the Cream-coloured Courser and Trumpeter Finch, penetrate further south in humid years. It is also certain that both rainfall and breeding along the southern edge of the desert are timed differently as Bates (1934 b, c) recorded Fulvous Babblers, Desert Sparrows and Black-crowned Finch Larks nesting in September. The House Bunting can nest in both the spring and autumn in the Tassili N'Azzer, corresponding to the barley and millet crops respectively (Heim de Balsac and Mayaud, 1962), and other species, particularly doves, may behave similarly. The Cricket Warbler breeds in both June and November in the southern Sahara.

It is not unexpected that many desert birds nest on the ground (Table 15.1). Nevertheless such nests are usually placed in the shade of a rock or bush and may, as in the case of several larks and wheatears, be supported on the outer side by small stones placed in position by the birds (Jourdain, 1927). In the case of the Thick-billed Lark, at least, pebbles are presented to the female by the male during courtship. Only a minority of species, mainly wheatears, place their nests in crevices or rodent burrows, whereas the Hoopoe Lark is unusual among larks in occasionally nesting somewhat conspicuously above ground level on a tuft of grass, perhaps because it is cooler (Heim de Balsac, 1936; Bannerman and Priestley, 1952). Certainly some species regularly nest in bushes or trees (Tables 15.1, 15.2). Nests can be completely isolated. A Brown-necked Raven's nest in the Arabian Desert was placed in a solitary *Calligonum comosum* bush, the highest plant within a radius of 50 miles. Another remarkable observation was made in August 1957 (Booth, 1961). A pair of Sooty Falcons were found to be breeding in a man-made camel cairn surrounded by the absolutely barren ground of the Calanscio Serir (Libyan Desert). There was no nest but three eggs were placed so that the incubating bird was continuously protected from the sun. The shade temperature was 42°C. Apparently this falcon, which also occurs on a few rocky islands in the Red Sea, resembles Eleonora's Falcon (*F. eleonorae*) in delaying its breeding until the returning Palaearctic migrants provide a plentiful supply of food. In fact many people have remarked upon the number of falcons present around rocky outcrops in barren desert (Moreau, 1934). Probably both Lanner and Barbary Falcons nest at such sites, but early in the year so as to exploit the spring migration. These raptors, obtaining water from their prey, are clearly less dependent on rainfall for successful breeding.

A family of birds that has received detailed study, even in the Sahara, is the Pteroclididae (George, 1969, 1970; Maclean, 1976). Spotted Sandgrouse select their nest site, to which no extra material is brought, among stones of roughly body size and of appropriate colour within 30 km of a source of water. Normally three eggs are laid at intervals of 1 to 2 days. The completed clutch is incubated by the female sandgrouse during the day and by the male sandgrouse at night, changeover being some 2 hours before sunset and after sunrise. The male remains near the nest by day and warns his mate of approaching danger. Once the highly precocial grey-brown young have hatched they leave the nest and hide among stones. The chicks become sandy yellow in colour after about 6 days when they take to crouching on small patches of windblown sand into which they burrow. Although not fed by either parent, they are shown food principally by the female. Water is provided by the male who collects it by soaking his belly feathers at a convenient source. On returning he adopts a position which enables the young to obtain the water with their beaks. After each daily watering, which usually occurs in the morning, both parents and chicks sandbathe. The female remains with the brood for about 4 days after hatching but thereafter accompanies the male to water. The chicks recognize their returning parents' calls but apparently do not respond to other adults. If surprised with young, adults exhibit an injury-feigning display. The Crowned Sandgrouse which inhabits more barren ground has been less studied but its behaviour certainly includes water transport by the male, while its breeding success is apparently less susceptible to adverse climatic factors (George, 1970). This contrasts with the Cream-coloured Courser for which the southern breeding limit is highly variable depending on the production of suitable conditions. The latter's two eggs are placed in a nest like a plover's, incubation is by both sexes, and the precocial young are fed with insects (Heim de Balsac and Mayaud, 1962).

The breeding behaviour of many desert species resembles that of comparable species in other environments; for example the elaborate display of the male Houbara Bustard with only the female incubating, the nuptial flights of the Hoopoe Lark, Thick-billed Lark and the Desert Warbler, or the use by House Buntings and White-crowned Black Wheatears of buildings for nesting. Both sexes of the Desert Warbler and Striated Scrub Warbler contribute to the building of nests, which are respectively cup-shaped and domed like those of closely related species. As expected Pale Crag Martins construct a mud half-cup while Desert Sparrows usually build globular nests, sometimes with an entrance spout. Clutch sizes vary widely according to the prevailing conditions (Tables 15.1, 15.2). Variation can be seasonal, particularly in potentially double-brooded species like the White-crowned Black Wheatear and Desert Sparrow, or spatial as nests are placed closer to the centre of the desert. Several species, including the Desert Warbler, Desert Lark and Fulvous Babbler, have extended breeding seasons, although each pair of babblers may only breed once each year. It appears that many birds that nest successfully in the desert subsequently move away. Such movements, which are more analogous to true migrations than to the local movements shown by graniverous species seeking out suitable localities in particular years, can be either northwards to the Maghreb or southwards. The species involved include the Cream-coloured Courser, Thick-billed Lark, Bar-tailed Desert Lark and Black-crowned Finch Lark.

With regard to oases, the number of nesting birds can vary markedly. At Kufra, almost exactly in the centre of the Libyan desert, the only certain breeding species in 1968 was a migrant, the Turtle Dove, although Cape Teal (*Anas capensis*) may also have bred (Cramp and Conder, 1970). At the other extreme some Algerian oases support a considerable number of species. These include such typical western palaearctic birds as the Blackbird (*Turdus merula*), Blue Tit (*Parus caerulens*), Reed Warbler (*Acrocephalus scirpaceus*) and Swallow (*Hirundo rustica*) as well as more southerly forms like the Hoopoe (*Upopa epops*), Cetti's Warbler (*Cettia cetti*) and the Spanish sparrow (*Passer hispaniolensis*) which is subject to hybridization with colonizing House Sparrows (*P. domesticus*). Even Coots (*Fulica atra*) nest at El-Goléa. Intermediate between these extremes are the oases in the Fezzan and Egypt which contain palmeries, cultivated fields and even surface water. The latter explains why Moorhens breed at El Jedid Lake (Guichard, 1955). Otherwise Rock Doves and Palm Doves, Little Owls and White-crowned Black Wheatears are commonly resident. In addition Pale Crag Martins, Turtle Doves, Fulvous Babblers and Desert Sparrows frequently breed while Rufous Bushchats, Olivaceous Warblers and even House Martins may do so. The Brown-necked Ravens which scavenge around the oases generally nest nearby in acacia trees.

15.3. BIOLOGICAL ADAPTATIONS OF DESERT BIRDS

Animals can show apparent adaptations to an environment through their structure, physiology and behaviour. However, at first sight, desert birds have no obvious structural adaptations, although many have long powerful legs corresponding to their largely terrestrial existence. Nevertheless, the feathers of the Desert Lark are arranged more loosely than those of a non-desert lark, presumably allowing ventilation of the body to occur more easily. A paradox exists with regard to physiology in that no bird has evolved that lives exclusively in a desert without water other than that resulting from metabolism, while several features of avian physiology make birds better prepared than mammals to establish themselves in arid environments. For example, the internal temperature maintained by most birds at rest (39–42°C) is higher than that characteristic of mammals (c. 37°C) and the initial response to heat stress is usually an increase of about 2 degrees. This reduces the period during which the thermal gradient from the body of a bird to its environment is reversed during a hot day and consequently the necessity for heat dissipation by increased evaporation. The excretion of nitrogen in the form of urates is a means of conserving water shared with reptiles. The solubility of uric acid is such that a semi-solid paste containing little water can be produced. Birds can also exploit their great mobility, for example in seeking water or the most favourable localities for breeding (nomadism). In fact several desert species usually run in preference to flying when disturbed. This is presumably to reduce predation. Similarly, unless breeding or very thirsty, Spotted Sandgrouse usually gather some distance from a waterhole and go to drink only when single prospecting birds consider it safe to do so (George, 1970). However, although soaring birds can escape excessive ground temperature by rising above 1000 m, the most obvious behavioural response to life in the desert is the restriction of activity to the morning and late afternoon thus avoiding the extreme heat of midday.

Although it may not always be possible for birds to avoid heat stress in deserts by behavioural means, the lack of any spectacularly adapted species does obscure the extent to which the physiology of desert birds differs from that of other birds. Only the Sand Patridge, which drinks frequently, and perhaps the Crowned Sandgrouse, resemble the Ostrich (Struthio camelus) and seabirds in being able to utilize saline water (Schmidt-Neilsen et al., 1964). However, two larks from the Namib Desert in South-west Africa had exceptionally low rates of evaporatory and excretory water loss at moderate temperatures and humidities which was interpreted as an adaptation to the desert environment (Willoughby, 1968). Similar results might be obtained with Saharan species like the Hoopoe Lark, which is frequently observed to be active during the day, or the nocturnal Egyptian Nightjar which apparently rests in the open desert. Nevertheless, even if such birds can exist under laboratory conditions on the water derived from their food, it is doubtful whether any desert bird refrains from drinking and utilizes a dry diet if water or succulent food is available. Thus tolerance of water deprivation if it occurs should be considered a reserve mechanism (Dawson, 1976). Further, the desert birds that have been examined do not appear to be more tolerant of increased body temperature than do non-desert birds. The initial response is to increase heat loss by radiation, for example by raising either the wings or the mantle feathers to expose the flanks or back respectively. Once the body reaches about 43°C, further increase to the lethal temperature around 47°C is counteracted by the evaporation of water. Either respiratory movements are increased (panting) or the soft skin under the throat is vibrated (gular fluttering) so that the passage of air over moist surfaces is increased and cooling results. In some cases both methods are used. Adult sandgrouse employ gular fluttering, while their chicks can only pant and seek the shade. Similarly larks are observed with open beaks in the Sahara during June and July (Heim de Balsac, 1936). Heat loss by convection, that is involving the movement of air molecules, can be increased by perching on bushes or rocks which may explain why owls and falcons roost in the open at certain times of the year. Nevertheless such birds also seek shade:

> On 6th November I was lucky to obtain a Desert Eagle-Owl as it sheltered from the sun standing on the sand under a lone Acacia tortilis in an otherwise bare little valley near the well of In Alahy (Bates, 1934a).

Larks and wheatears can also reduce heat stress during the day by keeping to the shade of stones, tufts of vegetation or the mouths of rodent burrows where the temperature may be 20°C less than the prevailing ground temperature. Similarly, although nesting tends to occur at the coolest time of the year, overheating of eggs is often more of a problem than keeping them warm. It is only necessary to shield eggs from the sun at air temperatures below 40°C but incubation is required at higher temperatures so that the brood patch can function to withdraw heat (Drent, 1975). Sandgrouse were incubating eggs in the Sahara when the soil surface temperature was 68–73°C and the shade temperature 48°C (George, 1970). This is interesting because temperature affected the breeding success of sandgrouse in the Namib Desert; above 50°C the birds were unable to keep the eggs cool enough and the embryos died (Dixon and Loew, 1978). It is probably the lack of development of burrow–dwelling and summer dormancy (aestivation) which explains why birds are less successful than mammals under extreme desert conditions. Indeed it is surprising that only Desert Wheatears may dig a hole in which to breed (Harrison, 1975), as the practice of placing nests directly on the ground in the open obviously represents a considerable challenge to bird physiology.

15.4. CONCLUSION

About 5000 years ago much of the Sahara was covered with a Mediterranean type of vegetation, although even then the Eastern Sahara was probably drier. Subsequently a dramatic spread of desert conditions occurred with the virtual elimination of plant and animal life, partly due in historical times to man's activity. Persecution certainly contributed to the decline of bustards and ostrich in the Sahara and it is now unclear whether the latter would otherwise still have been quite widely distributed in North Africa (Moreau, 1966). Nevertheless a number of birds have colonized the area, mainly from the east. It has been suggested that the existence of pairs of species with only slightly different ecological requirements, such as the Crowned and Spotted Sandgrouse and the Desert and Bar-tailed Desert Larks, indicates that at least two refuges for desert species existed during Neolithic times (Harrison, 1982). Nevertheless, as far as extreme desert conditions are concerned, birds have been less successful in adapting themselves than have other animals so that the resident population of mammals in ergs exceeds that of birds several times (Bartholomew and Cade, 1963). Further, desert bird populations are likely to fluctuate markedly from year to year depending on food supply; those of sandgrouse vary rather more than do those of the falcons which prey on them (McClean, 1976). However, quantitative data would be difficult to obtain; it has only recently become available for common British birds.

According to Heim de Balsac (1936) the 200-mm isohyet, rather than the 100-mm isohyet chosen as the geographical boundary of the Sahara, represents the breeding limit of the desert avifauna. Certainly the subdesert area of north-west Africa is sufficiently rich in desert species to be of conservation interest (Vernon, 1980). There is a comparable subdesert belt on the southern side of the Sahara in which many Afrotropical species including the White-throated Bee-eater (*Merops albicollis*), Yellow-breasted Barbet (*Trachyphorus margaritatus*) and Plain Nightjar (*Caprimulgus inornatus*) breed during the rainy season, that is between June and September (Bates, 1933; Zolotarevsky and Murat, 1938). These subdesert fringes, like the oases, are also used for wintering by some Palaearctic birds.

It is clear that additional study is required to provide much of the basic information about the breeding of Saharan birds and their physiology, particularly with regard to species with altricial young. However, large-scale reclamation of arid land for agricultural use would probably benefit species like the Palm Dove, Bulbul and Goldfinch (*Carduelis carduelis*) rather than the typical desert species. Similarly the oil industry does not seem to benefit the resident birds of the Libyan desert other than by providing water, although the installations do represent additional 'oases' trans-Saharan migrants and consequently prey for raptors (Hogg, 1974).

ACKNOWLEDGEMENT

The assistance of Jill Emery and Kerry Phillips of Birkbeck College with translation and typing respectively is gratefully acknowledged.

REFERENCES

BANNERMAN, D. and PRIESTLEY, J. (1952) An ornithological journey in Morocco in 1951. *Ibis*, **94**, 654–682.

BARTHOLOMEW, G. A. and CADE, T. J. (1963) The water economy of land birds. *Auk*, **80**, 504–539.

BATES, G. L. (1933) Birds of the southern Sahara and adjoining countries in French West Africa. Part 1. *Ibis*, **13** (3), 752–780.

BATES, G. L. (1934a) Birds of the southern Sahara and adjoining countries in French West Africa. Part III. *Ibis*, **13** (4), 213–239.

BATES, G. L. (1934b) Birds of the southern Sahara and adjoining countries in French West Africa. Part IV. *Ibis*, **13** (4), 439–466.

BATES, G. L. (1934c) Birds of the southern Sahara and adjoining countries in French West Africa. Part V. *Ibis*, **13** (4), 685–717.

BRUNEAU DE MIRÉ, P. (1957) Observation sur la faune avienne du Massif de l'Air. *Bull. Mus. Hist. nat., Paris*, **29**, 130–135.

BOOTH, B. D. McD. (1961) Breeding of the Sooty Falcon in the Libyan Desert. *Ibis*, **103A**, 129–130.

BUXTON, P. A. (1954). cited by Dorst (1971).

CRAMP, S. and CONDER, P. J. (1970) A visit to the oasis of Kufra, Spring 1969. *Ibis*, **112**, 261–263.

DAWSON, W. R. (1976) Physiological and behavioural adjustments of birds to heat and aridity. In *Proc. of the 16th Int. Ornithological Congress*, edited by H. J. FRITH and J. H. CALABY, pp. 455–467. Australian Academy of Science, Canberra.

DIXON, J. and LOUW, G. (1978) Seasonal effects on nutrition, reproduction and aspects of thermoregulation in the Namaqua Sandgrouse (*Pterocles namaqua*). *Madoqua* **11**, 19–29.

DORST, J. (1971) *The Life of Birds*, Vol. II. Weidenfeld & Nicolson, London.

DRENT, R. (1975) Incubation in *Avian Biology*, Vol. V, edited by D. S. FARNER and J. R. KING, pp. 333–420. Academic Press, New York.

ETCHÉCOPAR, R. D. and HÜE, F. (1964) *Les Oiseaux du Nord de L'Afrique*. N. Boubée & Cie, Paris.

GEORGE, U. (1969) Über das Tränken der Jungen und andere Lebensäuberungun des Senegal-Flughuhns, *Pterocles senegallus*, in Marokko. *J. Orn., Lpz*. **110**, 181–191.

GEORGE, U. (1970) Beobachtungen an *Pterocles senegallus* und *Pterocles coronatus* in der Nordwest-Sahara. *J. Orn., Lpz*. **111**, 175–188.

GUICHARD, K. M. (1955) The birds of Fezzan and Tibesti. *Ibis*, **97**, 393–424.

GUIRTCHITCH, G. (1937) Chronique ornithologique unisienne pour L'année 1936. *L'Oiseau*, **7**, 450–472.

HARRISON, C. (1975) *A Field Guide to the Nests, Eggs, and Nestlings of British and European Birds*. Collins, London.

HARRISON, C. (1982) *An Atlas of the Birds of the Western Palaearctic*. Collins, London.

HEIM DE BALSAC, H. (1936) Biogéographie des mammifères et des oiseaux de L'Afrique du nord. *Bull. biol. Fr. Belg.*, supplement 21.

HEIM DE BALSAC, H. and MAYAUD, N. (1962) *Les Oiseaux du Nord-ouest de L'Afrique*. Lechevalier, Paris.

HOGG, P. (1974) Trans-Saharan migration through Sarir, 1969–70. *Ibis*, **116**, 466–476.

JENNINGS, M. C. (1981) *Birds of the Arabian Gulf*. George Allen & Unwin, London.

JOURDAIN, F. C. R. (1927) Wall-building birds. *Brit. Birds*, **20**, 223–225.

MACLEAN, G. L. (1976) Adaptations of Sandgrouse for life in arid lands. In *Proc. of the 16th Int. Ornithological Congress*, edited by H. J. FRITH and J. H. CALABY, pp. 502–516. Australian Academy of Science, Canberra.

MEINERTZHAGEN, R. (1934) The biogeographical status of the Ahaggar plateau in the central Sahara, with special reference to birds. *Ibis*, **13** (4), 528–571.

MOREAU, R. E. (1934) A contribution to the ornithology of the Libyan desert. *Ibis*, **13** (4), 595–632.

MOREAU, R. E. (1966) *The Bird Faunas of Africa and its Islands*. Academic Press, London.

SCHMIDT-NIELSON, K., BORUT, A., LEE, P. and CRAWFORD, E. (1963) Nasal salt excretion and the possible function of the cloaca in water conservation. *Science, N.Y.* **142**, 1300–1301.

SERVENTY, D. L. (1971) Biology of desert birds. In *Avian Biology*, Vol. 1, edited by D. S. FARNER and J. R. KING, pp. 287–339. Academic Press, New York.

SIMON, P. (1965) Synthèse de l'avifaune du massif montagneux du Tibesti et distribution géographique de ces espèces en Afrique du nord et environs. *Gerfaut*, **55**, 26–69.

SMITH, K. D. (1965) On the birds of Morocco. *Ibis*, **107**, 493–526.

SMITH, K. D. (1968) Spring migration through southeast Morocco. *Ibis*, **110**, 452–492.

SUMMERS-SMITH, D. and VERNON, J. D. R. (1972) The distribution of *Passer* in northwest Africa. *Ibis*, **114**, 259–262.

SWIFT, J. (1975) *The Sahara*. Time-Life Books, Amsterdam.

TURK, R. F. (1959) Observations dans le massif du Tibesti en 1957. *Alauda* **17**, 199–204.

VALVERDE, J. A. (1957) *Aves del Sahara espanol*. Instituto de Estudios Africanos, Corsejo superior de Investigaciones cientificas, Madrid.

VERNON, J. D. R. (1980) 50 years of ornithology in north-west Africa 1930–80. *Bull. Br. Orn. Club*. **100**, 61–66.

WARD, P. (1972) The functional significance of mass drinking by sandgrouse: Pteroclididae. *Ibis*, **114**, 533–536.

WILLOUGHBY, E. J. (1968) Water economy of the Stark's lark and grey-backed finch-lark from the Namib desert of south-west Africa. *Comp. Biochem. Physiol.* **27**, 723–745.

ZOLOTAREVSKY, B. and MURAT, M. (1938) Divisions naturelles du Sahara et sa limite méridionale. In *La Vie dans la Région Desertique Nord tropicale de L'ancien Monde*. Mémoires de la Société de Biogéographie VI, pp. 335–350. Lechevalier, Paris.

CHAPTER 16

Migratory Birds

A. PETTET

23 Cole Park Road, Twickenham, London

CONTENTS

16.1. INTRODUCTION

In terms of their total biomass, migratory birds form one of the important, if transient and none too easily observed, components of the Saharan ecosystem. As is well known, the great seasonal movements occur during the spring and autumn months, but there is probably no time in the year when migration is not taking place somewhere in the desert as defined by the 100-mm isohyets.

16.2. TRANS-SAHARAN MIGRATION

Rather broadly, migratory birds found in the Sahara fall into three classes:

(i) *Trans-Saharan migrants* crossing the desert every year during their passage between breeding grounds in the Palaearctic and wintering grounds in Africa south of the Sahara. The southerly movement occurs mostly during July–October, and the return, northerly movement late February–early May. There is a related, but little-known, late movement into the Sudan and Egypt during December–January, associated with winter rains of the Red Sea area.

In terms of the number of individuals involved, the variety of species, and the regularity and ubiquity of movements, trans-Saharan migrants dominate the migratory system of the desert. Behaviourally and physiologically they show definite adaptations for the desert crossing and might be considered 'obligate-migrants'.

(ii) *Afrotropical migrants* penetrating beyond the southern 100-mm isohyet boundary in July and August when the rains associated with the Inter-Tropical Convergence Zone (ITCZ) are unusually vigorous. The species involved, and their numbers, tend to be correlated with the amount of

241

ephemeral vegetation produced by these rains, and with the resulting insect–life (particularly grasshoppers and locusts), and hence apt to vary greatly from year to year. Though sometimes important in limited areas, few species can be considered regular migrants in the desert; in a sense most are 'accidental migrants'.

In good years Abyssinian Rollers, *Coracias abyssinicus*, Grasshopper Buzzards, *Butastur rufipennis*, Kites, *Milvus migrans*, Pied Crows, *Corvus albus* and Nightjars, *Caprimulgus* spp., are the more conspicuous migrants, while Golden Sparrows, *Passer luteus*, and African Collared Doves, *Streptopelia roseogrisea*, tend to be pre–rains migrants. Other granivorous birds include the finch–larks, *Eremopterix nigriceps* and *E. leucotis*. Very occasionally, species such as the Painted Snipe, *Rostratula benghalensis*, Harlequin Quail, *Coturnix delegorguei*, Lesser Jacana, *Microparra capensis*, and Allen's Gallinule, *Gallinula alleni*, also get carried into the desert as vagrants by fast–moving storms.

(iii) *Breeding birds of steppe and desert* whose movements are often ascribed to nomadism rather than regular migration. In comparison with other classes of migrants the number of birds involved is small, but their movements are poorly documented and may well be more extensive than the limited data suggest. The Sooty Falcon, *Falco concolor*, is one of the unusual examples of this group known to be a long–distance migrant.

In view of their dominance in the Sahara the following discussion will be confined to *trans-Saharan migrants*. The name most closely associated with their study is that of Moreau (1961, 1972) who has summarized much of the early information on the group. More recently Curry-Lindahl (1981) has surveyed Palaearctic migrants in Africa and discussed aspects of trans-Saharan migration.

16.2.1. The migrants

Palaearctic migrants wintering in Africa south of the Sahara arrive from an area stretching from Iceland in the west, through Europe and eastwards to about 120°E in Siberian Russia. Some migrants are also drawn from the North Africa coastal area (the Mahgreb, and coastal regions of Libya and Egypt), Arabia, Iraq and Iran. This huge area falls fairly naturally into three zones: *West Palaearctic*, extending as far as 45°E; *Mid Palaearctic*, between 45°E and 90°E; and *East Palaearctic*, stretching east of 90°E. The 90°E boundary marks a faunal break for species migrating either to Africa or to Asia, which, as well as being related to the distances involved in their migrations, reflects topographical and vegetation changes. Most species of the East Palaearctic winter in southern Asia, but of those species migrating to Africa some breed as far as 120°E (Moreau, 1972).

Migrants from these different zones face very different conditions on passage. For those of the West Palaearctic the wintering grounds lie directly south and beyond the twin barriers of the Mediterranean Sea which varies in width up to 1100 km, and the Sahara whose average width is *c.* 1500 km between the 100-mm isohyets. For migrants of the Mid and East Palaearctic the passage to their African wintering grounds involves journeys over complicated, and sometimes adverse, topography which includes the Caspian and Black Seas, dry steppe, semi-desert or desert, and mountainous areas as well as the Sahara. The further east the breeding grounds, the greater the journey and its demands. Great circle routes from the more eastern of breeding grounds towards the south-eastern edge of the Sahara involve distances of 6000–9000 km. A few species of the Mid and East Palaearctic may bypass the Sahara altogether, of which the White-throated Robin *Irania gutturalis* may be an example, it possibly entering its wintering grounds in East Africa via Arabia, the Ethiopian Highlands and Somalia.

Of Palaearctic migrants 187 species winter regularly in Africa south of the Sahara (Table 16.1), while at least another thirty species are known to occur in smaller numbers, or as vagrants (Moreau, 1972). Most have been recorded somewhere in the Sahara or near its margins, at least infrequently, if not commonly. A few, for no obvious reason, remain unrecorded.

TABLE 16.1. Palaearctic species wintering in Africa south of the Sahara, excluding marine species (from Moreau, 1972). The number of species in a group is given in brackets

---

Passerines (74)
SYLVIINAE (29): warblers
Acrocephalus arundinaceus, A. melanopogon, A. paludicola, A. palustris, A. schoenobaenus, A. scirpaceus, Hippolais icterina, H. languida, H. pallida, H. polyglotta, Locustella fluviatilis, L. luscinioides, L. naevia, Phylloscopus bonelli, P. collybita, P. sibilatrix, P. trochilus, Sylvia atricapilla, S. borin, S. cantillans, S. communis, S. curruca, S. hortensis, S. melanocephala, S. mystacea, S. nana, S. nisoria, S. ruppelli
ALAUDIDAE (2): larks
Calandrella cinerea, Melanocorypha bimaculata
EMBERIZIDAE (3): buntings
Emberiza caesia, E. cineracea, E. hortulana
HIRUNDINIDAE (5): swallows and martins
Delichon urbica, Hirundo daurica, H. rupestris, H. rustica, Riparia riparia
LANIIDAE (6): shrikes
Lanius collurio, L. excubitor, L. isabellinus, L. minor, L. nubicus, L. senator
MOTACILLIDAE (6): pipits and wagtails
Anthus campestris, A. cervinus, A. trivialis, Motacilla alba, M. cinerea, M. flava
MUSCICAPIDAE–MUSCICAPINAE (3): flycatchers
Ficedula albicollis, F. hypoleuca, Muscicapa striata
MUSCICAPIDAE–TURDINAE (17): robins, thrushes, wheatears, chats
Cercotrichas galactotes, Irania gutturalis, Luscinia luscinia, L. megarhynchos, L. svecica, Monticola saxatilis, M. solitarius, Oenanthe deserti, O. hispanica, O. isabellina, O. oenanthe, O. pleschanka, O. xanthoprymna, Phoenicurus ochruros, P. phoenicurus, Saxicola rubetra, S. torquata, Turdus philomelos
ORIOLIDAE (1): orioles
Oriolus oriolus
PLOCEIDAE (1): sparrows
Petronia brachydactyla

Non-passerines (22) (other than raptors and water-birds)
APODIDAE (4): swifts
Apus apus, A. affinis, A. melba, A. pallidus
BURHINIDAE (1): stone-curlews
Burhinus oedicnemus
CAPRIMULGIDAE (3): nightjars
Caprimulgus aegyptius, C. europaeus, C. ruficollis
COLUMBIDAE (1): doves
Streptopelia turtur
CORACIDAE (1): rollers
Coracias garrulus
CUCULIDAE (3): cuckoos
Clamator glandarius, Cuculus canorus, C. poliocephalus
GLAREOLIDAE (2): pratincoles
Glareola nordmanni, G. pratincola
MEROPIDAE (2): bee-eaters
Merops apiaster, M. superciliosus
PHASIANIDAE (1): quails
Coturnix coturnix
PICIDAE–JYNGINAE (1): wrynecks
Jynx torquilla
STRIGIDAE (2): owls
Asio flammeus, Otus scops
UPUPIDAE (1): hoopoes
Upupa epops

Raptors (25)
ACCIPITRIDAE (16): hawks, eagles, buzzards, harriers, kites, vultures
Accipiter nisus, Aquila clanga, A. heliaca, A. pomarina, A. rapax, Buteo buteo, B. rufinus, Circaetus gallicus, Circus aeruginosus, C. macrourus, C. pygargus, Gyps fulvus, Hieraetus pennatus, Milvus migrans, Neophron percnopterus, Pernis apivorus
FALCONIDAE (8): falcons
Falco amurensis, F. cherrug, F. concolor, F. naumanni, F. peregrinus, F. subbuteo, F. tinnunculus, F. vespertinus
PANDIONIDAE (1): osprey
Pandion haliaetus

Table continued

TABLE 16.1—*continued*

Water-birds (66)
ANATIDAE (11): ducks
  Anas acuta, A. clypeata, A. crecca, A. penelope, A. platyrhynchos, A. querquedula, A. strepera, Aythya ferina, A. fuligula, A. nyroca, Tadorna ferruginea
ARDEIDAE (8): herons, egrets and bitterns
  Ardea cinerea, A. purpurea, A. ralloides, Botaurus stellaris, Egretta alba, E. garzetta, Ixobrychus minutus, Nycticorax nycticorax
CHARADRIIDAE–CHARADRIINAE (8): plovers
  Arenaria interpres, Charadrius alexandrinus, C. asiaticus, C. dubius, C. hiaticula, Pluvialis squatarola, Vanellus gregarius, V. leucurus
CHARADRIIDAE–RECURVIROSTRINAE (2): stilts and avocets
  Himantopus himantopus, Recurvirostra avosetta
CHARADRIIDAE–SCOLOPACINAE (20): waders
  Calidris alba, C. alpina, C. ferruginea, C. minuta, C. temminckii, Gallinago gallinago, G. media, Limosa limosa, Lymnocryptes minima, Numerius arquata, N. phaeopus, Philomachus pugnax, Tringa erythropus, T. glareola, T. hypoleucos, T. nebularia, T. ochropus, T. stagnatilis, T. totanus, Xenus cinereus
CICONIIDAE (2): storks
  Ciconia ciconia, C. nigra
CRUIDAE (2): cranes
  Anthropoides virgo, Grus grus
LARIDAE (5): gulls and terns
  Larus fuscus, Chlidonias hybrida, C. leucopterus, Gelochelidon nilotica, Hydroprogne tschegrava
RALLIDAE (5): rails, coots and gallinules
  Crex crex, Fulica atra, Porzana parva, P. porzana, P. pusilla
THRESKIORNITHIDAE (3): spoonbills and ibis
  Geronticus eremita, Platalea leucorodia, Plegadis falcinellus

Most trans-Saharan migrants are small passerines or non-passerines, weighing less than 100 g; the larger migrants, some more than 500 g in weight, are mostly raptors or water-birds. Raptorial birds, owls and water-birds apart, the majority are insectivorous, or largely so, and only the larks, *Alaudidae*, buntings, *Emberizidae*, Quail, *Coturnix coturnix*, Turtle Dove, *Streptopelia turtur*, and Pale Rock Sparrow, *Petronia brachydactyla*, are granivorous.

Outside the desert and sea crossings most passerines and non-passerines are nocturnal migrants, travelling during darkness and resting in cover during the day; others, such as the hirundines, *Hirundinidae*, rollers, *Coracidae*, and bee-eaters, *Meropidae*, are typically day migrants, travelling in the daylight hours and roosting at night. For obvious reasons this difference in behaviour breaks down when migrants start a sea or desert crossing, for once committed they are forced into a continuous flight, and in the case of the Sahara this can take many hours. In marked contrast with most trans-Saharan migrants, the soaring birds (mostly storks, *Ciconia* spp., and some of the larger raptorial birds, *Accipitridae*) migrate in quite a different manner, travelling only by day and taking far longer on journeys followed to avoid wide stretches of water. These birds will be discussed separately.

Some idea of the volume of trans-Saharan movement can be obtained from the number of species involved, and what is known of their breeding densities and distributions in the Palaearctic. On the somewhat conservative assumption that each breeding pair will provide two adults and two young for the journey to Africa, Moreau (1972) has estimated the number of landbirds leaving the West Palaearctic in autumn as 1300 millions, and as nearly 5000 millions for the whole of the Palaearctic. Of these landbirds, *c.* 90% will be passerines and *c.* 10% non-passerines with raptorial birds. With present knowledge the numbers of water-birds are very difficult to estimate with confidence, but *c.* 200 millions might be a reasonable guesstimate.

West Palaearctic landbirds entering the Sahara on a front stretching from *c.* 6°W (Straits of Gibraltar) to 34°E (eastern end of the North African coast), enter over a distance of *c.* 3900 km, giving a volume of *c.* 333,000 birds/km during the autumn, or *c.* 5500 birds/km/day for the central migration period of 60 days. Were all Palaearctic landbirds to enter on the same front, the volume would be nearer 1.28 million birds/km, a figure unlikely to be realized as some Mid and East Palaearctic birds enter the Sudan, Ethiopia and Somalia from the north-east. For the spring migration the volumes would be reduced through losses sustained during migration, and by mortality in the wintering grounds, but they would very unlikely be less than half of the estimates.

For the ground observer in the desert these estimates may seem unrealistically high. Away from the northern and southern boundaries very few birds are to be seen in the desert, even in times of the heaviest movements. Oases which might be expected to attract and concentrate migrants rarely shelter more than a handful, but of those present the unusual variety of species clearly represents the diversity of birds passing. That birds must be migrating at height, and out of sight, in large numbers is sometimes demonstrated by 'falls' of migrants brought down by contrary or difficult winds, especially in the north-eastern oases in spring.

16.2.2. Migrants and the desert environment

Although the spring passage may start, and the autumn passage end, in relatively cool weather, the major part of both migrations occurs during the hot months of the year when insolation is high, midday temperatures extreme, the ground burning, and the level of aridity severe. With conditions generally inimical to the survival of unadapted birds, for lack of shade or shelter, migrants all too easily die of heat-stress and dehydration if they remain grounded for long. The not infrequent periods of strong, often hot, winds can also be hazardous to flying birds, particularly where dust or sand has been raised; even the low-pressure systems along the North African coastal belt in spring will at times produce difficulties for migrants at the end of the desert crossing (e.g. Smith, 1967).

With practically the whole Sahara desolate, barren and waterless, very few places show any amelioration of conditions sufficient to encourage migrants. In the central desert the mountainous Ahaggar, Aïr and Tibesti have some grazing about them, and a few valleys with a little perennial vegetation, but too little to attract many migrants. In the extreme west, along a coastal belt c. 400 km wide, from Morocco to Mauritania, dewfall is sufficient to support scrubby growth, with food and shelter for migrants, but this produces no great concentration of passage. Likewise, on the eastern side of the desert the Nile valley supports a narrow thread of vegetation connecting the northern and southern boundaries, but as a feeding ground for migrants, and hence a migration route, it is relatively unimportant.

The oases scattered across the desert might owe their small numbers of migrants to their being mostly groves of date-palms with little other vegetation or undergrowth, although some of the larger ones have small pools and more suitable cover for migrants. But collectively their total area is so small compared with the vastness of the Sahara that their importance to trans-Saharan migrants is negligible. Their modern analogues, the oil-production wells with sewage effluents, and the small irrigation schemes fed by artesian water, are claimed to be rather more attractive to migrants, especially in times of unfavourable weather conditions (e.g. Sarir; Hogg, 1974), but they remain to be studied in detail.

For most migrants, then, the desert must appear as a foodless waste; although this can hardly be true for the bird-catching raptors (some falcons, *Falco* spp., harriers, *Circus* spp., and sparrowhawks, *Accipiter* spp.) for whom food, in the shape of other migrants, will be available generally throughout the passage. But in the case of insectivorous birds, the most numerous of migrants, the dearth will be pressing and absolute. On very rare occasions some food may be available in the form of migrant insects, locust swarms (Betts, 1976), or insects in patches of ephemeral vegetation brought up by the very infrequent desert storms, but these sources too unreliable to be of consequence. For the few granivorous species food will be limited, useful quantities only being found at the northern and southern margins where rains produce ephemeral grasses some time of the year. Suitable habitats for water-birds, hence food, is virtually absent over the whole of the desert outside of the Nile system, there being estimated 0.4 km^2, mostly in oases, in about 7.8×10^6 km^2 of desert (Moreau, 1967). Although some water-birds follow the Nile when migrating, the volume along this route is small.

Where migrants move south–north across the desert the northern and southern boundaries, as defined by the 100-mm isohyet, take on particular significance. The northern boundary, running along the North African coastal belt, is seen as a potential resting or refuelling 'stopover' between the Mediterranean and desert crossings; the southern boundary as the goal for autumn migrants, and adjacent to the 'take-off' zone of the spring crossing. Trans-Saharan migrants are certainly more numerous along these edges than in the desert but they are transients and numerically only a small proportion of the total on passage.

Vegetation about the 100-mm isohyet is sparse and patchy, and in terms of food supply for hard-pressed migrants has little to offer. Few birds, apart from steppe-adapted species such as wheatears *Oenanthe* spp., seem to stay more than a short time. But there are seasonal differences between the boundaries that are of some consequence to passing migrants.

Along the North African coastal belt rain falls in October–March, ameliorating the drought of the rest of the year, so that conditions for migrants are at their best at the start of the spring migration. Possibly as a consequence of this, but also because birds are arriving after a longer and more taxing desert crossing, more birds are seen in spring than autumn. This can be exemplified by past ringing records at Gabes, southern Tunisia, where more species were reported and ten times more migrants were ringed in spring than in autumn.

In contrast, at the southern boundary rain falls in July–August, brought by the northward movement of the ITCZ, so that trans-Saharan migrants arriving then find ecological conditions at their best, with ephemeral vegetation green, trees in leaf and flower, and insect life at its maximum. From September onwards the rains retreat southwards, the land and vegetation dry, and conditions worsen progressively until the next rains in the following year. At the time of spring migration conditions are severe and insect populations at a minimum, so that for insectivorous migrants, in particular, the effective desert edge at this time of year is much further south of the 100-mm isohyet. Even during the autumn migration the grounding zone of most birds after the desert crossing lies far south of the 100-mm isohyet. In many years the majority descend well past the 250-mm isohyet, probably not descending until they fly through the inter-tropical front (ITF) which, at the altitudes they are travelling, will be many km south of its surface position. For insectivorous birds there is considerable advantage in maintaining flight until this zone of wind convergence is reached, for here insect life is at its most concentrated.

In spring, with the effective desert edge far south of the southern 100-mm isohyet, the final 'take-off' for trans-Saharan fight is well into the savanna. For the different species the actual position for this seems to vary, probably controlled by where they find sufficient food for building up reserves needed for the long flight. One result of this is the virtual absence of migrants at the 100-mm isohyet in spring, in contrast with the autumn when numbers, if not large, are noticeable.

16.2.3. The broad-front crossing

With the exception of soaring birds, migration of trans-Saharan birds over both desert and Mediterranean is on a broad front. Although originally deduced from observations on the pattern of low-level landfalls, particularly along the northern and southern boundaries, the reality of broad-front migration has been amply confirmed by radar observations where actual passage of birds in the higher altitudes can be tracked directly. It follows, then, that overland migrants do not use narrow-front routes, there being no great concentrations along such obvious features as the Nile, or regions of the mountain masses of the central desert, as once surmised. Even over the Mediterranean small birds show no tendency to concentrate at the major narrows, although certain of the larger birds (herons, *Ardeidae*, some falcons, *Falco* spp., and harriers, *Circus* spp.) show some concentration at the Cape Bon–Sicily crossing. Radar traverses along the Mediterranean in spring and autumn indicate relatively uniform densities of migrants throughout, with a possible reduction over the Ionian Sea which is not only the longest sea crossing but also the position of a 'migratory divide' (Casement, 1966).

Although much of the wintering grounds of West Palaearctic birds lie more or less directly south of their breeding grounds, few migrants travel on direct N/S bearings, the great majority seeming to take diagonal routes, either NW/SE or NE/SW. Ringing studies in both Europe and Africa, together with fieldwork in tropical Africa, have shown movements of trans-Saharan migrants to be rather complex. For instance, some birds breeding in north-west Europe (Red-backed Shrike, *Lanius collurio*, Wood Warbler, *Phylloscopus sibilatrix*, Lesser Whitethroat, *Sylvia curruca*, Golden Oriole, *Oriolus oriolus*, and European Roller, *Coracias garrulus*) fly south-east in the autumn, to cross to Africa by way of the eastern Mediterranean, for wintering grounds in East Africa. Other birds, like the Blackcap, *Sylvia atricapilla*, Spotted Flycatcher, *Muscicapa striata*, and White Wagtail, *Motacilla alba*, have a 'migrational divide' about 5°E–15°E, where birds to the west migrate south-west to Iberia, through Morocco, and beyond the Senegal River, and those to the east migrate south-east to Egypt, Ethiopia and East Africa, with those few on the divide travelling directly southwards across the desert. Most migrants of north-west Europe on the NE/SW bearings are using the wintering grounds in tropical Africa lying up to 1000 km west of the European landmass. For birds of the Mid and East Palaearctic migrating on great circle routes between breeding and wintering grounds, the predominant bearings will be approximately NE/SW.

Migratory bearings deduced from ringing data tend to be those observed for low-level migrants crossing the North African coast in both seasons, but they have been confirmed by radar for migrants flying at higher altitudes. West of the Ionian Sea movement is NE/SW; east of the Ionian Sea it is much more variable but predominantly NNW/SSE. In Cyprus radar has also detected birds, presumably of the Mid and East Palaearctic, on west of south bearings in the autumn, and east of north bearings in the spring (Adams, 1962). Very broadly these bearings appear to be maintained over the desert. At In Abangharit, Niger, radar has shown *c.* 70% of autumn migrants on a SSW bearing, 15% between SSW and S, and 15% between S and SSE. For the many migrants making diagonal crossings the desert passage will be considerably longer than the nominal 1500 km for a direct route between the 100-mm isohyets.

Radar observations have also confirmed the generally high altitudes at which birds migrate, and demonstrated conclusively that large numbers are, indeed, moving at heights too great for them to be seen by ground observers. Although there may be a proportion travelling closer to the ground, the majority of small birds migrate at heights of 1500–2500 m, with some higher, to above 3000 m. For trans-Saharan migrants considerable benefits accrue from the use of high-altitude flight. Apart from avoiding the more severe manifestations of the desert environment, it reduces the chance of being killed by such predators as Eleonora's Falcon, *Falco eleonorae*, and Sooty Falcon, *F. concolor*, preying selectively on migrants crossing the Mediterranean and Sahara respectively. High-altitude flight also reduces air-drag to such an extent that, for a given fat (fuel) reserve, migrants experience a 20% improvement in air-speed at 3000 m. Water-loss, a critical aspect of survival on the crossing, is also reduced by increased altitude.

Most small passerines have air-speeds of *c.* 40 km/hr, while the larger birds such as waders, *Scolopacinae*, and ducks, *Anatidae*, are faster with air-speeds ranging up to 80 km/hr. With low air-speeds small birds are more likely to be adversely affected by unfavourable conditions such as contrary or difficult winds. Indeed, the stronger fliers are rarely seen in the desert as their quicker crossing leaves them less likely to be incommoded by difficult conditions. On the other hand, large slow-flying birds such as herons, *Ardeidae*, are more often grounded, presumably because, taking longer to cross, they are as vulnerable as small birds to being forced down by unfavourable conditions.

In still air the nominal direct crossing of 1500 km would take small passerines *c.* 40 hr of continuous flight but at the altitudes that most migrants cross the desert calms would not be common. Although the wind systems of the Sahara are far from being simple, and sometimes show dramatic reversals of direction, northerly winds generally prevail throughout the year with mean speeds of *c.* 15 km/hr. Thus in autumn migrants will tend to be assisted by tail-winds, and could make the direct crossing in less than 30 hr, and a diagonal crossing in *c.* 40 hr. In spring the crossing time is potentially longer because of the headwinds and the need for migrants to begin the crossing well south of the desert edge because of the unfavourable ecological conditions. If undertaken at lower altitudes where headwinds are weaker, the direct flight of 1500 km is likely to take *c.* 47 hr at this time of year, with an additional *c.* 12 hr for the extra 500 km (nominal) imposed by the more southerly start—a total of *c.* 60 hr for the full crossing. However, at this time of year a diagonal crossing is likely to take less time. In spring the wind system is such that, except for in the extreme western desert (where ground conditions are favourable anyway) and in the Northern Sudan, tailwinds of *c.* 25 km/hr are common at 2000 m for migrants travelling on a north-east bearing, as, indeed, are many from the West Palaearctic, and all from the Mid and East Palaearctic. Flying time for the north-east flight at the right altitude would be *c.* 32 hr from the 100-mm isohyet, or *c.* 45 hr for the more southerly start, i.e. not much longer than the autumn crossing, but liable to disruption by *khamsin*-type winds in the north-eastern quarter of the desert. At the end of the desert crossing most West Palaearctic migrants continue across the Mediterranean, to complete both crossings in one continuous flight (Moreau, 1961, 1972).

Energy for these long continuous flights is supplied by the metabolism of fat (lipid) laid down during a period of 2–3 weeks immediately prior to the crossing. At the start of the journey birds are conspicuously fat; at its end they are more or less fatless or lean, and may even be emaciated should difficulties on the journey have forced the mobilization of the additional reserves in the musculature.

Such changes can be monitored in living birds by examining weight changes statistically at the different stages of migration: lean weight during the non-migratory phase, premigratory fattening, and stages during the journey (e.g. Ash, 1969; Fry *et al.*, 1970; Moreau *et al.*, 1970; Moreau, 1969; Dowsett *et al.*, 1971; Ward, 1963). Such information often provides useful ancillary facts about trans-Saharan migrants, e.g. migratory state of readiness; probable distances already covered; degree of refuelling at intermediate stages. For any detailed understanding of the energetics involved in weight changes laboratory analysis of body components has to be undertaken.

Carcase analysis of migrants collected prior to, or during, the spring migration has shown the heaviest passerines about to migrate to have lipid contents of 35–55% of live weight (warblers, *Sylvia* spp. & *Acrocephalus* spp., Wheatear, *Oenanthe oenanthe*). Similar amounts have been found in several waders, whereas hirundines, *Hirundinidae*, have been found to have lower lipid levels, at about 20–30% of live weight. Calculations have shown that the heaviest birds have more than adequate supplies for the desert crossing, and usually for the Mediterranean crossing as well. For example, a Sedge Warbler, *Acrocephalus schoenobaenus*, leaving the edge of Lake Chad where it had been fattening, with a lean (fat-free) weight of 10.3 g, and carrying 5.9 g of fat, would need 3.1 g of fat for 44 hr direct flight across the desert), or 4.5 g for a 59 hr north-east flight to the Egyptian coast (Fry *et al.*, 1970).

It is still uncertain to what extent all birds leaving for the desert crossing are carrying the higher amounts of fat recorded; some may leave underweight, and with insufficient fuel for the journey, to succumb *en route*. In this connection it is possibly relevant to note that in autumn, at the end of the desert crossing, it is mostly the young birds of the year that come to ground first, while the more experienced adults, and possibly birds heavier with fat at the outset of the journey, fly on to come to ground further south. The lower fat reserves of hirundines is thought to reflect their greater aerodynamic efficiency and consequent lower energy requirements for the journey, though it has also been suggested they may be more efficient food gatherers on migration as well, especially at the desert edges.

Premigratory rates of lipid deposition can be high where food is plentiful, and rates of 0.3–0.6 (−1.0) g/day have been recorded for a number of insectivorous passerines feeding in favoured places in the Sahel and Guinea Savanna south of the desert edge. Under pressure of hyperphagia many insectivorous birds, and most notably the *Sylvia* warblers, will supplement their intake with quantities of succulent fruits where these are available. Birds breaking their journey will normally feed if food is available but even at the desert edges where some food exists they are not always capable of replenishing fat reserves and rarely, if ever, are able to attain high rates of deposition.

Excessive dehydration is an ever-present threat for trans-Saharan migrants, and it is common to find that birds dying *in extremis* on migration were severely dehydrated as well as emaciated. Body water content seems as much an essential reserve as fat, and possibly equally critical in limiting migratory range.

Evaporative losses from respiration are largely responsible for dehydration during the flight across the desert, only being replenished from metabolic water produced by oxidation of fat which releases its own weight in water. Respiratory losses are greatly affected by the temperature and saturation deficit of the air, and, under the conditions of the journey, are very likely to exceed production of metabolic water, so that some degree of dehydration is probably characteristic of trans-Saharan migrants. Of a sample of migrants taken in autumn when temporarily grounded on the Egyptian coast after passage across the Mediterranean, 78% were shown to be dehydrated to some degree, and 12% severely dehydrated with losses greater than 40% of normal water content (Moreau *et al.*, 1970).

The full significance of dehydration in trans-Saharan migrants is yet to be evaluated, especially as it may appear before migration has started. In the Reed Warbler, *Acrocephalus scirpaceus* , it has been shown that premigratory fattening in the savanna under hot (33–35°C) and dry conditions can induce a marked premigratory dehydration, even where fat deposits were apparently adequate for the desert crossing (Fogden, 1972). This might imply migrants risk incurring more serious dehydratioin by breaking journeys at intermediate stages where temperatures were high and food, though present, was scarce, than making one continuous flight. In these circumstances the migratory restlessness so evident in many migrants in the boundary zones, driving them onwards to further destinations, would be biologically advantageous.

16.2.4. The soaring birds

The soaring birds form a fairly well-defined group, if small in number of species, whose migratory behaviour is conspicuously different from that of the other, more numerous, trans-Saharan migrants. On migration they use upcurrents of thermal or topographical origins for soaring and gliding, rather than using the straight, flapping flight typical of the others. Forward progression is obtained by gliding after height has been obtained by circling in upcurrents. As most upcurrents are thermally generated, actual passage is restricted to the daylight hours, and to country where upcurrents are reliable—a feature concentrating movement through hilly or mountainous land and across the narrowest of sea passages. Movement is, therefore, on narrow fronts, with some routes being well defined and very localized.

Being diurnal migrants, large in size, and often moving majestically at a leisurely pace in groups or flocks at heights usually less than 1000 m and along well-defined routes, soaring birds are more frequently noticed by ground observers, and consequently their movements better known. The White Stork, *Ciconia ciconia*, is undoubtedly the best known but not, however, the most numerous of the soaring birds to be seen on passage.

Soaring birds leaving the West Palaearctic take routes crossing the Straits of Gibraltar (Lathbury, 1970; Evans *et al.*, 1973) or the Bosphorus (Collman *et al.*, 1967; Porter *et al.*, 1968), and thus enter the Sahara on the extreme west or the extreme east. Of the passage of raptorial birds across the narrows between Sicily and Tunisia only one soaring bird, Honey Buzzard, *Pernis apivorus*, is common, the rest being narrow-winged species, mostly *Falco* spp. and *Circus* spp. which cross the Sahara on a broad-front movement (Beaman *et al.*, 1974). Of the two major routes the Bosphorus is used by the great majority of soaring birds, not only from the Mid and East Palaearctic but also from the West.

From the Bosphorus the European birds cross Asia Minor to the Taurus where they turn south, down the Levant coast to Suez, or down the Rift Valley to Sinai (Safriel, 1968). The commonest species are White Stork, Honey Buzzard, Lesser Spotted Eagle, *Aquila pomarina*, Booted Eagle, *Hieraetus pennatus*, Black Kite, *Milvus migrans*, and sparrowhawks, *Accipiter* spp. From the Gulf of Iskendem southwards these are joined by migrants from Russia and Turkey which include more storks, harriers, *Circus* spp., and large numbers of buzzards, *Buteo* spp., together with vultures and Short-toed Eagles, *Circaetus gallicus*. From Suez or Sinai migrants move either along the Red Sea coast and the eastern side of the Ethiopian Highlands into East Africa (most of the raptors), or across the desert to the Nile, picking up the river between Qena and Luxor, and following it southwards, using thermals generated by the river system (White Storks and White Pelicans, *Pelicanus onocrotalus*). Whether storks following the Nile bypass the Dongola loop of the river by going over the Nubian desert is unknown, but once past this they start dispersing east and west.

In the case of the crossing at the Straits of Gibraltar, not only are numbers fewer compared with the Bosphorus, but certain species do not occur other than as vagrants, notably the *Aquila* eagles, Long-legged Buzzard, *Buteo rufinus*, and the Pallid Harrier, *Circus macrourus*. On autumn passage south of Gibraltar and Tangiers numbers are augmented by birds breeding in Morocco, especially White Storks, but little is known of their subsequent movements across the desert to West Africa where some at least winter.

Spring passage is largely the reverse of the autumn movement although there seem to be small differences in routes followed and the main passage of some species less protracted. Compared with the autumn, in spring White Storks tend to move in larger flocks and pass any one location over a smaller number of days.

As is to be expected, the energetic relations of migration in soaring birds is quite different from those of other trans-Saharan migrants, and related to the constraints imposed by the size of the birds involved. Soaring birds are quite large birds, e.g. 3–4 kg for *Ciconia* spp., 2 kg for *Accipitrinae*, and for aerodynamic reasons could not carry extra weight equivalent to 50% of live weight as can smaller birds fattened for the desert crossing. Though the soaring-and-glide mode of progression is a slower means of migration than straight, flapping flight, it consumes very little energy, even if many days are needed to complete the journey. In other words, soaring birds are 'economizing on fuel at the expense of making slower progress' (Pennycuick, 1969). The difference in the case of the White Stork has been calculated as reducing fuel consumption by a factor of 23 compared with straight, flapping flight (Pennycuick, 1972). Unfortunately little work has been done on weight changes in this type of migrant but, since there can be little opportunity for some species to feed over much of the journey, the limited fat reserves they carry must be sufficient for the long migratory routes, which in the case of the eastern route is *c.* 30–40% longer than the more direct route taken by smaller migrants.

16.3. REFERENCES

ADAMS, D. W. H. (1962) Radar observations of bird migration in Cyprus. *Ibis*, **104**, 113–146.
ASH, J. S. (1969) Spring weights of trans-Saharan migrants in Morocco. *Ibis*, **111**, 1–10.
BEAMAN, M. and GALEA, C. (1974) The visible migration of raptors over the Maltese Islands. *Ibis*, **116**, 419–431.
CASEMENT, M. B. (1966) Migration across the Mediterranean observed by radar. *Ibis*, **108**, 46–91.
COLLMAN, J. R. and CROXALL, J. P. (1967) Spring migration at the Bosphorus. *Ibis*, **109**, 359–372.
CURRY-LINDAHL, K. (1981) *Bird Migration in Africa. Movements between six continents.* Academic Press, London & New York. 2 vols.
DOWSETT, R. J. and FRY, C. H. (1971) Weight losses of trans-Saharan migrants. *Ibis*, **113**, 531–533.
EVANS, R. R. and LATHBURY, G. W. (1973) Raptor migration across the Straits of Gibraltar. *Ibis*, **115**, 572–585.
FOGDEN, M. P. L. (1972) Premigratory dehydration in the Reed Warbler *Acrocephalus scirpaceus* and water as a factor limiting migratory range. *Ibis*, **114**, 548–552.
FRY, C. H., ASH, J. S. and FERGUSON-LEES, I. J. (1970) Spring weights of some palaearctic migrants at Lake Chad. *Ibis*, **112**, 58–82.
HOGG, P. (1974) Trans-Saharan migration through Sarir, 1969–1970. *Ibis*, **116**, 466–476.
LATHBURY, Sir G. (1970) A review of the birds of Gibraltar and its surrounding waters. *Ibis*, **112**, 25–43.
MOREAU, R. E. (1961) Problems of Mediterranean–Saharan migration. *Ibis*, **103a**, 373–427, 580–623.
MOREAU, R. E. (1967) Water birds over the Sahara. *Ibis*, **109**, 232–259.
MOREAU, R. E. (1969) Comparative weights of some trans-Saharan migrants at intermediate points. *Ibis*, **111**, 621–624.
MOREAU, R. E. (1972) *The Palaearctic–African Bird Migration Systems.* Academic Press, London & New York.
MOREAU, R. E. and DOLP, R. M. (1970) Fat, water, weights and wing-lengths of autumn migrants in transit on the northwest coast of Egypt. *Ibis*, **112**, 209–228.
PENNYCUICK, C. J. (1969) The mechanics of bird migration. *Ibis*, **111**, 525–556.
PENNYCUICK, C. J. (1972) Soaring behaviour and performance of some East African birds observed from a motor-glider. *Ibis*, **114**, 178–218.
PORTER, R. and WILLIS, I. (1968) The autumn migration of soaring birds at the Bosphorus. *Ibis*, **110**, 520–536.
SAFRIEL, U. (1968) Migration at Elat, Israel. *Ibis*, **110**, 283–320.
SMITH, K. D. (1968) Spring migration through southeast Morocco. *Ibis*, **110**, 452–492.
WARD, P. (1963) Lipid weights in birds preparing to cross the Sahara. *Ibis*, **105**, 109–111.

CHAPTER 17

Small Mammals

D. C. D. HAPPOLD*

Department of Zoology, Australian National University, Canberra, Australia

CONTENTS

17.1. INTRODUCTION

It comes as rather a surprise to learn that most hot deserts of the world have a rich and diverse fauna of small mammals. These small mammals (adult weight up to 3 kg) are rarely seen by the casual observer as they are mainly nocturnal and widely dispersed. Some unfavourable parts of the Sahara are completely devoid of mammals while other favourable habitats support flourishing populations of several species. The most successful orders of small mammals are the rodents, carnivores and bats and, to a lesser extent, the hares and hyraxes. All of them are adapted to the harshness and unpredictability of the desert environment. Many of these adaptations, such as nocturnal activity, efficient water conservation, behavioural control of body temperature, and rapid initiation of reproduction after drought, are shared by similar species of small mammals in other hot deserts of the world. Hence, there appear to be only a limited number of evolutionary options for small mammals in deserts.

---

* Formerly University of Khartoum, Sudan and University of Ibadan, Nigeria.

The principal species of small mammals in the Sahara are relatively well known, although there are many differences of opinion regarding the taxonomic status of some species and subspecies. Most taxonomic studies have been on a regional or country basis (see Table 17.2) and there are virtually no studies which encompass the whole Sahara. Biological and ecological information on Saharan small mammals, with a few exceptions (e.g. Heim de Balsac, 1936; Petter, 1961; Daly and Daly, 1975) is very scanty and is not nearly as comprehensive as for the American, South African and Australian deserts. However, there are many similarities in the biology of desert small mammals regardless of locality, and consequently many of the general ideas in this chapter still need detailed investigation in the Sahara.

17.2. ORIGINS AND FORMATION OF THE FAUNA

The distribution patterns of Saharan small mammals has changed frequently in the past, as a result of the climatic and vegetational changes of past periods. During the pluvial periods, when the extent of the Sahara diminished, species which were unable to cope with wetter conditions became locally extinct and savanna species became more numerous and widespread. Although there are few data for small mammals, many African savanna species probably reached what is now the central Sahara as did hippopotamuses, elephants and many large antelopes. During the dry interpluvials, the opposite trend occurred, and species which favoured drier conditions extended their geographical ranges; some of these species may have occurred as far south as the present Guinea savannas of West and Central Africa (6–8°N). During the process of desert contraction, some species of small mammals from Eurasia gained access to the African continent; some of these are still present as isolated relic populations along the Mediterranean coast of North Africa (Table 17.1). Similarly, other species which are more typical of the arid zones of the Middle East (extending into India and southern USSR) were able to recolonize Africa when the interpluvial periods re-established arid conditions. The chronology of these events is uncertain, although the common heritage of the Sahara–Middle East small mammal fauna is well established.

One effect of these past changes is that the present fauna of the Sahara is derived from several sources (Table 17.1). Five distinct origins may be recognized:

1. Saharan–Middle Eastern: species and genera which are widespread in the arid areas of the Sahara, Middle East, India and southern USSR.
2. Temperate Eurasian: species and genera which occur in Europe and which gained access to North Africa via the Iberian peninsula or the Levant during the pluvial periods, or when the Ice Ages resulted in southward movements in distribution. These species now occur only as isolated relic populations in the Mediterranean coastal region.
3. African: species and genera which occur widely in Africa south of the Sahara and which reach the northern limit of their distribution along the southern edge of the Sahara. These species may be considered as two subgroups: (a) widespread species which occur throughout the savannas of Africa, and (b) northern savanna species which occur only north of the tropical rainforest zone.
4. Cosmopolitan: species which are widespread and ecologically very tolerant. These include species which occur throughout Africa and the Middle East (not just in arid regions), or throughout Africa and Eurasia.
5. Endemic: species which are found only in the Sahara and are assumed to have evolved there.

The present distribution pattern of each species is closely linked to its origins and affinities and, as indicated above, is always liable to change. The current increase in man-made desertification along the southern edge of the Sahara will gradually enable many desert species to extend their geographical ranges to the south.

TABLE 17.1. The origins of the genera of Saharan small mammals

Order	Cosmopolitan	Temperate Eurasian	Saharan Middle Eastern	African		
				Widespread	Northern savannas	Endemic
Insectivora	Crocidura		Paraechinus	Elephantulus		
	Suncus		Hemiechinus			
	Erinaceus					
Chiroptera	Rousettus	Plecotus	Rhinopoma	Eidolon		
	Taphozous		Asellia	Nycticeius		
	Nycteris		Otonycteris			
	Rhinolophus					
	Eptesicus					
	Pipistrellus					
	Tadarida					
Lagomorpha	Lepus					
Rodentia	Tatera	Microtus	Dipodillus	Xerus	Taterillus	Pachyuromys
	Mus	Apodemus	Gerbillus	Acomys	Desmodilliscus	Ctenodactylus
	Hystrix	Spalax	Sekeetamys	Arvicanthis		Massoutiera
		Eliomys	Meriones	Mastomys		Felovia
			Psammomys	Graphiurus		
			Jaculus			
			Allactaga			
Carnivora	Vulpes		Fennecus	Ictonyx	Poecilictis	
	Genetta					
	Herpestes					
	Felis					
Hyracoidea				Procavia		

17.3. THE SMALL MAMMAL FAUNA

As the margins and extremities of the Sahara are imprecise, it is only possible to give an approximate number of species of small mammals in the Sahara. Six orders, twenty-four families and eighty-three species are recorded (Table 17.2) when eight countries or localities are considered. The rodents are by far the most successful order in terms of species (40), and within the rodents the Cricetidae contributes the greatest number of species (22). The gerbils and jirds (*Gerbillus, Dipodillus, Desmodillicus, Meriones, Pachyuromys, Psammomys, Sekeetamys, Tatera* and *Taterillus*) are all characterized by brown or sandy dorsal pelage, whitish ventral pelage, long tails often with a tuft of hairs, large eyes, and inflated tympanic bullae. The Muridae, although very numerous in sub-Saharan Africa, appear to be less successful in deserts except for the widespread spiny mice (*Acomys* spp.). The jerboas (Dipodidae) are the only bipedal rodents of the Sahara. Only one species, *Jaculus jaculus*, is widespread; the other species are confined to parts of the Mediterranean coastal region. The other families of rodents comprise only a few species, often with limited and disjunct distributions. Dormice (Muscardinidae) and mole rats (Spalacidae) are not really desert rodents and they survive as relic populations in a few confined localities. The gundis (Ctenodactylidae) and hyraxes (Procaviidae) are rock-dwelling mammals which occur as isolated populations on massifs (Aïr, Hoggar, Tibesti) and other rocky habitats. The only other group of small herbivores in the Sahara are the hares (Leporidae) which maintain dispersed populations where grass is available.

Insectivores and carnivores form an interesting and important group of small predators. The Insectivora are represented by the hedgehogs (*Erinaceus, Hemiechinus* and *Paraechinus*), shrews (*Crocidura*), and elephant-shrews (*Elephantulus*). Although rarely seen, hedgehogs are fairly widespread where insect food is available whereas shrews appear to be very rare and confined to rocky or mesic habitats. Elephant-shrews are known only from a few localities in the coastal and Pre-Saharan regions. The Carnivores comprise three species of foxes (*Fennecus, Vulpes*), two weasels (*Ictonyx, Poecilictis*), a genet (*Genetta*), a

TABLE 17.2. Small terrestrial mammals of the Sahara

+ = present; (+) = probably present, but no definite record; — = absent for geographical reasons, or because the habitat is unsuitable.

Order/Family/Species	Mauritania and Rio de Oro (1)	Aïr (2)	Hoggar (3)	Tunisia and NE Algeria (4)	Libya (5)	Egypt (6)	Northern Sudan (7)	Lake Chad and Southern Niger (8)
INSECTIVORA								
Soricidae								
Suncus etruscus	—	—	—	+	—	—	—	—
Crocidura anthonyi	—	—	—	+	—	—	—	—
Crocidura fulvastra	(+)	—	—	—	—	—	+	+
Crocidura lusitania	+	—	—	—	—	—	—	—
Crocidura pasha	—	—	—	—	—	—	+	+
Crocidura suaveolens	—	—	—	+	—	+	—	—
Crocidura flavescens	—	—	—	—	—	—	+	+
Crocidura whitakeri	+	—	—	+	—	—	—	—
Macroscelididae								
Elephantulus rozeti	+	—	—	+	+	—	—	—
Erinaceidae								
Erinaceus algirus	—	—	—	+	+	+	—	—
Erinaceus albiventris	—	(+)	—	—	—	—	+	+
Hemiechinus auritus	—	—	—	—	+	+	—	—
Paraechinus aethiopicus	(+)	+	+	+	+	+	+	
CHIROPTERA								
Rhinopomatidae								
Rhinopoma microphyllum	+	(+)	—	—	—	(+)	—	(+)
Rhinopoma hardwickei	+	+	+	+	(+)	+	+	+
Emballonuridae								
Taphozous nudiventris	+	—	—	—	—	+	—	+
Taphozous perforatus	+	—	—	—	—	+	+	(+)
Nycteridae								
Nycteris thebaica	(+)	—	—	—	+	+	+	+
Nycteris hispida	+	—	—	—	—	—	+	+
Rhinolophidae								
Rhinolophus clivosus	—	—	+	—	+	+	+	(+)
Hipposideridae								
Asellia tridens	+	+	(+)	+	+	+	+	+
Hipposideros caffer	+	+	+	—	—	—	+	—
Vespertilionidae								
Pipistrellus kuhli	—	—	+	+	+	+	+	(+)
Pipistrellus deserti	—	—	(+)	—	+	+	+	—
Pipistrellus rueppelli	+	(+)	—	—	—	+	+	+
Pipistrellus ariel	—	—	—	—	—	+	+	—
Plecotus austriacus	(+)	—	—	+	—	+	+	—
Otonycteris hemprichi	—	+	(+)	+	+	+	+	—
Nycticeius schlieffeni	+	+	—	—	—	+	+	—
Scotophilus sp.	—	—	—	—	—	—	+	+
Molossidae								
Tadarida aegyptiaca	—	+	+	—	—	+	—	(+)
Tadarida pumila	—	—	—	—	—	—	—	+
Tadarida nigeriae	—	—	—	—	—	—	—	+
LAGOMORPHA								
Leporidae								
Lepus capensis	+	+	+	+	+	+	+	+

Table continued

TABLE 17.2—*continued*

	Localities							
Order/Family/Species	Mauritania and Rio de Oro (1)	Aïr (2)	Hoggar (3)	Tunisia and NE Algeria (4)	Libya (5)	Egypt (6)	Northern Sudan (7)	Lake Chad and Southern Niger (8)
RODENTIA								
Sciuridae								
Xerus erythropus	+	+	(+)	—	—	—	+	+
Cricetidae								
Dipodillus simoni	—	—	—	+	+	+	—	—
Desmodilliscus braueri	+	+	(—)	—	—	—	(+)	(+)
Gerbillus amoenus	—	—	—	—	+	+	—	—
Gerbillus andersoni	—	—	—	—	—	+	—	—
Gerbillus campestris	(+)	+	+	+	+	+	+	—
Gerbillus gerbillus	+	+	(+)	+	+	+	+	+
Gerbillus henleyi	—	—	—	+	+	—	—	—
Gerbillus nanus	+	+	+	+	—	+	+	—
Gerbillus pyramidum	+	+	+	+	+	+	+	—
Gerbillus watersi	—	—	—	—	—	—	+	—
Meriones crassus	+	—	+	+	+	+	+	—
Meriones libycus	+	+	+	+	+	+	—	—
Meriones shawi	—	—	—	+	(+)	+	—	—
Pachyuromys duprasi	+	—	+	+	+	+	—	—
Psammomys obesus	+	—	+	+	+	+	—	—
Sekeetamys calurus	—	—	—	—	—	+	—	—
Tatera valida	—	—	—	—	—	—	—	+
Tatera hopkinsoni	—	—	—	—	—	—	—	+
Tatera robusta	—	—	—	—	—	—	+	—
Taterillus sp.	+	—	—	—	—	—	—	+
Microtus socialis	—	—	—	—	+	—	—	—
Muridae								
Acomys cahirinus	(+)	+	+	—	+	+	+	+
Acomys russatus	—	—	—	—	—	+	—	—
Apodemus sylvaticus	—	—	—	+	—	—	—	—
Arvicanthis niloticus	—	+	—	—	—	+	+	+
Mastomys sp.	—	+	—	—	—	—	—	+
Mus musculus	—	—	—	+	+	+	+	—
Muscardinidae								
Eliomys quercinus	—	—	—	+	+	—	—	—
Graphiurus murinus	—	+	—	—	—	—	—	(+)
Dipodidae								
Jaculus jaculus	+	+	+	+	+	+	+	+
Jaculus orientalis	—	—	—	(+)	+	+	—	—
Allactaga tetradactyla	—	—	—	—	+	+	—	—
Spalacidae								
Spalax ehrenbergi	—	—	—	—	+	+	—	—
Hystricidae								
Hystrix cristata	—	+	—	+	+	—	—	(+)
Ctenodactylidae								
Ctenodactylus gundi	—	—	—	+	+	—	—	—
Ctenodactylus vali	—	—	—	+	+	—	—	—
Felovia vae	+	—	—	—	—	—	—	—
Massoutiera mzabi	—	+	+	(+)	—	—	—	—
CARNIVORA								
Canidae								
Fennecus zerda	+	+	(+)	+	+	+	+	—
Vulpes pallida	+	+	—	—	—	—	+	+
Vulpes rueppelli	+	+	—	+	+	+	+	—

Table continued

TABLE 17.2—*continued*

Order/Family/Species	Mauritania and Rio de Oro (1)	Aïr (2)	Hoggar (3)	Tunisia and NE Algeria (4)	Libya (5)	Egypt (6)	Northern Sudan (7)	Lake Chad and Southern Niger (8)
CARNIVORA (*continued*)								
Mustelidae								
Ictonyx striatus	+	+	—	—	—	+	+	+
Poecilictis libyca	+	(+)	—	+	+	+	+	(+)
Viverridae								
Genetta felina	—	—	—	—	—	—	+	+
Genetta genetta	+	+	—	(+)	—	—	—	—
Herpestes ichneumon	+	+	—	+	—	—	+	(+)
Felidae								
Felis libyca	+	+	(+)	+	+	—	+	+
Felis margarita	+	+	(+)	+	(+)	+	(+)	—
Felis chaus	—	—	—	—	—	+	—	—
HYRACOIDEA								
Procaviidae								
Procavia ruficeps	+	+	(+)	—	+	+	+	(+)

Sources: (1) Mauritania and Rio de Oro—Dekeyser and Villiers, 1956; Valverde, 1957; Jullien and Petter, 1970; Poulet, 1974;
     (2) Aïr—Dekeyser, 1950; Niethammer, 1963; Fairon, 1980; Newby and Jones (n.d.);
     (3) Hoggar—Regnier, 1960; Niethammer, 1963; Meester and Setzer, 1971–77;
     (4) Tunisia and NE Algeria—Schomber and Kock, 1960; Niethammer, 1963; Kock, 1980;
     (5) Libya—Setzer, 1957; Ranck, 1968; Hufnagl, 1972;
     (6) Egypt (excluding Sinai and Nile delta)—Hoogstraal, 1962, 1963, 1964; Gaisler *et al.*, 1972; Osborn and Helmy, 1980;
     (7) Northern Sudan—Setzer, 1956; Bauer, 1963; Happold, 1967 a, b, c;
     (8) Lake Chad and Southern Niger—Kock, 1978; Happold, in press;
Additional sources: Ctenodactylidae—George, 1974; *Felis margarita*—Hemmer *et al.*, 1976; *Crocidura*—Hoogstraal, 1962; Heim de
     Balsac, 1968; Chiroptera—D. Kock, pers. comm., 1982 (using date in Heim de Balsac, 1936; Valverde, 1957;
     Rosevear, 1965; Kock, 1969; Viellard, 1974; Koopman, 1975); General—D. Kock, pers. comm., 1982; F. Petter,
     pers. comm., 1982.
Nomenclature follows Meester and Setzer, 1971–77.

TABLE 17.3. The numbers of species of rodents and bats in desert and non-desert countries
Numbers in brackets indicate number of species/km^2 × 10^5

	Sahara Desert		Non-desert		
	Egypt[a]	Libya[b]	Nigeria[c]	East Africa[d]	Europe[e]
Rodentia	31 (3.1)	20 (1.0)	53 (5.8)	101 (5.8)	48 (1.1)
Chiroptera	21 (2.1)	n.d.	70 (7.7)	98 (5.7)	31 (0.7)
Area (km^2 × 10^3)	989	2073	911	1737	c. 4530

Sources: (a) Gaisler *et al.*, 1972 (bats); Osborn and Helmy, 1981 (rodents);
     (b) Ranck, 1968;
     (c) Happold, in press;
     (d) Kingdon, 1971;
     (e) van den Brink, 1967.

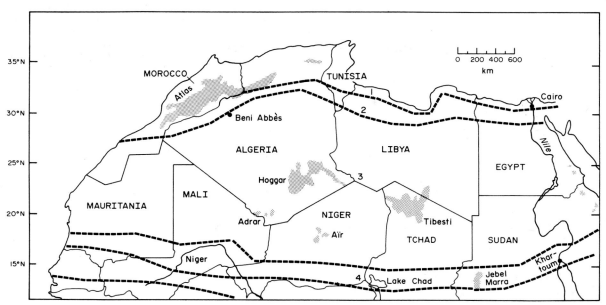

FIG. 17.1. The faunal regions of small mammals in the Sahara. 1 = Mediterranean coastal region; 2 = pre-Saharan semi-desert region; 3 = Saharan desert region; 4 = sub-Saharan or Sahel savanna region. Stippled areas indicate massifs and jebels above 1500 m. See text for details.

mongoose (*Herpestes*), and three cats (*Felis* spp.). The populations of all these predators are small and dispersed, mainly because of the scarcity of their food.

Compared with the rodents, bats are not an important component of the Saharan small mammal fauna. Although nineteen species are listed in Table 17.2, only about five species are true desert bats (Table 17.10); additional species live on the margins of the Sahara where they are associated with mesic conditions.

The richness of the small mammal fauna may be put into perspective by comparison with some non-desert countries (Table 17.3). Although such a comparison is difficult due to differences in area and latitude, species richness is lower than in sub-Saharan Africa, but relatively higher than in Europe.

17.4. PATTERNS OF DISTRIBUTION

The topography and rainfall of the Sahara is extremely varied and has an important effect on the geographical distribution of small mammals. Each species only occurs in regions which it has been able to colonize in the past, and where the present environment (rainfall, soil and rock characteristics, and food availability) is suitable for survival. Analysis of the distribution patterns suggests five faunal regions, each characterized by its own small mammal communities; these regions run in essentially parallel lines from east to west:

1. Mediterranean coastal region: this thin strip along the Mediterranean coast has a mild climate, rainfall up to 400 mm/yr, and comparatively dense vegetation often associated with salt marshes and saline lakes. In Libya, for example (Table 17.4), more species occur here than elsewhere in the country (18 spp. out of a total of 20 spp.), and some (*Gerbillus henleyi*, *Dipodillus simoni*, *Meriones libycus*, *Psammomys obesus*, *Microtus socialis*, *Spalax ehrenbergi*, *Jaculus orientalis* and *Allactaga tetradactyla*) are confined to this region.

TABLE 17.4. Geographical distribution, and number of subspecies, of rodents in Libya (modified from Ranck, 1968, using nomenclature of Meester and Setzer, 1971–77). + = present; (+) = sometimes present; — = absent.

Species	Geographical distribution			Number of subspecies
	Mediterranean Coastal Region	Pre-Saharan Semi-desert	Sahara	
Microtus socialis	+	—	—	1
Gerbillus gerbillus	(+)	+	+	5
Gerbillus pyramidum[1]	+	+	+	8
Gerbillus amoenus	+	+	+	1
Gerbillus campestris	+	+	+	5
Gerbillus henleyi	+	—	—	1
Gerbillus simoni	+	—	—	1
Pachyuromys duprasi	+	+	—	1
Meriones crassus[2]	—	+	+	1
Meriones libycus	+	+	+	5
Psammomys obesus[3]	+	(+)	—	3
Spalax ehrenbergi	+	—	—	1
Mus musculus	+	+	+	1
Acomys cahirinus	—	+	+	1
Eliomys quercinus	+	+	+ (oases)	3
Jaculus jaculus[4]	(+)	+	+	9
Jaculus orientalis	+	—	—	1
Allactaga tetradactyla	+	—	—	1
Hystrix cristata	+	—	—	1
Ctenodactylus gundi	+	+	—	2
Totals	18	13	10	

Some of Ranck's (1968) species are considered as synonyms by Meester and Setzer (1971–77) as follows:

[1] Includes *G. aureus* (3 subspecies) and *G. eatoni* (3 subspecies).

[2] Includes *M. caudatus* (4 subspecies).

[3] Includes *P. vexillaris* (1 subspecies).

[4] Includes *J. deserti* (4 subspecies).

2. Pre-Saharan semi–desert region: this region is situated between the coastal region and the Sahara itself, and is characterized by lower predictable rainfall, sparser and more scattered vegetation, and greater aridity. The number of species falls dramatically compared with the coastal region; in Libya from 23 spp. to 14 spp., and in Egypt from 15 spp. to 11 spp. In Libya (Ranck, 1968) the rodent fauna of the pre-Saharan region is composed of species which occur in the coastal region, Saharan species which reach their northern limit in the pre-Saharan (*Acomys, Meriones crassus*), and species with widespread distributions (*Gerbillus gerbillus, G. pyramidum, G. amoenus, G. campestris, Meriones libycus* and *Jaculus jaculus*).

3. Saharan desert region: this region is the most arid of all the regions; rainfall is low or non-existent and unpredictable, and vegetation is extremely sparse and scattered. Consequently, this region supports the fewest number of species of rodents in any region (11 spp. in Libya, 8 spp. in Egypt). Most of them are widespread species occurring in other regions, but because of their ability to survive in the most arid environments, their geographical ranges are larger than the less adapted and geographically more confined species.

4. Sub-Saharan region: along the southern fringe of the Sahara is the Sahel savanna where an increase in the predictable rainfall results in less aridity and denser vegetation. This situation is similar in some ways to the pre-Saharan region to the north, except that the small mammals are derived from the African savanna fauna (e.g. *Taterillus, Tatera, Arvicanthis, Mastomys, Desmodillicus*).

5. Massif and jebels: several large massifs (Aïr, Hoggar, Tibesti, Adrar) and numerous large jebels provide rocky environments which contrast greatly with the hammadas, ergs and sand plain habitats. Several species are confined to these rocky habitats (*Procavia, Massoutiera, Felovia*), and other species which occur elsewhere may be found in larger numbers (*Acomys* spp., *Gerbillus compestris, Sekeetamys calurus, Vulpes* spp.).

A secondary pattern of distribution occurs from east to west. The distribution of rodents in Egypt indicates how the Nile river has influenced the geographical ranges of some species. Discounting the coastal species, many of which only occur to the west of the Nile, six species occur on both sides of the river, two species only on the west, and four species only on the east. An apparently similar trend, which cannot be substantiated due to inadequate data, is that species richness is highest in the eastern Sahara (Egypt, Libya, northern Sudan) than in the western Sahara (Mauritania, Rio de Oro and western Algeria).

17.5. ENDEMISM

Endemic taxa, whether at the genus, species or subspecies level, are ones which are confined to a discrete geographical region. In the Sahara only four genera and five species are endemic (Tables 17.1 and 17.2), whereas many endemic subspecies, often of doubtful validity, have been described.

Subspecific variation is most likely to occur where populations of the same species are geographically isolated from each other or when populations of a widespread species occur in different habitats at the extremes of its geographical range. Under these situations the total range of the species is composed of a series of discrete populations, each isolated and morphologically distinct from each other. The Sahara desert, with its mosaic of varied, unconnected habitats such as massifs, jebels, sandy plains separated by stony hammadas, and isolated oases, is the perfect situation for the evolution of morphological variations. In Libya, for example (Table 17.4), many species of rodents show morphological variation which are accorded subspecific status. The jerboa, *Jaculus jaculus*, illustrates the difficulties of interpreting morphological variation: Ranck (1968) lists five subspecies for Libya alone, whereas Misonne (1974) does not recognize any subspecies throughout the whole Sahara. Although there are differences of interpretation of what constitutes a subspecies, many Saharan small mammals show morphological differences throughout their geographical range.

In contrast, some species are morphologically very similar throughout their geographical ranges. Species with small geographical ranges, such as *Gerbillus henleyi, Dipodillus simoni* and *Spalax ehrenbergi*, do not show sufficient variation to warrant subspecific distinctions. Perhaps more interesting are species with large and sometimes disjunct ranges which remain remarkably constant in form wherever they occur (e.g. *Acomys cahirinus, Pachuryomys duprasi, Meriones crassus* and all species of Ctenodactylidae). This suggests that these species have not been geographically isolated for sufficient time to form subspecies, or they may have less tendency to genetically alter in isolation. Alternatively, there may be adequate gene flow between populations to suppress any subspecific variation. As might be expected, the widespread Saharan bats (*Rhinopoma hardwickei, Asellia tridens* and *Otonycteris hemprichi*) do not form recognizable subspecies.

17.6. SMALL MAMMALS AND THE DESERT ENVIRONMENT

There are many problems associated with the survival of mammals in the Sahara, especially those which are small and totally dependent on local resources. The most important of these problems concern regulation of body temperature, conservation of water, and finding an adequate supply of food. Small mammals solve these problems by careful habitat selection, physiological and behavioural adaptations, and by having flexible life-histories. Most problems of desert life may be solved in several ways, and hence a number of life-history strategies may be exemplified by sympatric species. One advantage of being small is that a whole community of small mammals require substantially less food and water than, for example, a single addax or oryx; and consequently small mammals can flourish in regions which are unable to support larger species. Some of the small mammals which flourish in the Sahara and the ways in which they solve the problems of desert life will be considered in the following pages.

17.7. HABITAT SELECTION

The varied topography and soil characteristics of the Sahara result in a series of habitats and microhabitats. Habitat selection by species with similar habitat requirements results in the typical communities of rodents associated with each habitat. For example, rocky areas are preferred by *Acomys* spp., *Sekeetamys calurus, Gerbillus campestris, Ctenodactylus* spp., *Massouetiera mzabi, Felovia vae* and *Procavia ruficeps*. Similarly, saline marshes in the coastal region are inhabited only by species which can tolerate salinity such as *Gerbillus andersoni, G. gerbillus, G. campestris, G. amoenus, Psammomys obesus* and *Mus musculus*. The vast sand plains, hammadas and sand dunes are colonized by species, such as *Gerbillus gerbillus, G. pyramidum* and *Jaculus jaculus*, which can tolerate (when necessary) the severe desert conditions. If a habitat is sufficiently varied, several species may occur in close proximity (Table 17.5).

TABLE 17.5. The rodents of Beni Abbès, Algeria, and their habitats. ++ = preferred habitat; + = other habitats
(data from Petter, 1961)

Species	Sand dunes	Sand patches in valleys and bases of jebels	Wadis, rarely flooded	Wadis, regularly flooded	Hammadas with vegetation	Hammadas no vegetation	Rocky valleys in hammadas and jebels
Meriones libycus			++	+	++		
Meriones crassus						++	
Psammomys obesus			+	+			
Pachyuromys duprasi					?+	?+	
Gerbillus pyramidum	+	++			++		
Gerbillus gerbillus	++						
Gerbillus campestris							+
Gerbillus nanus			+	+	+		
Jaculus jaculus	+	+			+	+	
Ctenodactylus vali							++

Many carnivores use massifs and jebels as their domiciles although they frequently forage in the surrounding plains and wadis. Fennec foxes, by contrast, live among sand dunes where they make their burrows. Bats are extremely dependent on caves and crevices in jebels, and trees in oases, for suitable roosting sites during the day.

Few species are found throughout the whole Sahara. The geographical range of each species has been determined by past geographical and climatic events (p. 252). However, within its geographical range, each species is found in a particular habitat or series of habitats which suit its own physiology, behaviour and ecological tolerance. The ability to live in a variety of habitats is probably advantageous in times of drought. Conversely, species which exhibit limited ecological tolerance are probably more prone to local extinction during adverse conditions. Habitat selection, resulting in the association of each species with particular habitats, enables the special physiological and behavioural adaptations of each species to be utilized to the best advantage and with minimal competition from sympatric species.

17.8. TEMPERATURE REGULATION AND WATER CONSERVATION

There are many different ways of regulating body temperature and ensuring that the body obtains adequate water. The methods adopted by desert mammals are related to body size, activity patterns, the availability of water and the ability to eliminate waste products with the minimum loss of water.

There are two major problems for small mammals. Firstly, small mammals gain heat rapidly when exposed to high temperatures. In the Sahara, the daytime air and substrate temperatures increase rapidly after sunrise, and quickly exceed the limits of tolerance during most days of the year; hence small mammals must not expose themselves on the surface during the day. In this respect, small mammals are very different to large mammals, such as the oryx, which allow their body temperature to rise during the day from the normal 37°C to about 42°C. When this happens, water is not lost by evaporation in an attempt to cool the body; instead, the excess 'heat load' is gradually lost by radiation during the evening and early night, and the body temperature falls to 37°C again (Taylor, 1969). Secondly, the scarcity of water necessitates very careful conservation of body fluids. Loss of water in the urine and faeces, and through the lungs, must be kept to a minimum as there may be no way of replacing these losses. Consequently, water

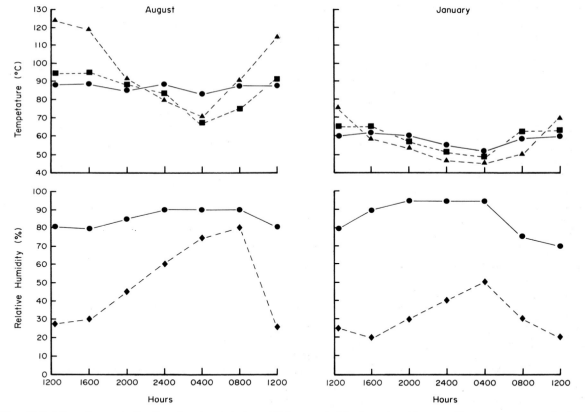

FIG. 17.2. Microclimates in the Egyptian desert during a 24-hour period in August and January. ● = temperature and relative humidity in burrows; ■ = air temperature; ▲ = sand surface temperature; ◆ = relative humidity of air. Burrow temperatures are lower than air and sand surface temperatures during the day, but are higher during the night. All temperatures are higher in August than in January. Relative humidity of burrows remains at 80–90% despite the fluctuating relative humidity of the air at both times of the year. Measurements on burrow and air temperatures near Khartoum (see Delany and Happold, 1979, p. 159) show a similar trend. Source: Yunker and Guirgis (1963).

cannot be used, except in an emergency, as a means of cooling the body by evaporation. Thus, the ability not to become overheated is a very necessary attribute to successful water conservation. Some species are much better at conserving water than others, but there are many ways of balancing the heat and water budgets in small mammals. Five species exemplify the ways in which Saharan small mammals cope with these temperature and water problems; how they do this has an important bearing on their life histories.

Sand rats, *Psammomys obesus* (adult wt 100 g), live in the coastal and pre-Saharan region. Their preferred habitats are saline marshes and wadis where halophytic (salty) plants (family Chenopodiaceae) are abundant. They are also found in gardens and farmlands where moist conditions and green plants are available. They build a network of fairly shallow burrows and, because they are frequently diurnal, they may be seen feeding on halophytic plants and climbing shrubs during the day. This kind of activity is only possible provided sand rats possess special mechanisms for staying cool, or for obtaining water which may be used for evaporative cooling. The key to their success seems to be their ability to eat the succulent leaves of halphytic plants and then eliminate the excess salt in a concentrated urine. Sand rats eat up to their own weight each day of salty plants, obtaining adequate nourishment and water for thermoregulation but, in addition, a lot of salt. The salt is eliminated in a fairly copious yet concentrated urine (up to 25 ml/day). Measurements of their urine show that it is up to four times as salty as sea water although urea concentration is lower than in jerboas (Table 17.6). The concentration of this urine is one of the highest ever recorded in mammals. No measurements have been made on the body temperature of sand rats while foraging during the day, nor on their precise method of thermoregulation. It is unlikely (but has not been tested) that sand rats can survive on dry foods as can *Meriones crassus, Jaculus jaculus* and many other species of Saharan rodents.

Libyan jirds, *Meriones crassus*, are similar in size to sand rats, but are more widely distributed in coastal and pre-Saharan regions and are not confined to saltbush habitats. In contrast to sand rats, Libyan jirds survive on a diet containing very little water and therefore they have a totally different set of problems. They are mostly nocturnal and feed on seeds and other non-halophytic vegetation if available. Their water intake is low and they maintain their water balance by producing small volumes of concentrated urine. Again in contrast to sand rats, the osmotic concentration of the urine is mostly due to metabolic waste products and not to salts (Table 17.6). They have the capacity to vary the volume and concentration of the urine in relation to water availability, and are able to maintain their body weight without drinking provided the food contains a small percentage of water. Furthermore, metabolic (or oxidation) water formed during the metabolism of food may be an important additional supply of water in Libyan jirds (as it is in some North American and Australian rodents), especially during times of drought. These characteristics enable jirds to live in habitats which are much more arid than those required by sand rats. Their nocturnal activity is no doubt related to their very tight water budget which does not allow any surplus water for thermoregulation.

Jerboas, *Jaculus jaculus*, live on sand plains and hammadas where they hop rapidly with their long hindlegs, and forage widely for seeds and grasses. They are strictly nocturnal, and during the day they rest in deep burrows where they are well insulated from daytime air and substrate temperatures. Like jirds, they produce a very concentrated urine (Table 17.6) and probably also make good use of metabolic water. However, they do not appear to be as well adapted to very dry conditions as they lose weight when deprived of water in captivity, and drink more water than *Meriones crassus* when water is available again (Shkolnik, in Schmidt-Nielsen, 1964). In the Sudan, jerboas may supplement their generally dry food by eating bulbs and corms (50–73% water) when these are available (Ghobrial and Hodieb, 1973). They do not hoard food, and therefore do not have a supply of food (and water) which could be used when conditions deteriorate. In many of these characteristics, jerboas are less adapted to really arid conditions than many jirds and gerbils; their success and widespread distribution is due mainly to their ability to minimize water loss and to increase water intake by eating seeds and bulbs.

A rather different method of thermoregulation in herbivorous small mammals is found in rock hyraxes

TABLE 17.6. Characteristics of food, activity and urine in four Saharan small mammals

Species	Adult weight (g)	Volume of water in food	Concentration of salt in food	Activity Pattern*	Volume of urine/day (ml)	Volume of urine/100 g body weight/ day (ml)	Total osmotic concentration of urine (osm/1)	Urea concentration in urine (mM/1)	Electrolyte concentration in urine (mEq/1)
Meriones crassus	80[f]	low	low	N	7–9	9–11[b]	n.d.	n.d.	1793[a]
Psammomys obesus	147[f]	high (80–90%)	high	D	25[a]	17	6.34[a]	2850[a]	1920[a]
Jaculus jaculus	50($\male$) 60($\female$)	low	low	N	2.3[e]	4.6	6.5[a]	4320[a]	1530[a]
Fennecus zerda	1200[d]	high	low	N	32[d]	2,6	2,6[d] (1.4–3.8) 4.02[d] (on restricted water)	1991[d] −2620[a]	n.d.

* N = nocturnal; D = diurnal.
Sources: (a) Schmidt-Nielsen, 1964;
    (b) Kulzer, 1972;
    (c) Daly, 1979;
    (d) Noll-Banholzer, 1978;
    (e) Ghobrial and Nour, 1975 (dry dura seeds only);
    (f) Osborn and Helmy, 1980.
n.d. = no data.

and gundis. These mammals make their domiciles in caves and crevices among rocks where the temperature remains cool. Paradoxically, these domiciles remain too cool for hyraxes and gundis, and so they sunbathe on the rocks in the early morning. Observations of gundis (George, 1974) show that they are completely diurnal; the highest levels of activity are in the first 4 hours after dawn when the air and rock surfaces are warming after the cool night. Sunbathing occurs when the air temperature is 22–29°C and the rocks are at about the same temperature. When the air reaches 31°C, and rock temperatures increase rapidly, gundis retreat into their domiciles. In the evening, gundis may sunbathe on a hot rock while the air temperature is 35–38°C, but not in direct sunlight. By alternating between sunbathing on rocks and retiring into rock crevices, gundis are able to take advantage of radiated and convected heat when available, prevent overheating during the day, and reduce heat loss at night. Measurements of body temperatures in relation to these behavioural activities are not available, but it seems likely gundis have poorly developed physiological mechanisms for temperature regulation and consequently have developed a highly refined behavioural mechanism. Gundis feed on herbs and leaves, but their water balance has not been investigated. As they feed in the early morning when the relative humidity is comparatively high and when dew may be found at some times of the year, they may not be exposed to severe water stress. No information is available on Saharan rock hyraxes, but their behaviour patterns in sub-Saharan Africa suggest that in the Sahara they are rather similar to gundis.

The final example is a desert carnivore, even though much less is known about these animals than for rodents. The only carnivore studied in this respect is the fennec fox. These foxes live in sand dune habitats where, at night, they hunt for small mammals, lizards and insects, and perhaps eat some vegetable foods. Fennec foxes occur in very arid regions and they are probably better adapted than any other carnivore for desert life. The relatively high water content of their diet means that water conservation is not as crucial as in rodents. This is reflected in the lower total osmotic concentration and lower urea concentration of their urine (Table 17.4); urine volume and concentration may be altered depending on diet and water availability. The body temperature of fennecs normally shows daily fluctuations between 37.4°C and 38.7°C, but under heat stress, it may rise to nearly 40°C. During heat stress, fennecs pant and lose 3–4 times more water than normal. Because of their water-rich diet, this method of thermoregulation is possible for fennecs whereas it is not possible for the majority of Saharan rodents.

These Saharan mammals exemplify four methods of maintaining a constant body temperature and remaining in water balance. The differences between them are primarily related to activity patterns, water availability and ability to produce concentrated urine. As a stable body temperature is of paramount importance, all other features are related to this. The four methods may be summarized as follows:

1. When the food is dry, water loss must be kept to a minimum by producing a concentrated urine and minimizing evaporative water loss during respiration; nocturnal activity is essential to avoid overheating (*Meriones crassus* and *Jaculus jaculus*).
2. If halophytic plants can be eaten, due to the ability to excrete the excess salt ingested in the food, diurnal activity is possible and water may be used for thermoregulation (*Psammomys obesus*).
3. Carnivores obtain more water in their food than most rodents, but more water is required to maintain the water balance as urine is not particularly concentrated and panting occurs during heat stress; nocturnal activity is essential except during the coolest part of the year (*Fennecus*).
4. Temperature regulation is attained by alternating between sunbathing to gain heat, and resting in domiciles to prevent overheating and heat loss at night; maintenance of a constant body temperature in this way minimizes water requirements for thermoregulation (gundis and rock hyrax).

17.9. ACTIVITY PATTERNS

The majority of small mammals are nocturnal, for the reasons given above. Many species emerge from their domiciles soon after sunset while the air and substrate temperatures are cooling but have not reached their lowest levels. It is highly likely that Saharan mammals show peaks of activity at different times of the night (as do many other species of non-Saharan small mammals) and vary their times of nocturnal activity according to seasonal changes in temperature and food availability. Limited information on Sudanese mammals (D. C. D. Happold, unpublished) suggests that some species are less active in the winter months when air temperatures at night are less than 10°C. In all these instances, activity patterns are closely related to the central problem of temperature regulation; this is usually thought to mean the prevention of overheating, but the problem of keeping warm may be of equal importance to some species during the cold winter nights, especially if food resources are scarce.

Sand rats and gundis do not fit this general pattern. Sand rats are frequently diurnal but probably gain some relief from radiated and convected heat amongst the shrubs, salt bushes and grass clumps where they live. Even so, their heat load must be greater than for nocturnal species. Similarly gundis are diurnal, but they shelter in rock crevices during the hottest times of the day. Diurnal activity in sand rats and gundis is facilitated by their special methods of temperature regulation and water conservation (pp. 262–263).

17.10. BURROWS AND CREVICES

Burrows are excavated by many species of rodents around the bases of bushes, on sand plains and hammadas, and in wadis. Although the sand surface is very hot and dry during the day, the temperature decreases and relative humidity increases with increasing depth below the surface. At about one metre, both temperature and relative humidity remain almost constant throughout the day and night. Measurements in burrows in the Egyptian desert at different times of the year (Fig. 17.2) show that burrow temperatures are lower than air or surface temperatures during the day, but higher at night. The relative humidity remains higher in burrows at all times compared with the air above ground. By varying the

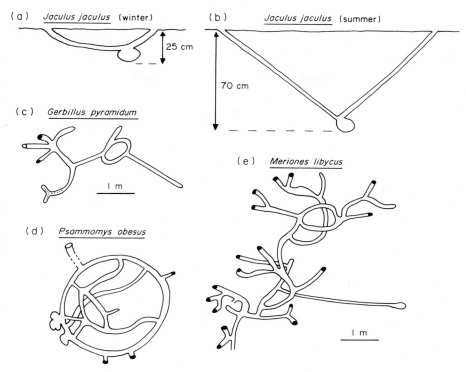

FIG. 17.3. Burrows of some Saharan small mammals. (a) *Jaculus jaculus* during the winter and (b) during the summer at Khartoum; (c) *Gerbillus pyramidum*, (d) *Psammomys obesus* and (e) *Meriones libycus* at Benni Abbès, Algeria. Sources: (a) and (b) Ghobrial and Hobieb (1973); (c), (d) and (e) Petter (1961).

depth of the burrow small mammals can select the optimum burrow temperature, and there is some evidence that deeper, and therefore cooler, burrows are dug in the summer when the surface temperature is especially hot (Ghobrial and Hodieb, 1973). When in their burrows, small mammals experience a constant environment and avoid the extremes of air and surface temperatures. Similarly, the higher constant relative humidity of the burrows reduces evaporative water loss. At night, burrow temperatures are warmer than outside and in the winter provide some protection from the cold winter air (see also p. 261). Many species of rodents plug the entrance(s) of their burrows with sand after entering and leaving; this probably results in an increase in the relative humidity of the burrow and may also conceal the burrow from other animals. The burrows of rodents (Fig. 17.3) vary greatly in size, complexity, number of entrances, depth, and number of individuals in each burrow (Petter, 1961). Simple burrows consisting of one or two shafts and a nestchamber are built by most *Gerbillus* species and *Jaculus jaculus*; these are usually used by only one or two individuals. In contrast, large complex burrows with many shafts and entrances are built by *Meriones crassus*, *M. libycus*, *Psammomys obesus* and *Pachyuromys duprasi*, and may be used by many individuals.

Species which live in rocky habitats (e.g. *Acomys* spp., *Gerbillus campestris*, *Sekeetamys calurus*, *Vulpes* spp. and gundis) live in crevices in the rocks and under boulders. No climatic measurements have been made in these domiciles but, undoubtedly, they provide a cool non–fluctuating microclimate similar to that of burrows. All species obtain excellent insulation from the outside climatic conditions in their burrows and rock crevices.

17.11. FOOD

A wide selection of foods is available for Saharan herbivores. However, the abundance of food is extremely variable and unpredictable and is largely determined by the frequency and amount of rainfall. The coastal, pre-Saharan region and Sahel savanna have a higher and more predictable rainfall and a higher primary production than the central Sahara which may have no rainfall for several consecutive years and an extremely low primary production. Localities within the same rainfall region may vary in their primary production and available water; rocky hillsides and sand plains have fewer trees, shrubs and grasses than do wadis, depressions and sheltered valleys. Food availability for herbivores therefore depends on locality, topography and rainfall, and is very low in the central Sahara compared to the regions to the north and south.

Many species of small rodents are grass-, stem- and seed-eaters. After rain, annual herbs and grasses germinate, grow and fruit quickly, providing a large source of energy for small herbivores. After these plants have died, their dead stems and seeds are blown into the bases of shrubs where they form 'food stores'. However, such stems and seeds have a low percentage of water although this may be adequate for species which are efficient at water conservation. Seeds are generally good sources of carbohydrate and fat, with the additional benefit that both provide metabolic water. It is not surprising that many small rodents are seed-eaters, except when green vegetation is available. Some herbivores eat specialized foods which are rich in water; *Psammomys obesus* eats salt bushes (p. 262) and *Jaculus jaculus* sometimes eats bulbs (p. 262). *Sekeetamys calurus* supplements its herbivorous diet with insects, and *Acomys* spp. with insects and snails. The species which eat water-rich foods do not have really efficient water-conservation abilities. Their survival in the Sahara is dependent on these water-rich foods, and they are more susceptible to very dry conditions than are other species.

Small mammals which eat animal foods are also indirectly dependent on rainfall and plant production for their food. Insects, lizards and small mammals are the main food of insectivores and small carnivores and, like plant foods, ar subject to fluctuations in abundance. Animal foods are less abundant than plant foods, and can only support a much lower biomass of predators.

Although certain foods are often preferred, many species are opportunistic and at times may be forced to eat less preferred foods if only these are available. Recent studies in other parts of the world on food selection and foraging behaviour have shown that in most mammalian communities, each species selects different types of food or different proportions of the same food. In this way, several species can feed in the same area without any (or undue) competition for the same food resource. This partitioning of food resources also enables more species and individuals to co-exist and results in a better utilization of the available plant biomass.

If Saharan small mammals are classified according to their feeding niches, five niche types may be distinguished (Table 17.7). The small granivores are the most numerous and successful in terms of genera and species; herbivores are much less successful probably because moist vegetation is uncommon except after rain. The insect-eaters and carnivores contain a surprisingly large number of genera which suggests that, although the population numbers of these species are low, the 'predator niche' is an important and successful one in the Sahara. If more was known about the feeding habits of all these mammals, a much finer distinction between species within a single feeding niche might be apparent. If so, this would indicate very precise food selection, less competition for food, and the possibility of more co-existing species. However, in times of food shortage, any available food may be acceptable and food competition is likely to be intensified.

One of the principal characteristics of the Sahara is the disjunct abundance of food in time and space. Although small mammals can select their preferred microenvironment, they have little control over food availability. Long periods of reduced food availability are frequent in the central Sahara, but less frequent around the margins. Nevertheless, all species may be subject to food shortages for several months, or

TABLE 17.7. Food niches of the genera of Saharan small mammals

	Granivores (seed-eaters)	Herbivores (grass; root-eaters)	Omnivores (seeds, grasses, insects, roots, animal foods)	Insectivores	Carnivores
INSECTIVORA	—	—	—	*Crocidura* *Suncus* *Elephantulus* *Erinaceus* *Paraechinus* *Hemiechinus*	
LAGOMORPHA	—	*Lepus*	—	—	—
RODENTIA					
Sciuridae	—	—	*Xerus*	—	—
Cricetidae	*Dipodillus (?)* *Desmodilliscus (?)* *Gerbillus* *Pachyuromys* *Tatera* *Taterillus*	*Meriones* *Psammomys*	*Sekeetamys*	—	—
Muridae	—	—	*Acomys*	—	—
Muscardinidae	*Eliomys Graphiurus*	—	—	—	—
Dipodidae	*Jaculus*	*Allactaga*	—	—	—
Spalacidae	—	*Spalax*	—	—	—
Hystricidae	—	*Hystrix*	—	—	—
Ctenodactylidae	—	*Ctenodactylus* *Felovia* *Massoutiera*	—	—	—
CARNIVORA					
Canidae	—	—	—	—	*Fennecus* *Vulpes*
Mustelidae	—	—	—	—	*Ictonyx* *Poecilictus*
Viverridae	—	—	—	—	*Genetta* *Herpestes*
Felidae	—	—	—	—	*Felis*
HYRACOIDEA	—	*Procavia*	—	—	—

years, at a time. No food at all results in population extinction, but any species, community or individual which can find food, or reduce its daily food requirements, is more likely to survive until conditions improve again. The principal ways of doing this are by hoarding food, utilizing body fat reserves, and aestivation.

Storing or hoarding food when conditions are good appears an ideal way of overcoming periods of shortage, provided the food can be located and eaten by the individual who hoarded the food. This option is only available to granivores and some omnivores, but not to mammalian predators and species requiring fresh moist vegetation. Somewhat surprisingly, hoarding is known only in several *Gerbillus* spp. and *Meriones crassus* and perhaps *Allactaga*. The other granivores and omnivores are not known to hoard and evidently rely on the blown 'food stores' and standing vegetation.

In times of abundant food, some species of gerbils (and no doubt other species as well) increase their body fat. At these times, too, much energy is used for reproduction and lactation. In any individual or species, a compromise is necessary between laying down fat (to assist that individual to survive a period of food shortage) and reproduction. In both instances, available food is used to perpetuate the population. Deposits of body fat are known in *Gerbillus pyramidum* and *Jaculus jaculus* (D. C. D. Happold, unpublished), and deposits of fat occur in the tails of *Pachyuromys duprasi* (Petter, 1961) and around the lower back and abdomen of *Rhinopoma* spp. (D. Kock, J. E. Hill, pers. comm.). There are no records of fat deposits in other species, nor on the importance of fat as an energy reserve in other Saharan small mammals.

A third way of reducing energy expenditure (and water loss) is by aestivation, when the body temperature, energy and oxygen consumption are lowered, and metabolic rate declines. Aestivation, or torpor, is recorded in some North American desert rodents during the coolest months of the year (Schmidt–Nielsen, 1964), but in the Sahara only *Gerbillus pyramidum* and *Jaculus jaculus* are known to reduce their activity (p. 264), huddle and become torpid when it is cool. The ability to aestivate is clearly advantageous in the winter when food is becoming less abundant and water (except dew) is scarce. Aestivation also occurs in some species of bats (p. 273).

17.12. REPRODUCTION AND POPULATION DYNAMICS

The characteristics of reproduction (fecundity, number of young in a litter, and percentage of females actively breeding) are important determinants of population numbers and population density.

Reproduction in desert mammals is usually closely related to rainfall and the length of the reproductive season may be adjusted according to the timing of the rainfall and subsequent plant growth. Reproduction ceases during very adverse environmental conditions. Species vary in their responses to the environmental cues which initiate and maintain reproduction; consequently, different species in a community breed at different times.

Although reproductive studies are a very topical aspect of mammalian research, there is little comprehensive data for Saharan small mammals. However, a few studies illustrate the principles of desert reproduction. As climatic regimes vary greatly throughout the Sahara, there is no generalized reproductive season. Figure 17.4 shows when pregnancies have been observed in *Gerbillus andersoni*, *G. pyramidum*, *G. gerbillus* and *Jaculus jaculus*. For these species, at the localities where the observations were made, there are well-defined periods of reproduction during and just after the rainy season, whether this is in the winter (Egypt) or summer (Sudan). In contrast, *Psammomys obesus*, which eats halophytic plants and is less dependent on rainfall and a flush of annual plants to initiate reproduction, can breed quite prolifically for many months of the year (Daly, 1979). Unfortunately, there are inadequate data on reproductive patterns of several sympatric species in a community to illustrate species variations in this respect; however, at Khartoum, *Jaculus jaculus* appeared to have a more confined reproductive season than *Gerbillus pyramidum* (Happold, 1967, 1968).

Population studies on small mammals in the Sahara are hampered by the scarcity of individuals except in optimal habitats. As population increases are mainly due to recruitment during the reproductive season, and as these seasons are (for most species) determined by the irregular and unpredictable rainfall, population increases are neither regular nor predictable. No generalizations can be made about the timing and magnitude of population increases as they are dependent on local conditions and the responses of each species to these conditions. The population numbers of *Gerbillus pyramidum* at Khartoum, as indicated by the number of individuals caught (Happold, 1968), fluctuated during the course of the year, with maximum numbers in October and November (at the end of the rains). Population numbers probably oscillate greatly between high numbers and densities after good prolonged rainstorms (perhaps several years apart) and very low numbers and densities during droughts. In localities with no rainfall for several years, populations fluctuate more widely and erratically than in the pre- and sub-Saharan regions. Detailed population studies on single species, and on a community of species, are required before the population dynamics of Saharan small mammals are properly understood.

The population densities of small mammals are generally very low, and on bare sand plains and

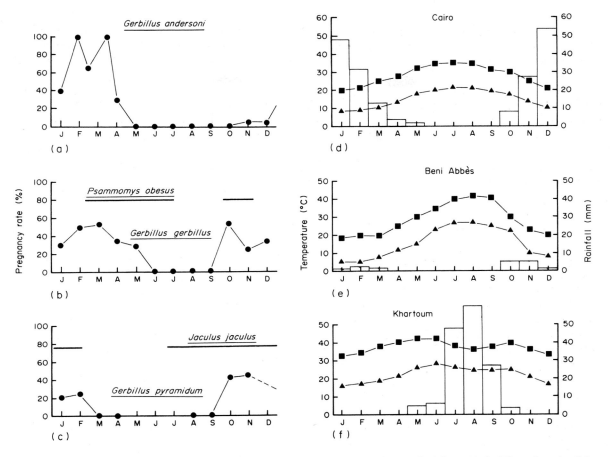

FIG. 17.4. The relationship between reproductive activity and climate at three localities in the Sahara. (a) *Gerbillus andersoni* at Cairo, Egypt) (b) *Psammomys obesus* (—) and *Gerbillus gerbillus* (●) at Beni Abbès, Algeria) (c) *Jaculus jaculus* (—) and *Gerbillus pyramidum* (●) at Khartoum, Sudan.

$$\text{Pregnancy rate} = \left(\frac{\text{number of pregnant females}}{\text{number of adult females}} \right) \times 100;$$

● = monthly pregnancy rate (*Gerbillus andersoni*, *G. gerbillus* and *G. pyramidum* only); ■ = mean monthly maximum temperature; ▲ = mean monthly minimum temperature; histogram = mean monthly rainfall. See text for details. Sources: (a) Wassif and Soliman (1980); (b) Amirat *et al.* (1977); (c) Happold (1967, 1968); (d) Griffiths (1972); (e) Petter (1961); (f) Griffiths (1972).

hammadas, especially in drought years, small mammals may be non-existent. However, small isolated populations often occur in clumps of bushes or grasses well separated from adjacent populations. In a study area near Beni Abbès, the average density of *Meriones crassus* was one male and one female per km^2 (Daly and Daly, 1975); in addition, there were four other, less abundant, species. At Khartoum, *Gerbillus pyramidum* and *G. watersi* were rarely, if ever, found in the open sand plains as all populations centred around small clumps of *Acacia* and *Capparis* bushes (D. C. D. Happold, unpublished). At these centres, population numbers were relatively high although in the habitat as a whole, numbers were extremely low.

The rapid development of plants and widespread abundance of food after rainfall allows the populations of many small mammals to expand their ranges into nearby habitats. Expansion, followed by contraction

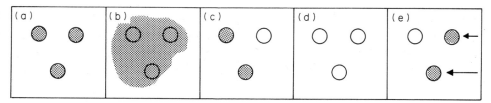

FIG. 17.5. Sequential patterns of distribution of a hypothetical species during climatic change in the Sahara. (a) Normal conditions: distribution in optimum habitats only. (b) Heavy rainfall and increased primary production: expansion into surrounding habitats. (c) Drought: contraction into optimum habitats with local extinction in one optimum habitat. (d) Prolonged drought: local extinction in all habitats. (e) Normal conditions resumed: immigration from outside regions into two of the optimum habitats. Circles—areas of optimum habitat; stippled—area of distribution; arrows—immigration from other regions.

as conditions deteriorate again, appears to be a normal feature of the populations of Saharan small mammals (Fig. 17.5). This is a similar process, although on a smaller scale, to the changes in geographical distribution that have occurred as arid conditions have expanded and contracted in the past.

The ability to travel from one optimum habitat to another is probably advantageous during both normal and drought conditions. Studies on five species of rodents near Beni Abbès showed that during a 6-month period, some individuals could be found as much as 4500 m from their original sites of capture (Table 17.8). Males normally moved further than females. When individuals of *Meriones crassus* were transported 350–3800 m (Petter 1968), most returned to their original site of capture within a single night. These studies suggest that some species have a detailed knowledge of extensive areas which enable them to move from one optimum patch to another.

In *Meriones libycus, Gerbillus pyramidum* and *G. nanus*, females occupy relatively small non-overlapping home ranges, whereas males have large overlapping ranges, especially in the reproductive season. Consequently, females are spaced out in patches of apparently good habitat, and the males move from one patch to another. The result is that the population density is low and dispersed.

Differences in food requirements and mobility are associated with differences in social organization. It is obviously advantageous for an individual to defend a patch of habitat which contains food. *Psammomys obesus* exemplifies a form of territorial system in which the saltbush patch where an individual feeds and has its burrow is defended; aggressive interactions between individuals are not uncommon, and home ranges are small (Daly, 1979). *Meriones crassus* and *M. libycus* feed more widely and on a greater variety of foods, and show less intraspecific aggression. These studies suggest that there are numerous correlations between population structure, social organization, movements and food habits in the few Saharan

TABLE 17.8. Maximum distances between capture sites for six species of rodents near Beni Abbès between November and April (after Daley and Daly, 1975)

Species		Number of individuals recaptured	Maximum distance (m)
Meriones libycus	♀	13	2280
	♂	16	2420
Meriones crassus	♀	9	960
	♂	8	1870
Gerbillus pyramidum	♀	4	910
	♂	7	4500
Gerbillus gerbillus	♀	1	1250
	♂	2	0
Psammomys obesus	♀	1	470
	♂	1	476
Jaculus jaculus	♀	2	1020
	♂	3	2420

mammals which have been studied. The general impression from these and other studies is that low population numbers are primarily a result of low and fluctuating food availability, and that the other characteristics of social organization and behaviour have been determined by these fluctuations in food resources.

17.13. RESOURCE PARTITIONING

When several species of mammals live in the same habitat, they tend to differ from each other in a whole range of characteristics. This widespread principle is known as 'resource partitioning'. Resources such as food and domiciles are utilized in different ways and at different times by each of the species in the community, so that competition for a particular resource is minimized. Some form of resource partitioning probably occurs in every mammalian community. The ecological and behavioural characteristics of any species may vary according to the numbers and characteristics of the species with which it co-exists.

TABLE 17.9. Resource partitioning in two species of Saharan rodents at Beni-Abbès, Algeria (data from Daly, 1979)

	Libyan jird (*Meriones libycus*)	Fat sand rat (*Psammomys obesus*)
Habitat	Hammadas and wadis	Sandy wadis, gardens and farmlands in oases
Home range	♀ large, exclusive, no overlap	♀ small, exclusive, no overlap
	♂ very large, not exclusive, overlap	♂ large, not exclusive, overlap
Aggressiveness	Low (related to dispersed social organization)	Very high (helps protect special feeding areas)
Activity	Mostly nocturnal	Mostly diurnal
Burrow occupancy	♂♂ and ♀♀ separate	♂♂ and ♀♀ separate
Home range	♀ 1.5–10 acres	0.01 acre
Diet	Seeds, grasses	Saltbush (Chenopodiaceae spp.). Eats about 100% body weight of leaves/day
Kidney function	Small volumes of concentrated, non-salty urine	Large volumes of salty urine
Foraging time	Food eaten as found; not more than 1 hr/day	Food collected and taken to burrow (15 min/day); eating time about 4 hr/day
Population density	Low; individual home ranges scattered in suitable habitat	High in suitable saltbush habitat

The only documented example for Saharan mammals is for Libyan jirds and sand rats (Table 17.9). These two co-existing species show many differences in habitat preferences, social organization, activity times and diet, so it is unlikely that there is much overlap or competition between them. Many morphological differences between species, such as size, length of limbs, type and size of teeth and characteristics of the kidneys, suggest niche differences. Many other examples may be cited, such as general habitat selection (p. 260), different activity times (p. 264), food selection (p. 266), or even differences in times of reproduction (p. 268), all of which result in each co-existing species occupying a niche which is not *exactly* the same as the niche of any other species. Resource partitioning may be particularly important in desert environments so that a species may exploit a particular resource solely to its own advantages; conversely, if a resource becomes scarce, it may be more advantageous to be a generalist. Probably both situations occur among Saharan small mammals; *Psammomys obesus* (p. 262) is a specialist with a narrow niche whereas *Gerbillus gerbillus* and *G. pyramidum* are generalists and more adaptable.

17.14. BATS

Bats are a very successful order of mammals, and in many tropical non-desert habitats they contribute the greatest number of species to the total mammalian fauna. Bats also occur in deserts, but they are generally less successful than the rodents or carnivores. In some deserts, such as those of North America

TABLE 17.10. Bats of the Sahara

(a) Species which occur in suitable habitats in the most arid regions of the Sahara
      Rhinopoma hardwickei
      Asellia tridens
      Pipistrellus deserti
      Otonycteris hemprichi
      Tadarida aegyptiaca

(b) Species which occur around the edges of the Sahara, make seasonal movements into the desert when conditions are suitable, and
      are associated with mesic habitats in the desert (e.g. the Nile Valley)
      Eidolon helvum
      Rousettus aegyptiaca
      Rhinopoma microphyllum
      Taphozous nudiventris
      Taphozous perforatus
      Nycteris thebiaca
      Nycteris hispida
      Lavia frons (?)
      Rhinolophus clivosus
      Hipposideros caffer
      Nycticeius schlieffeni
      Plecotus austriacus
      Pipistrellus kuhlii
      Pipistrellus rueppelli
      Scotophilus sp.*
      Tadarida pumila

   * *Scotophilus dingani* (=*nigrita*) or *S. leucogaster*.

where bats have been well studied, certain cave-dwelling species are extremely numerous but this situation does seem to be true for the Sahara. The presence of bats is dependent on suitable roosting sites, such as caves, rock crevices, trees or bushes, and an adequate supply of flying insects. Each species, as for the other mammals already considered, has a very precise resource requirement and many of these resources are very localized (or absent) in desert environments. The success of some species is due to the general characteristics of bats, such as flight, migration, nocturnal activity, and the ability to lower the body temperature if necessary, and not due to any special characteristics essential for living in deserts.

   Saharan bats are poorly known compared with the rodents, and there is virtually no information on their biology and ecology. They may be divided into two groups (Table 17.10): firstly, those which may occur in the most arid, central part of the Sahara; of these *Rhinopoma hardwickei* and *Otonycteris hemprichi* are perhaps the best adapted for arid habitats and are found only in desert environments. The second group comprises species which are known from semi-arid regions on the margins of the Sahara and have been recorded from the Sahel savanna, the Nile valley and parts of northern Africa. All these bats are insect-eating species, except for two fruit-eating bats (*Eidolon helvum* and *Rousettus aegyptiacus*) which migrate into semi-arid regions when fruits are available. In some habitats in the Nile valley, the planting of fruiting trees has enabled some populations to become resident.

   The aerial habits of bats enable them to search for food over larger areas than is possible for any other small mammals. Even so, their presence is dependent on aerial insects which, in turn, are dependent on a minimum amount of water and vegetation; consequently, some regions of the central Sahara are unsuitable for bats at any time. Most records of bats are from rocky massifs, wadis with scattered trees, oases, houses and near permanent water holes. These habitats provide domiciles where bats may roost, and from where they can hunt in the surrounding desert.

   The ability to migrate to other locations when resources are scarce is a method of desert survival unavailable to other Saharan small mammals. Bats can therefore utilize the food resources in habitats where they are unable to maintain permanent populations. Probably many Saharan bats undertake extensive and migratory movements, although there are few definite records. Colonies of *Rhinopoma* spp. and *Asellia tridens* which roosted in a cave near Khartoum were found there only in the winter, 3 to 6 months after the end of the rains (D. C. D. Happold, unpublished). At this time, they fed on insects near the Nile river. During and after the summer rains, they vacated the cave, and it is presumed they moved further away from the Nile into the semi-arid desert where there is an abundance of insects after the local rains.

The nocturnal activity of insect-eating bats minimizes the chances of excessive increases in body temperature. Very small mammals, such as bats, are particularly vulnerable to large fluctuations in environmental temperature. Bats evade the high daytime temperatures by selecting roosting sites (just as rodents retire into burrows or rock crevices), and may eliminate excess body heat by gentle flapping of the large wing membranes. Perhaps of more importance to bats are the cool temperatures of the Saharan winter. Very small mammals lose heat rapidly in the cold, and it is energetically expensive to maintain a constant body temperature. Rodents have a similar, though less severe problem, but as their food is relatively plentiful, they can eat more to obtain adequate energy. In contrast, bats often experience a shortage of food in the winter, a time when more energy is required for the maintenance of a constant normal body temperature. Bats overcome this problem by being heterothermic and allowing their body temperature to fall so it is slightly above the air temperature; in this way, energy is not expended or needed, in trying to maintain a body temperature of about 37°C when the air temperature may be as cold as 5–10°C. Bats vary in their ability to let their body temperature fall, and in their ability to warm up again. When the body temperature falls, all the Saharan species in which this has been recorded (such as *Rhinopoma* spp., *Tadarida* spp. and *Asellia tridens*) become lethargic and rather inactive, but they have the ability to rewarm themselves (by expending energy) even though the air temperature may still be cool (Kulzer, 1965). There are limits to this method of survival, but it is obviously a very successful strategy for small insectivorous mammals in a cool climate.

The presence of any species of bats in the Sahara is therefore a function of body-temperature characteristics, the availability of food and domiciles and the ability to migrate. Only a few species (Table 17.10a) have the necessary combination of characteristics to survive in the really arid regions, although many others are able to live on the margins of the desert.

17.15. EPILOGUE

There is no doubt that small mammals are a most successful group of desert animals, even though most species are rarely seen. Despite their success, population numbers are necessarily small because of the low primary production of the desert ecosystem. However, their ability to reproduce quickly whenever there are local temporary flushes of new resources, and to evade the usually extreme conditions, enables many species of small mammals to survive and flourish in this seemingly hostile environment.

Most studies on Saharan small mammals have been on their geographical distribution, taxonomy and physiology. Much less is known about their ecology in the natural environment. Detailed information on food requirements, foraging strategies, diversity patterns, habitat selection and seasonal changes is still required for many species and communities to give a more complete 'natural history' of these fascinating animals.

Small mammals in all deserts of the world tend to be rather similar in their range of physiological and morphological adaptations and community structure (Mares, 1980), regardless of phylogeny and taxonomic relationships. Thus, desert mammals exhibit parallel and convergent evolution resulting from selection pressures imposed by the severe environment in which they live. Small mammals of the Sahara and Kalahari deserts illustrate many of these evolutionary similarities (Delany and Happold, 1979).

However, there is still a lot to learn about Saharan and other African desert mammals; one can surmise that many of the as yet unknown characteristics of these small mammals will show even more clearly the general ecological similarities of small mammals in different deserts of the world.

ACKNOWLEDGEMENTS

My grateful thanks are due to Dr D. Kock (Senckenberg Museum, Frankfurt) and Dr F. Petter (Natural History Museum, Paris) for their comments and corrections on Table 17.2; to my wife, Dr Meredith Happold, and to C. R. Dickman for their constructive comments on the drafts of this chapter; and to Doreen Kjeldsen and Wendy Lees for their willing secretarial assistance.

REFERENCES

AMIRAT, Z., KHAMMAR, F. and BRUDIEUX, R. (1977) Variations saisonnieres comparées de l'activité sexuelle (données pondérales) chez deux espèces de Rongeurs (*Psammomys obesus* et *Gerbillus gerbillus*) du Sahara occidental algerien. *Mammalia*, **41**, 341–356.

BAUER, K. (1963) Ergebnisse der zoologischen Nubien-Expedition 1962. XIX. Saugetiere. *Ann. naturhistor. Mus. Wien*, **66**, 495–506.

DALY, M. (1979) Of Libyan jirds and fat sand rats. *Nat. Hist.* **88**, 64–71.

DALY, M. and DALY, S. (1975) Socio-ecology of Saharan gerbils, especially *Meriones libycus*. *Mammalia*, **39**, 289–311.

DEKEYSER, P. L. (1950) Contribution à l'étude de l'Aïr: Mammifères. *Mem. I.F.A.N.* **10**, 389–425.

DEKEYSER, P. L. and VILLIERS, A. (1956) Contribution à l'étude du peuplement de la Mauretanie. Notations écologiques et biogeographiques sur la faune de l'Adrar. *Mem. I.F.A.N.* **44**, 1–222.

DELANY, M. J. and HAPPOLD, D. C. D. (1979) *Ecology of African Mammals*. Longman, London. 434 pp.

FAIRON, J. (1980) Deux nouvelles espèces de Cheiropteres pour la faune du massif de l'Aïr (Niger): *Otonycteris hemprichi* Peters 1859 et *Pipistrellus nanus* (Peters 1852). *Bull. Inst. r. Sci. nat. Belg.* **52** (17), 1–7.

GAISLER, J., MADKOUR, G. and PELIKAN, J. (1972) On the bats (Chiroptera) of Egypt. *Acta Sci. Nat. Brno*, **6** (8), 1–40.

GEORGE, W. (1974) Notes on the ecology of gundis (Ctenodactylidae). *Symp. zool. Soc. Lond.* **34**, 143–160.

GHOBRIAL, L. I. and HODIEB, A. S. K. (1973) Climate and seasonal variations in the breeding of the desert jerboa, *Jaculus jaculus*, in the Sudan. *J. Reprod. Fert.*, Suppl. **19**, 221–233.

GHOBRIAL, L. I. and NOUR, T. A. (1975) The physiological adaptations of desert rodents. In *Rodents in Desert Environments*, edited by I. PRAKASH and P. H. GHOSH, pp. 413–444. Junk, The Hague.

GRIFFITHS, J. F. (1972) (Ed.) *Climates of Africa*. (*World Survey of Climatology*, Vol. 10.) Elsevier, Amsterdam. 604 pp.

HAPPOLD, D. C. D. (1967a) Biology of the jerboa, *Jaculus jaculus butleri* (Rodentia, Dipodidae), in the Sudan. *J. Zool.* (Lond.) **151**, 257–275.

HAPPOLD, D. C. D. (1967b) The natural history of Khartoum Province. III. Mammals. *Sudan Notes Rec.* **48**, 111–132.

HAPPOLD, D. C. D. (1967c) Additional information on the mammalian fauna of the Sudan. *Mammalia*, **31**, 605–610.

HAPPOLD, D. C. D. (1968) Observations of *Gerbillus pyramidum* (Gerbillinae, Rodentia) at Khartoum, Sudan. *Mammalia*, **32**, 43–53.

HAPPOLD, D. C. D. (in press) *Mammals of Nigeria*. Oxford University Press.

HEIM DE BALSAC, H. (1936) *Biogeographie des mammifères et des oiseaux de l'Afrique du Nord*. *Bull. Biol. Franc Belg.* **21** (Suppl.), 1–446.

HEIM DE BALSAC, H. (1968) Les Soricidae dans le milieu désertique saharien. *Bonn. zool. Beitr.* **3/4**, 181–188.

HEMMER, H., GRUBB, P. and GROVES, C. P. (1976) Notes on the sand cat, *Felis margarita* Loche, 1858. *Z. Saugetierk.* **41**, 286–303.

HOOGSTRAAL, H. (1962) A brief review of the contemporary land mammals of Egypt (including Sinai). 1. Insectivora and Chiroptera. *J. Egypt Publ. Hlth. Assn.* **38**, 143–162.

HOOGSTRAAL, H. (1963) A brief review of the contemporary land mammals of Egypt (including Sinai). 2. Lagomorpha and Rodentia. *J. Egypt Publ. Hlth. Assn.* **38**, 1–35.

HOOGSTRAAL, H. (1964) A review of the contemporary land mammals of Egypt (including Sinai). 3. Carnivora, Hyrracoidea, Perissodactyla and Artiodactyla. *J. Egypt Publ. Hlth. Assn.* **39** (4), 205–239.

HUFNAGL, E. (1972) *Libyan Mammals*. Oleander Press.

JULLIEN, R. and PETTER, F. (1970) La faune du gisement d'Akjoujt (Mauretanie). *Bull. Mus. natl. Hist. nat.* (Paris), **41** (5), 1290–1291.

KINGDON, J. (1971) *East African Mammals. An Atlas of Evolution in Africa*. I. Academic Press, London. 446 pp.

KOCK, D. (1969) Die fledermaus-fauna des Sudan. *Abh. senckenberg. naturforsch Ges.* **521**, 1–238.

KOCK, D. (1978) Vergleichende untersuchung einiger saugetiere im sudlichen Niger. *Senckenbergiana biol.* **58**, 113–136.

KOCK, D. (1980) Distribution of hedgehogs in Tunisia corrected. *Afr. small Mamm. Newsletter*, **5**, 1–6.

KOOPMAN, K. (1975) Bats of the Sudan. *Bull. Amer. Mus. nat. Hist.* **154**, 353–444.

KULZER, E. (1965) Temperaturregulationen bei fledermausen (Chiroptera) aus verschiedenen Klimazonen. *Z. vergl. Physiol.* **50**, 1–34.

KULZER, E. (1972) Temperaturregulation und Wasserhaushalt der Sandratte *Meriones crassus* Sund., 1842. *Z. Saugetierk.* **37**, 162–177.

MARES, M. A. (1980) Convergent evolution among desert rodents: a global perspective. *Bull. Carnegie Mus. nat. Hist.* **16**, 1–51.

MEESTER, J. and SETZER, H. W. (Eds.) (1971–1977) *The Mammals of Africa: an Identification Manual*. Smithsonian Institution Press, Washington, D.C.

MISONNE, X. (1974) Part 6: Rodentia. In *The Mammals of Africa: an identification manual*, edited by J. MEESTER and H. W. SETZER, pp. 1–38. Smithsonian Institution Press, Washington, D.C.

NEWBY, J. E. and JONES, D. M. (n.d.) An ecological survey of the Takolokouzet Massif and surrounding area in the eastern Aïr Mountains, Republic of Niger. Unpublished report, Government of the Republic of Niger.

NIETHAMMER, J. (1963) Nagetiere und Hasen aus der zentralen Sahara (Hoggar). *Z. Saugetierk.* **28**, 350–369.

NOLL-BANHOLZER. U. (1978) Thermoregulation, heart rate and water balance in the fennec, *Fennecus zerda* (Zimm.). Ph.D. dissertation, University of Tubingen.

OSBORN, D. J. and HELMY, I. (1980) The contemporary land mammals of Egypt (including Sinai). *Fieldiana (Zool.)* n.s. **5**, 1–579.

PETTER, F. (1961) Repartition geographique et écologie des rongeurs désertiques du Sahara occidental a l'Iran oriental. *Mammalia*, **25** (spec. number), 1–222.

PETTER, F. (1968) Retour au gite et nomadisme chez un rongeur à bulles tympaniques hypertrophiees. *Mammalia*, **32** (4), 537–549.

POULET, A. R. (1974) Rongeurs et insectivores dans des pelotes d'effraie en Mauritanie. *Mammalia*, **38**, 145–146.

RANCK, G. L. (1968) The rodents of Libya—taxonomy, ecology, and zoogeographical relationships. *Bull. U.S. nat. Mus.* **275**, 1–264.

REGNIER, J. (1960) Les mammifères au Hoggar. *Bull. Liaison Saharienne*, **11** (40), 300–320.

ROSEVEAR, D. R. (1965) *The Bats of West Africa*. Brit. Mus. (Nat. Hist.), London. 418 pp.

SCHMIDT-NIELSEN, K. (1964) *Desert Animals. Physiological Problems of Heat and Water*. Oxford University Press, Oxford. 277 pp.

SCHOMBER, H. W. and KOCK, D. (1960) The wildlife of Tunisia. Part 2. *Afr. Wildl.* **14** (2), 146–151.

SETZER, H. W. (1956) Mammals of the Anglo-Egyptian Sudan. *Proc. U.S. nat. Mus.* **106**, 447–587.

SETZER, H. W. (1957) A review of Libyan mammals. *J. Egypt. Publ. Hlth. Assn.* **32** (2), 41–82.

TAYLOR, C. R. (1969) The eland and the oryx. *Sci. Amer.* **220**, 88–95.

VALVERDE, J. A. (1957) Aves del Sahara espanol. *Inst. Estudios afr.* (Madrid).

VAN DEN BRINK, F. H. (1967) *A Field Guide to the Mammals of Britain and Europe*. Collins, London. 221 pp.

VIELLARD, J. (1974) Les chiropteres du Tchad. *Rev. suisse Zool.* **81**, 975–991.

WASSIF, K. and SOLIMAN, S. (1980) Population studies on gerbils of the western desert of Egypt, with special reference to *Gerbillus andersoni* de Winton. *Proc. 9th Vert. Pest. Conf.*, pp. 154–160. Fresno, Calif.

YUNKER, C. E. and GUIRGIS, S. S. (1969) Studies of rodent burrows and their ectoparasites in the Egyptian desert. 1. Environment and microenvironment; some factors influencing acarine distribution. *J. Egypt. Publ. Hlth. Assn.* **44**, 498–542.

CHAPTER 18

Large Mammals

J. E. NEWBY

Wildlife Consultant, WWF/IUCN

CONTENTS

18.1. INTRODUCTION

Few of the many spectacular sights the Sahara has to offer are more thought-provoking than its superb rock paintings and petroglyphs. What a thrill it is to find elephant (*Loxodonta africana*), giraffe (*Giraffa camelopardalis*) and rhinoceros (*Diceros bicornis*) emblazoned on the walls of sandstone shelters or along the rocky banks of long since redundant and sand-choked river valleys. It is a fact that these familiar savanna animals once inhabited the Sahara, a place many people today think harbours no life at all.

The recent climatic history of the Sahara has been characterized by a humid phase, optimal about 6000–7000 years ago, followed by a drying-up period that became marked some 2500 years ago and continues to the present day. During the humid phase, the Sahara probably enjoyed a rainfall in the order of 500–700 mm annually, and under such conditions, would have taken on the appearance of a wooded savanna supporting a varied fauna. Evidence left by Stone Age man in the form of his exquisite art-work and in his middens shows that the Sahara was within the last 10,000 years home to not only elephant and rhino but also to buffalo (*Syncerus caffer*), reedbuck (*Redunca redunca*), roan (*Hiooptragus equinus*), greater kudu (*Tragelaphus strepsiceros*) and hippopotamus (*Hippopotamus amphibius*) (Lhote, 1973; Smith, 1979).

As drier conditions brought about a change in the composition and abundance of the Sahara's flora, the less-well-adapted mammals retreated towards the periphery of the expanding desert, leaving behind them hippopotamus and crocodile (*Crocodylus* sp.) to perish in dwindling pools and marshes. In the mountains, change was slower and the elephant and giraffe were probably able to survive in wooded valleys long after the plains had been depopulated. Though we have no reliable records of elephant or giraffe, we do know that other species managed to survive in the mountains until fairly recently. During the 1850s, the German explorer Barth (1860) found the lion (*Panthera leo*) common in the Aïr, and it was apparently not until 1932 that the Aïr's last lion was killed in the region of Aouderas an area that still harbours some of Africa's most northerly occurring warthogs (*Phacochoerus aethiopicus*).

The anubis baboon (*Papio anubis*) is another typical savanna mammal that manages to survive in some of the Saharan massifs, though confined to small areas around permanent springs and rock-pools. Another primate, the patas monkey (*Cercopithecus patas*), also occurs in the mountains but unlike the baboon, is somewhat less water-dependent.

Two to three thousand years ago, the species composition of the Sahara's fauna was probably much as it is today. Doubtless a few less-demanding species were able to hang on in isolated enclaves, but these were gradually reduced by increasing desertification or by hunting. Large predators like the lion and the leopard (*Panthera pardus*) were restricted to areas with permanent water. Out on the plains, where the vegetation was becoming dominated by quick-growing annuals and hardy xerophytes, a small number of well-adapted herbivores took advantage of a vast and virtually predator-free habitat.

18.2. THE PRESENT-DAY SAHARAN FAUNA

18.2.1. Aoudad and asses

Mountains and rocky areas make up a large proportion of the Sahara's total surface area. The larger massifs like the Hoggar, Aïr or Tibesti all harbour considerable water resources in the form of springs and rain-fed rock-pools. There is plenty of water for any animal that can get to it and the large mammal most capable of this is the aoudad or Barbary sheep (*Capra lervia*), the undisputed king of the Sahara's mountains. Morphologically and behaviourly more goat than sheep, the aoudad has solved its water problems by rarely being without it. Its ability to climb gives it year-round access to permanent springs and pools.

Standing about a metre high, an adult male aoudad may weight up to 120 kg. The female is smaller and both sexes bear sturdy, backward and laterally curving horns (Fig. 18.1). Fur colour is reddish-brown and the most noticeable feature of the adult male is the long, honey-coloured beard that adorns his chin, neck and chest. Aoudad are extremely wary and seldom leave their mountain strongholds except during the wet season, when they come down into uninhabited valleys to rut. Rams spar fiercely and a dominant male will mate with several females. Young are born in the cold season after a gestation period of 6 months.

Aoudad feed on a large variety of plants and amongst their favourites are the thorn-trees *Acacia ehrenbergiana* and *Acacia tortilis*. There is little of them the animals do not eat: twigs, shoots, bark, leaves, flowers and, above all, the curly, green fruit that the trees produce (Fig. 18.2). After having eaten the fallen pods and those within reach on the tree, the aoudad will often ram the trunk in an effort to dislodge more.

Although quite able to satisfy its water requirements most of the time by feeding on green plants, the aoudad prefers to drink during the hot season and it is whilst at waterholes that they are most vulnerable to hunters. Nets, foot-nooses and specially trained dogs are traditionally employed in the hunt and the meat obtained is dried for local consumption. The skins are highly prized and until recently, the tents of many noble nomads were made of aoudad leather.

Apart from the occasional cheetah (*Acinonyx jubatus*) (Fig. 18.3), the aoudad has few predators to worry about other than man. It enjoys a scattered distribution and major populations exist in the Tibesti, Ennedi, Aïr and Termit mountains.

FIG. 18.1. Female aoudad (*Capra lervia*). (Photo by A. Dragesco.)

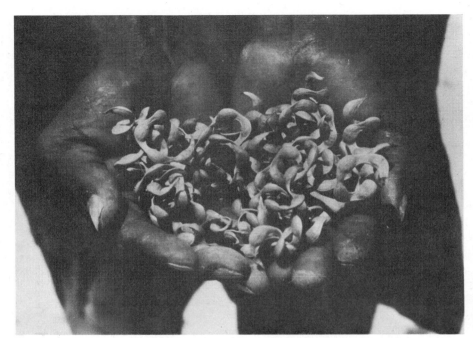

FIG. 18.2. Fruits of *Acacia tortilis*. (Photo by. J. Newby/WWF.)

FIG. 18.3. Saharan cheetah (*Acinonyx jubatus*). (Photo by. J. Newby/WWF.)

One characteristic group of Old World desert mammals is the wild asses. In the Sahara, the status of the Nubian wild ass (*Equus a. africanus*) is surrounded by considerable controversy. There is a strong feeling that, considering its widespread and ancient domestication, the free-living asses one sees are either feral or cross-bred animals (Sidney, 1965). Other authors staunchly believe that pure-bred asses still exist in the Sahara (Dupuy, 1966). Whatever their lineage, the asses living in the less inhabited parts of the Hoggar and Tibesti bear a striking resemblance to the wild ass of classical descriptions. Standing about 1.20 m tall, they are buff to bluish-grey in colour and sport well-defined pale eye-rings and muzzle, as well as the distinctive 'cross' on the back and shoulders. Herds are wily and contain up to twenty individuals. The asses' sure-footedness over rock and scree allows them access to the permanent waterholes of the larger massifs and oases.

18.2.2. Saharan antelopes: Addax and scimitar-horned oryx

Just as the aoudad reigns supreme in the mountains, the desert sands are the realm of the addax (*Addax nasomaculatus*), an antelope so remarkable and so well adapted that it is able to exploit the very heart of the Sahara in a way no other large mammal can. The addax is a relatively heavy-bodies animal and weighs up to 130 kg. Short limbs give it a dumpy appearance and although not without stamina, it is not a fast runner. Its predominantly white coat is offset by a thick wig-like fringe of black hair on the top of its head. Both sexes bear long and heavily annulated, spiral horns.

The meagre desert pastures severely limit the formation of large herds and addax live in small groups of from four to ten individuals. A dominant male mates with a hierarchically ordered number of mature females and young are usually born once a year between wet and cold seasons.

The addax is a highly endangered species and has disappeared from large parts of its former range. Table 18.1 outlines the country-by-country status of the addax and some of the Sahara's other large mammals.

TABLE 18.1. The status of some large Saharan mammals

Country	Scimitar horned-oryx	Addax	Dama gazelle	Dorcas gazelle	Slender-horned gazelle	Aoudad	References
Ex-Spanish Sahara	Extinct 1963	Very rare	Extinct	Very rare	Very rare	Extinct	Valverde, 1968 Gillet, 1969
Mauritania	Extinct early 1960s	Rare	Very rare	Local	Rare	Very rare	Trotignon, 1975 Lamarche, pers. comm.
Morocco	Extinct 1932	Extinct	Very rare	Rare	Very rare	Extinct	Cabrera, 1932 Chapuis, 1961
Algeria	Extinct	Very rare	Very rare	Rare	Rare	Rare	Grenot, 1974 de Smet, pers. comm.
Tunisia	Extinct 1906	Extinct 1885	Extinct	Rare	Rare	Very rare	Lavauden, 1924 Schomber & Koch, 1961
Libya	Extinct 1940s	Extinct late 1960s	Extinct	Rare	Rare	Rare	Osborn & Krombein, 1969 Hufnagl, 1972
Mali	Extinct	Rare	Local	Local	Very rare	Local	Sayer, 1977 Lamarche, 1980
Niger	Very rare	Very rare	Local	Local	Very rare	Local	Newby, 1975a Newby, 1982
Chad	Rare	Very rare	Local	Local	Very rare	Local	Newby, 1975b Newby, 1981
Sudan	Extinct mid-1970s	Very rare	Very rare	Local	Very rare	Very rare	Brocklehurst, 1931 Wilson, 1980
Egypt	Extinct 1850	Extinct *c.* 1900	Extinct	Rare	Rare	Extinct	Flower, 1932 Jones, pers. comm.

FIG. 18.4. Trophy of slaughtered oryx (*Oryx dammah*). (Photo by. J. Newby/WWF.)

Addax populations in the northern parts of the Sahara were all but exterminated during the last century and although the situation is somewhat better to the south, the addax's only remaining strongholds are the large sand-seas where hunters cannot penetrate. Unfortunately, when conditions become too harsh during the hot season, addax are obliged to leave the desert and migrate to the more inhabited desert fringes and into the domain of the Sahara's other large antelope, the scimitar-horned oryx (*Oryx dammah*).

The scimitar-horned oryx, although a predominantly sub-desert species, exploits Saharan pastures during the wet and cold seasons. Oryx are a little heavier than addax and stand about 1.25 m at the shoulder. Coat colour is café-au-lait to off-white, with the neck, upper chest and shoulders russet. As might be expected, the horns are long and gracefully scimitar-shaped. Oryx form larger herds than addax and where the species has not been decimated, groups of up to thirty are not unusual. The oryx's life-style is nomadic and, like the addax, appears to exhibit spatially-reduced territoriality. Honey-coloured calves are born after a pregnancy of $8\frac{1}{2}$–9 months and are weaned some 6 months later.

Until a century ago, the oryx population of Africa must have been enormous. Here was a species exploiting the sub-Saharan grasslands much as the bison exploited the Great Plains of North America. Oryx ranged from Tunisia to the Atlantic coast and then eastwards across the Sahel to the Nile Valley. The story of the oryx's decline from an abundant to a critically endangered species is representative of the fate that has befallen most of the Sahara's larger mammals. Although the oryx populations of North Africa were mostly eliminated by the turn of the century, those to the south remained abundant until the Second World War. The advent of automatic weapons and, more pertinently, vehicles has had a devastating effect on wildlife numbers. Being large and gregarious, oryx have always attracted the attention of hunters and some of the Sahara's rock-art graphically depicts their flight before armed men. Oryx meat is good to eat and has excellent drying qualities. The hide is especially sought after, that of the neck, shoulders and chest being unusually thick; doubtless an adaptation to minimize the damage caused to sparring males from the needle-sharp horns. Touaregs and Berbers traditionally used the hide to make the shields with which they went into battle. The nineteenth-century explorer, Gustav Nachtigal (1881), even describes the use of oryx hide for shoeing horses.

It is unlikely that the traditional oryx hunts practised by such peoples as the Nemadi of Mauritania or the Azza of Niger and Chad, significantly affected oryx numbers. These hunters were experts and killed their fair share of animals, but in the oryx's waterless domain, the logistics imposed upon them were restrictive. The motor car solved these problems but it was no longer the pastoralists doing the hunting but soldiers, prospectors and bush functionaries. On the open plains the oryx were no match for powerful engines and vast numbers were killed annually (Fig. 18.4). The oryx is today one of the world's most endangered species and has disappeared from over 90% of its former range (Table 18.1). Desperate efforts are being made by the World Wildlife Fund to save the remaining 1500 that survive in Chad and Niger (Newby, 1981).

18.2.3. Gazelles

Although both the addax and the oryx exploit the desert habitat well, it is probably fair to say, that as a group, the gazelles are the most successful herbivores inhabiting the Sahara. Five species can be found within its limits: red-fronted (*Gazella rufifrons*), Cuvier's (*Gazella cuvieri*), dama (*Gazella dama*), slender-horned (*Gazella leptoceros*) and dorcas (*Gazella dorcas*). Of these, the red-fronted gazelle is a sudanbo-sahelian species and can only be found in the Sahara where suitable wooded habitat occurs. Another atypical Saharan gazelle is the edmi or Cuvier's gazelle. This species is restricted to the Atlas and Anti Atlas mountains of Morocco, Algeria and Tunisia. Mostly living at altitudes above 2000 m, Cuvier's gazelle is rare but seemingly well protected in a number of special reserves.

The dama gazelle, although a predominantly subdesert species, may spend much of its time in the Sahara and there are many purely Saharan populations. Known colloquially as the mohor or adrar, the dama stands over a metre tall and can weigh up to 80 kg when adult. Coat colour is white with, according to geographic location, a varying amount of reddish or rusty-brown on the neck, cape, back and flanks. The head is white and there is a distinct pale spot in the middle of the gazelle's unusually long and graceful neck. Horns are strongly curved, somewhat like a flattened 'S', and although borne by both sexes, the male's are longer and heavier. Dama gazelles are gregarious, and although migratory wet season herds may number over a hundred, they normally live in small groups of up to ten individuals.

Of all the gazelles, the two most well adapted to life in the Sahara are the dorcas and slender-horned. The latter, also known as the rhim or Loder's gazelle, occurs mainly in the large ergs of North Africa. It is the palest gazelle and although a little larger, can be mistaken for a dorcas. Apart from its isabelline colour, careful observation will reveal longer, thinner and much straighter horns and a considerably larger hoof-print.

The dorcas gazelle is the most adaptable of the Sahara's larger mammals and is at home over virtually the whole range of arid habitats. Its adaptability makes it the Sahara's commonest, most widespread and most successful ungulate. Coat colour is fawn on the upper parts and white below, with a more or less distinct chestnut-coloured side-stripe dividing the two. Both sexes bear horns, although those of the

Fig. 18.5. Dorcas gazelle (*Gazella dorcas*). (Photo by Don Miller.)

female compare poorly with the male's handsome lyrate trophy. Adult dorcas gazelles weigh up to 20 kg and stand about 0.55 m tall at the shoulder (Fig. 18.5).

Dorcas gazelles form herds which vary greatly in number and where pasture is good, aggregations of up to fifty can be seen. Where resources are limited, smaller groups of up to fifteen are encountered. Herd size and structure are also determined by social behaviour. A dominant male may live with up to six or seven reproductive females and their offspring. Bachelor herds of non-reproductive males also exist. Young dorcas gazelles are born throughout the year but Sahara populations show a marked peak in births during the wet season. Gestation takes about 6 months and although twins are sometimes born, a single birth is usual. Just before giving birth, the pregnant female goes into seclusion, rejoining the herd or its mate after a day or two. Of all the larger Saharan mammals, the dorcas gazelle is the most vulnerable to predators and kids especially may be lost to common jackals (*Canis aureus*), striped hyaena (*Hyaena hyaena*), African wild cat (*Felis libyca*), caracal (*Felis caracal*) and the now extremely rar cheetah. During the first few days of its life, the young gazelle is zealously protected by its mother and I have seen both jackals and hyaena sent soundly packing on approaching new-born kids. Attack may also come from the air in the form of both tawny and martial eagles (*Aquila rapax, Polemaetus bellicosus*), secretary birds (*Sagittarius serpentarius*) and white-headed vultures (*Trigonoceps occipitalis*). Because of the gazelle's relative vulnerability, warning signals are well developed. Tail-wagging is often observed and made highly conspicuous by the contrast between the black tail and the brilliant-white rump. If tail-wagging is a sign of mild alarm, stotting is more demonstrative and consists of making a number of comical, stiff-legged jumps into the air.

18.3. ADAPTATIONS TO LIFE IN THE DESERT

In the words of the respected physiologist, Knut Schmidt-Nielsen (1979), '. . . there are three ways in which animals can arrange their life in the hot desert: they can evade the heat, they can passively put up with it, or they can actively combat it by the evaporation of water.' Notwithstanding the destructive effects of modern man, the larger mammals of the Sahara not only manage to survive in a hot, waterless desert but to exploit it most successfully. To overcome the problems posed by high temperatures and lack of water, they have evolved a number of highly effective adaptations. Unlike the smaller mammals, the larger ones cannot evade the heat by burrowing. Although heat evasion plays an important role in their lives, emphasis is put on other behavioural, morphological and physiological mechanisms that enable them to reduce their heat loads whilst conserving enough body water to satisfy basic life-functions and to cool themselves when necessary by evaporation.

18.3.1. Morphological adaptations

In a hot environment, the heat load on an animal consists of two major components: metabolic heat generated from within and environmental heat acquired passively from the surroundings. Mammals with large bodies are better adapted than those with smaller ones because they have a relatively smaller surface area through which heat can penetrate. The addax is the Sahara's largest indigenous mammal and although the introduced dromedary (*Camelus dromedarius*) is larger, it does not approach the addax's ability to live almost permanently in a waterless and meagrely pastured environment. The addax's bulky body has probably been retained through evolutionary selection processes at the expense of the rapidity that becomes less important in a largely predator-free habitat.

All of the Sahara's larger mammals are pale in colour. The multiplicity of thought surrounding the question of coloration in desert animals has been reviewed by Cloudsley-Thompson (1979), and whilst it is evident that colour serves a cryptic function in some species, this may not be the case for those that face few threats from predation. Adult addax, oryx or dama gazelle have little to fear from predators unless sick or aged. Their predominantly white coats tend to be conspicuous rather than cryptic and the fact that only the calves of these species are cryptically-coloured may well indicate that a change in the function of coat colour exists. Although white is a good reflector of energy in the visible wavelengths, it is of little use against the infrared and far-infrared radiation from the sun and heat from the ground. In a habitat where pastures are far-flung and herd sizes are small, the conspicuous white colour of the addax or oryx could have an advertising function to facilitate herd cohesion or encounter.

Fur is a good insulator against both heat and cold and ideally needs to be long enough to act as a barrier whilst still allowing efficient evaporation of sweat from the skin's surface. Most large Saharan mammals grow a winter coat that is shed as the need to sweat increases.

Hoof size is also an important adaptation to life in the sandy desert. It is to an animal's advantage to be as efficient as possible in moving from place to place. Exertion ultimately means more body heat to dissipate by sweating away valuable water and, as might be expected, the addax shows excellent adaptation to walking in sand. Its hoof-print is large and round (Fig. 18.6) and progression is facilitated by the large surface area of contact that reduces sinking. Size for size, the addax's spoor matches that of the buffalo, an animal six or seven times as heavy. Although not as exaggerated, the slender-horned gazelle's hoof is also splayed, as are those of the oryx and the dama gazelle.

18.3.2. Physiological adaptations

To survive in the Sahara, the larger mammals have come to terms with two important limiting factors:

FIG. 18.6. Addax spoor. (Photo by. J. Newby/WWF.)

abnormally high temperatures and acute aridity. Radical heat evasion is not possible and so to a certain degree, most large mammals have developed tolerances to both heat and drought. The gazelles and the antelopes can withstand abnormally high body temperatures, up to 46°C (Cloudsley-Thompson, 1977), and by doing so store heat until it can be lost passively to a cooler environment during the night. One of the major disadvantages of storing heat is the risk of brain damage and in recent years research has discovered a protective mechanism that keeps the brains of ungulates up to 3°C lower in temperature than the rest of their bodies (Taylor and Lyman, 1972). Before passing to the brain, arterial blood comes into contact with a dense network of small vessels in the cavernous sinus, the carotid rete. Evaporation of moisture from the exposed mucous membranes in the nasal passages cools the carotid rete, which in turn cools the incoming arterial blood.

Sweating, although a highly effective way of reducing body temperature, is costly in terms of limited and often difficult-to-replenish body water. Desert mammals do sweat but maximize on its effectiveness through the presence of an insulating layer of fur that enables evaporation to take place at the skin's surface where the cooling effect is the greatest. They also have a relatively high 'switch-on' temperature at which sweating begins. By supporting a certain degree of hyperthermia, desert mammals reduce the temperature gradient between their bodies and the environment, and by doing so reduce the amount of sweating required to combat further rises in temperature.

Just as the addax and dorcas gazelle tolerate hyperthermia, they can also withstand high degrees of dehydration. Ghobrial (1974) has shown that gazelles can sustain and recuperate a 20% loss in body weight when deprived of water. Further dehydration leads to a serious increase in blood viscosity and failure of the vascular system to evacuate hot blood from the animal's core to its periphery. The result is explosive heat death.

The advanced water economy of desert mammals is aided by their ability to conserve water through the production of highly concentrated urine and faecal pellets. The absorptive faculties of the kidney and gut are highly developed, and the production of concentrated urine made partly possibly by the animals' low-protein diet.

18.3.3. Behavioural adaptations

Although morphologically and physiologically well adapted to desert life, most experts would agree that the large Saharan mammals are not biologically so radically different from related species in less arid climes. Behavioural adaptations such as heat evasion and migration play an important role in making life in the Sahara possible at all.

During the hot season, the large mammals spend much of the day in shade, conserving valuable energy and water. As temperatures creep into the 40s, the wildlife becomes virtually nocturnal, feeding and migrating during the hours of darkness. The hot season distribution of oryx and addax is often dictated by the presence or absence of shade, and even the addax is obliged to quit his treeless domain for shadier climes during the hottest part of the year.

If burrowing is the best way of evading heat, the larger mammals have little choice but to try and shade themselves from the sun. The gazelles are versatile in exploiting all manner of rocky crannies, thickets and even subterranean workings. Where there are aardvarks (*Orycteropus afer*), their abandoned burrows become prime hot season real estate. On one occasion, three gazelles were seen in a single hole, but the prize for variety must go to the dorcas gazelle, the desert monitor lizard (*Varanus griseus*) and the desert eagle-owl (*Bubo bubo*) flushed from a burrow in northern Chad. In areas where shade is scarce, wildlife will use any cover the topography affords them: dunes, wadi banks, gullies. In sandy country, addax, oryx and dama gazelles often excavate shallow scrapes in which to lie on the cooler exposed sand.

Scimitar-horned oryx, addax and to a lesser extent the gazelles, undertake seasonal migrations in response to the availability of pasture and water. Because of the erratic nature of Saharan rainfall, pasture is widely dispersed and variable in quantity and quality. To exploit these resources the wildlife must be mobile. Whilst many local movements take place throughout the year, wildlife generally migrates into the desert during the wet and cold seasons and towards its periphery during the hot. In Chad, oryx annually exploit wet- and dry-season pastures as much as 300 km apart (Newby, 1974). The most impressive migrations take place in response to rainfall at the end of the hot season. The wildlife is driven by thirst to reach water and new pasture and in days gone by, thousands of animals might be seen on the move.

Although large mammals migrate in response to shade during the hot season, the principal function of migration is ultimately tied up with their need to feed on vegetation that not only satisfies their hunger but also their thirst. Apart from species like the aoudad or wild ass, which have or require permanent access to water, the other large mammals are capable of going without drinking for many months at a time. Some authors (Ghobrial, 1974; Delany and Happold, 1979) have questioned this ability in dorcas gazelles and whilst some animals may have access to water, the vast majority of Saharan gazelles do not drink during the hot season. Some have even suggested that the addax can spend its whole life without ever drinking. Under ideal conditions, this is possible but during the recent Sahelian drought, addax mortality was high in areas devoid of moist pasture.

Desert ungulates feed on a large variety of plants and while the gazelles browse a great deal, the addax and the oryx are predominantly grazers. Major items eaten include grasses of the genera *Aristida*, *Stipagrostis*, *Panicum* and *Lasirus*, many annual forbs (*Tephrosia*, *Indigofera*, *Monsonia*, *Euphorbia*, *Fagonia*)

Fig. 18.7. Jizzu pasture in the Sahara. (Photo by. J. Newby/WWF.)

and perennials like *Pulicaria*, *Chrozophora* and *Aerva*. Important browse species include the trees *Acacia tortilis*, *A. ehrenbergiana*, *Maerua crasifolia* and *Leptadenia pyrotechnica*.

One of the most important pastures occurring in the Sahara is 'jizzu', and it is unlikely that the addax would be able to survive without it. Jizzu is the colloquial arabic word for the ephemeral pastures that grow following the rains that do occasionally fall in the Sahara's interior (Fig. 18.7). The water-retentive nature of the soil and the cold winter nights keep the pasture green and succulent until desiccated during the hot season. Animals grazing on jizzu can remain independent of water almost indefinitely. Whilst many plants have been variously recorded from jizzu pastures, some of the principal elements are *Indigofera berhautina*, *I. hochstetteri*, *Neurada procumbens*, *Tribulus longipetallus*, *Fagonia bruguieri*, *Cyperus conglomeratus* and *Stipagrostis acutiflora* (Newby, 1974; Wilson, 1978).

As the need to satisfy higher water requirements becomes more compelling, so the proportion of moist plants in the diet increases. The presence of succulents like *Schouwia thebaïca*, *Chrozophora brocchiana*, the wild melon *Citrullus colocynthis* (Fig. 18.8) and the xerophyte, *Cornulaca monacantha*, often dictates hot-season wildlife distribution patterns and similarly, the green-stemmed shrubs *Leptadenia pyrotechnica* and *Capparis decidua*.

18.4. CONCLUSION

From the high massif to the chaos of the sand-seas, the Sahara is inhabited by a highly adapted large mammal fauna. Its ability to survive in a hot, arid environment is founded not only upon refined

FIG. 18.8. Wild melon, *Citrullus colocynthis*, and oryx spoor. (Photo by. J. Newby/WWF.)

physiological capabilities and habitat-adapted morphological features, but also on behavioural traits that allow it to exploit the meagre desert resources to their fullest. The Sahara's large mammal fauna is by any standards successful and although few in terms of species was, considering the aridity of its range, productive in terms of biomass. Wherever pasture was available there are wildlife to exploit it and the Sahara's explorers were unanimous when describing its fauna as abundant. But today, in spite of its apparent success, the Sahara's fauna is critically endangered. Many of the species already described have disappeared from vast parts of their former ranges and a few are actually on the verge of extinction. In a frighteningly short space of time, over-hunting has reduced once populous and widespread animals like the oryx and addax, to relics confined to the most isolated and inaccessible parts of the Sahara.

18.4.1. Conservation for development: The wildlife resource

Organizations like the World Wildlife Fund and the International Union for the Conservation of Nature and Natural Resources believe that, apart from the necessity of maintaining the genetic diversity of our planet, desert wildlife can play an important role in the wise development of some of the world's poorest nations. Through activities such as culling, sport hunting and tourism, wildlife could bolster the poverty-level rural economies of the Sahelo-Saharan countries. Apart from inhabiting otherwise virtually unusable terrain, the wildlife resource requires little management and unlike mining, can be exploited indefinitely. Already twice this century, the inhabited parts of the southern Sahara have experienced major droughts. The existence of a healthy wildlife resource could greatly help alleviate famine during periods of natural disaster.

To return once again to the wonderful pictorial record left by Stone Age man, one can see friezes depicting oryx and addax alongside elephant, rhino and giraffe. The measure of the oryx's and addax's success is clearly demonstrated by the fact that, several thousands of years later, they alone still roam the Sahara. But for how much longer?

ACKNOWLEDGEMENTS

I wish to thank Messrs A. R. Dupuy, K. de Smet, B. Lamarche and D. M. Jones for valuable information concerning the distribution of some Saharan mammals.

To Mr John Grettenberger goes my gratitude for the constructive criticism made throughout the preparation of this manuscript.

REFERENCES

BARTH, H. (1860) *Travels and Discoveries in North and Central Africa*. Longmans & Green, London.

BROCKLEHURST, H. C. (1931) *Game Animals of the Sudan. Their Habits and Distribution*. Gurney & Jackson, London.

CABRERA, A. (1932) *Los mammiferos de Marruecos*. Madrid.

CHAPUIS, M. (1961) Evolutioin and protection of the wild life of Morocco. *Afr. wild Life*, **15** (2), 107–112.

CLOUDSLEY-THOMPSON, J. L. (1977) *Man and the Biology of Arid Zones*. Arnold, London.

CLOUDSLEY-THOMPSON, J. L. (1979) Adaptive functions of the colours of desert animals. *Journal of Arid Environments*, **2**, 95–104.

DELANY, M. J. and HAPPOLD, D. C. D. (1979) *Ecology of African Mammals*. Longman, London.

DUPY, A. R. (1966) Especes manacees du territoire algerien. *Trav. Inst. Rech. sahar.* **25**, 29–56.

FLOWER, S. S. (1932) Notes on the recent mammals of Egypt, with a list of the species recorded from that kingdom. *Proc. Zool. Soc. London*, **2**, 369–450.

GHOBRIAL, L. I. (1974) Water relation and requirement of the dorcas gazelle in the Sudan. *Mammalia*, **38**, 88–107.

GILLET, H. (1969) L'oryx algazelle et l'addax. Distribution geographique. Chances de survie. *C. R. Biogeogr.* **405**, 117–189.

GRENOT, C. (1974) Ecologie appliquee a la conservation et a l'elevage des ongules sauvages au Sahara algerien. Paris, *C.N.R.S.*

HUFNAGL, E. (1972) *Libyan Mammals*. Oleander Press, Cambridge.

LAVAUDEN, L. (1924) *La Chasse et la Faune Cynegetiaue en Tunisie*. Tunis.

LAMARCHE, B. (1980) L'addax *Addax nasomaculatus* (Blainville): I. Biologie. Report to WWF/IUCN, Gland, 66 pages mimeo.

LHOTE, H. (1973) *A la devouverte des fresques du Tassili*. Arthaud, Paris.

NACHTIGAL, G. (1881) *Sahara und Sudan*. Weidmannsche Buchhandlung, Berlin.

NEWBY, J. E. (1974) The ecological resources of the Ouadi Rime–Ouadi Achim Faunal Reserve, Chad. Report to F.A.O., Rome.

NEWBY, J. E. (1975a) The addax and the scimitar-horned oryx in Niger. Report to WWF/IUCN/UNEP, Morges.

NEWBY, J. E. (1975b) The addax and the scimitar-horned oryx in Chad. Report to WWF/IUCN/UNEP, Morges.

NEWBY, J. E. (1981) Desert antelopes in retreat. *World Wildlife News*, 1981 (Summer), 14–18.

NEWBY, J. E. (1982) *Avant-projet de classement d'une aire protegee dans l'Air et le Tenere (Republique du Niger)*. Gland, WWF/IUCN.

OSBORN, D. J. and KROMBEIN, K. V. (1969) Habitats, flora, mammals and wasps of Gebel 'Uweinat, Libyan Desert. *Smithsonian Contrib. Zool.* **11**, 1–18.

SAYER, J. A. (1977) Conservation of large mammals in the Republic of Mali. *Biol. Conserv.* **12**, 245–263.

SCHMIDT-NIELSEN, K. (1979) *Desert Animals. Physiological Problems of Heat and Water*. Dover, New York.

SCHOMBER, H. W. and KOCH, D. (1961) Wild life protection and hunting in Tunisia. *Afr. wild Life* **15**, 137–150.

SIDNEY, O. J. (1965) The past and present distribution of some African ungulates. *Trans. zool. Soc. Lond.* **30**, 1–397.

SMITH, A. B. (1977) Biogeographical considerations of colonization of the lower Tilemsi valley in the second millenium B.C. *Journal of Arid Environments*, **2**, 355–361.

TAYLOR, C. R. and LYMAN, C. P. (1972) Heat storage in running antelopes: independence of brain and body temperatures. *Amer. J. Physiol.* **222**, 114–117.

TROTIGNON, J. (1975) Le status et la conservation de l'addax et de l'oryx et de la faune associee en Mauritanie. Report to IUCN, Morges.

VALVERDE, J. (1968) Ecological bases for fauna conservation in western Sahara. *Proc. IBT/CT Tech. Meeting Cons. Nat.*, Hammamet.

WILSON, R. T. (1978) The 'gizu': Winter grazing in the South Libyan desert. *J. Arid Environments*, **1**, 327–344.

WILSON, R. T. (1980) Wildlife in northern Darfur, Sudan: A review of its distribution and status in the recent past and at present. *Biol. Conserv.* **17**, 85–101.

CHAPTER 19

Archaeology and Prehistory

M. MILBURN

c/o Society of Antiquaries of London, Burlington House, Piccadilly, London

CONTENTS

19.1. THE AREA AND THE PERIOD

The limits of the greatest desert on earth are hard to define, though it runs broadly from the Sahel (or southern 'shore') to the lower limits of the fertile Mediterranean zone. From east to west, it stretches between the Atlantic and the Red Sea, with a total area of almost 9 million km². Here we encounter problems of terminology, however, since several recent studies make clear differentiation between the Sahara and the Nile. Another brings in the Nile as part of the Eastern Sahara, which in turn is clearly applied, *inter alia*, to the Western Desert of Egypt.

Bowing to this current trend only limited reference will here be made to the Nile. Next comes the identification of 'North Africa', in which context British archaeologists, consciously or no, seem more at home either in 'Africa south of the Sahara' or 'North Africa', which latter they see as the region bordering the Mediterranean. French writers on the other hand sometimes apply the term to the whole northern half of Africa. Considerable confusion can arise between North Africa, Northern Africa and the Sahara, dependent on just who is writing.

Maghreb is an Arabic term meaning 'west' and can cause muddles where 'West Africa' is involved. One contemporary journal sees fit to include Algeria, Libya, Morocco and Mauritania, along with Tunisia, in the Maghreb, which thus comprises desert, non-desert and even sahelian terrain. In terms of what actually constitutes West Africa, two modern studies hover around the equator as southern limit, while reaching up to between *c.* 22–25° in the north: the zone's eastern boundary is placed between 10 and 18° east, quite a large variation. Clearly, if we accept such definitions and terminology, parts of the Maghreb, West Africa and the Sahara overlap.

The above are offered in the context of archaeology and prehistory only. As a further hint to increasing the chances of mutual comprehension between specialists, notably those with some grasp of geography, it is wise to avoid such generalizations as 'the Adrar', 'the Tassili' or 'the Tenere'; various examples of each do exist.

In terms of the period, few may have stopped to wonder when prehistory began. But a number of differing ideas clearly exist as to when it ended. Meanwhile archaeology will be understood throughout to mean prehistoric archaeology. Dates will be written as B.C. when calibrated to calendar years, b.c. when uncalibrated (in both cases where this is evident from the text) or B.P., meaning before present.

The separation of prehistory, proto-history—a term beloved of French authors—and history is not easy. This can be shown by taking a trip back in time, imagining a caravan passing from Classical Roman or Greek 'North Africa' into the comparatively unknown lands to south: it would have left a zone already in the historic period, to enter another still in its own prehistoric setting. Perhaps the same is particularly true for adjacent zones with and without metal technology, if it should be that metal did not spread in a predictable and uniform manner across the Sahara.

It is important to realize that the Classical Period in North Africa is also a pre-Islamic epoch, though some so-called pre-Islamic stone structures look very much like pre-dating not only Greeks, Romans and Garamantes, but also the advent of the horse, which takes them firmly back into prehistory. Much depends, as ever, upon who is talking and upon how other cultures, objects and eras relate to his or her own subject. The height of pseudo-academic lunacy was attained by a Viennese enthusiast who insisted that certain stone monuments of Ahaggar (Hoggar) were 'Caballine' (Horse Period) rather than prehistoric.

In one study the proto-historic era is seen as about 3000 B.C., though it is fair to say that most French writers would probably opt for *c.* 2000 B.C. to the time of Christ. Another recent idea is that the prehistory of West Africa ended only with the Arab arrival around A.D. 1000; no proto-history is mentioned.

I have tried to slant references cited towards English-language works. This in itself is a Herculean task, due to an unaccountable lack of British interest in, or knowledge of, Saharan prehistory. Against this the last few years have seen the appearance of a prodigious quantity of continental publications, many of these profusely illustrated, at both popular and scientific level. The 1978 Cologne work, from which many references are cited below, is a fine example.

A word of warning about English-language texts resulting from translation, notably from the French. A bad description, subsequently translated by a linguist who has never seen the material described, can herald disaster, as evidenced by some of the stone monuments mentioned by Salama (1981). One must also be careful about words like 'Soudan', rendered as 'Sudan': area names like Mouydir can become place-names: and one description of the Neolithic (New Stone Age) states clearly that 'the first inhabitants of the Sahara . . . made adzes, hoes, grinders . . . polished stone axes'. To add to all the above, a French lady specialist becomes male in an English translation.

19.2. SOME PRACTICAL PROBLEMS

Before setting out to see just how little is really known about our subject, let us look at the circumstances and especially the environment, which today is hardly conducive to a prolonged stay in any one area. Granted that much of the territory was French-administered at the time prehistoric studies were still in their infancy, it is difficult not to marvel at the interest which quite junior officials showed in what they found on their travels. Hence a good many French reports are by non-specialists: this practice continues even today, with parts of the desert being criss-crossed by the pounding wheels of a myriad powerful vehicles.

Often spectacular finds are made by those whose main interest lies elsewhere and who are thus simply incapable of reporting as thoroughly as an expert, even should time allow. I have been in this situation more times than I care to recall. It seems to be the hardest thing in the world to get the right man to the right spot. Massive 'expeditions', assuming time, money and the potential host country permit, are one way around the problem. The French Berliet Missions around the start of the 1960s and the British Expedition to the Aïr Mountains in 1970 both achieved a great deal.

Frequently all on which one has to go is a laconic though painstaking report by some long-vanished NCO, who did his very best, probably under extremely adverse circumstances, to describe what he found. His site(s) may well lie in an area which, for a variety of reasons, it is hard to reach today, had never previously been reported and may never be so again in the foreseeable future. Even professionals had their difficulties, as witnessed by the inability of Monod (1932) to bring out skeletal material from the central Saharan Ahnet.

Survey work is still all-important; knowing more or less even what does not occur in a given zone all helps towards establishing, or improving, the archaeological knowledge of the area. Huge distances may be traversed without coming upon a trace of what one is after, far less what one confidently expects. Hence the humour of the situation when a non-Saharan specialist complained a few years ago that parts of the Sahara had not been properly searched to turn up the bone and ceramic artefacts on which his heart was set.

The administration required to keep a party in the field for even a few weeks may be considerable, especially where water is in really short supply. A safety limit has to be set and adhered to: knowing the location of a well only 50 km distant is quite different to being sure that it contains water. Locals often know of supplies marked on no map, always assuming that one finds people in the area being studied. It does seem to be the fate of the small survey team to run short of water or fuel, or both, and to have to abandon, in the interests of safety to human life, a search that was just starting to pay dividends.

In addition to the above, some technical difficulties have been summarized by Balout (1981). While the Maghreb, for instance, got away to an early start in investigation of prehistory, excavation methods and lack of C14 dates caused this head start to be lost. Lack of human fossils from the Lower Pleistocene and of dates for Palaeolithic (Old Stone Age) occupation sites have meant that settlements in the Maghreb and Sahara can only be dated by hypothetical correlations of typology of stone industries and of fauna. Although the continuity of human occupation seems probable, lack of stratigraphical evidence hinders progress in determining this. Large-scale excavations, carried out in accordance with modern scientific requirements, are vital in future if we are to gain full understanding. In a nutshell, Man has inflicted what mostly amounts to a number of pin-pricks over a huge area of desert, generally not stopping to delve beneath the surface: hence much of a current uncertainty or even ignorance.

With the above reservations constantly before us, all that it seems possible to attempt in these few pages is a cautious and far from comprehensive or conclusive survey.

THE PALAEOLITHIC (OLD STONE AGE)

Subject as ever to the limited evidence and to currently-unceasing reconstruction of global theories, only a brief description of some of the industries involved will be given. Attempts to define the various Ages, specifically in terms of time, artefact assemblages, technological and socio-economic levels or on the basis of associated human physical types, result in extensive divergence of views (Clark, 1982).

Distribution zones and nomenclature are also controversial. For instance, the Middle Palaeolithic of the Maghreb and Sahara is usually split into two broad complexes, Aterian and Mousterian, though it is not always certain which is present. Time after time we are referred back to comparatively well-known sites like Haua Fteah (outside the Sahara) and Adrar Bous, Niger, demonstrating the lack of detailed study and publication in general.

Using where possible the terminology of Coppens and Balout (1981) and Wymer (1982), an attempt will be made at some sort of chronological account, following to some extent the views of Hugot (1981).

TABLE 19.1. Simplified chronology (after Kuper, 1978a).

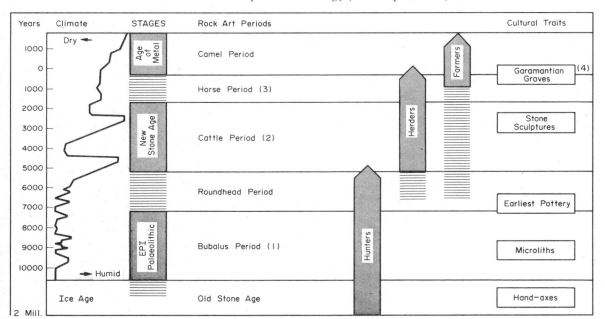

Notes: (1) Also known as the Great Wild Animal Period and limited to carvings. (2) Also called Bovidian or Pastoralist Period. (3) Sometimes known as Equiolian or Caballine Period. (4) Preferably termed protohistoric stone monuments.

19.3.1. Pebble culture

This is the oldest known industry, whereby a cutting edge was created on a pebble by striking off one or more flakes. Very little is known about Man's habitat or way of life in the Sahara during this period or indeed in the succeeding Palaeolithic periods. It has been suggested that such pebble-tools belonged to *Homo habilis*, being widely found throughout the very humid region of those times. They seem to be the implements manufactured and used by our most remote forbears.

19.3.2. The Acheulian

Following on from the above came a far more sophisticated tool shaped like an almond and trimmed on both edges, known as a handaxe or biface and used for cutting, scraping and digging. These are the typical tools of the Lower Palaeolithic and may today be found in thousands lying on the desert surface, whether in dunes or open reg. Their number may not perforce indicate a huge population, in view of studies by Ziegert (1978) postulating a large number of bifaces being rapidly produced by a small party. The people may have belonged to specialized hunting groups and seem not to have continued even as late as 100 000 B.P.

Two conflicting views of Tihodaine, a later Acheulian site in Algeria, make interesting comparison. The occurrence of handaxes and cleavers among dunes in a now totally desiccated area is indicative to Wymer (1982) of periods of wetter climate, with some water available and with movement being possible from one water-hole to another. The account of Hugot (1981) includes mention of remains of elephant, hippo and crocodile: elephants are prolific feeders and need abundant vegetation, while hippo and crocodile require water aplenty.

19.3.3. The Mousterian

Here we start the Middle Palaeolithic, not to be confused with the Middle Stone Age, or Mesolithic, which current opinion is less and less willing to accept as ever having existed in the Sahara.

The industry is characterized by the prodigious quantity of points and scrapers obtainable by trimming flakes from one of their faces. If the handaxe is the characteristic tool of the Lower Palaeolithic (once the technique was evolved), there exist reasons for thinking that flakes were used too, as well as chips off the core. Flake tools emerge as a development, rather than a discovery, as presumably was the handaxe that replaced the pebble-tool. A revolutionary feature, however, is the Levallois technique, a process enabling large flakes of predetermined form to be obtained from a prepared core, with little or no trimming being necessary.

To complicate the picture, the Mousterian is seen by some as being absent, other than in the Eastern Sahara, where tentative dates of c. 60000–40000 B.P. are suggested, with its place in the Sahara being taken by the Aterian.

19.3.4. The Aterian

The type site is Bir el Ater in eastern Algeria and the industry comes between the Mousterian and the Capsian (of North Africa); it consists of tanged points, scrapers and a few foliate points, being considered as the tools of a migratory people who possessed implements much lighter than those used by preceding cultures and essentially useful for hunting. The presence of a tanged point implies that the concept of shafting was known, while that of the spear and the bolas have been suggested. The Aterian has some Mousterian features and uses the Levallois technique. Two 'provinces' are apparent, one to north of Ahaggar and one to south.

It seems to have been a North African industry that forged south into the Sahara, with a clear lower limit slightly south of a line running between Ahaggar and Tibesti. Change is apparent during this southwards progression, culminating in the superb tool-kit known at Adrar Bous. Chronology has been suggested as being between c. 45000–38000 B.P., though another view puts the upper limit at c. 35000 B.P. with sites as late as c. 11000 B.P. occurring in Chad. From these widely-divergent opinions we strike another obstacle:

when agreement cannot be reached even about positive identification of an industry of the Neolithic (New Stone Age), then acceptance of even C14 dates becomes a hazard, in so far as they relate to an industry whose exact nature and composing elements are disputed.

19.3.5. The Hiatus

If the Aterian can be dubbed Terminal Palaeolithic, it is becoming less and less fashionable to speak of a Middle Stone Age in the Sahara. In fact there is a blank period in prehistory, seemingly indicating very harsh climatic conditions which may have rendered certain zones uninhabitable, probably from *c*. 23,000 to *c*. 10,000 years ago. Cultural assemblages dating to this time, contemporary to the main advance of the last Glacial in Europe, are reported lacking. Certainly in the Tenere Tafassasset severe deflation of Aterian sites and wind-polishing of the artefacts illustrates the untenable conditions. The Sahara is further thought of as being for the most part lightly settled between 16000 B.P. and the arrival of food-producers. We shall see that this lower end of the time bracket is fraught with uncertainty and there is no firm evidence of contact between the Aterian and the earliest Epi-Palaeolithic (P. Smith, 1982).

19.3.6. The Epi-Palaeolithic (see Table 19.2)

This much-debated and controversial epoch, whose time-span will be subject to revision as new finds occur, can be explained as encompassing a hunter/gatherer way of life. After the long millenia of development from handaxe industries of specialized hunting groups it is perhaps about the time of the Hiatus that the leptolithic industries of advanced hunting communities were evolving, possibly via transitional industrial industries like the Aterian. From now on lithic assemblages are more and more typified by tools made on blades owing their production to the indirect-percussion or punch technique of separation from the nucleus. Such tools as end-of-blade scrapers, burins and backed knives were gaining in popularity, as were complex tools combining several types all on one piece. Small tools approaching microlithic proportions became frequent around 15000 B.C.: some were mounted in wooden or bone hafts as cutting or piercing instruments and the bow was probably in common use. Industries based on blades and bladelets are seemingly not found until around *c*. 8000 B.C. (P. Smith, 1982).

In conclusion, one should ask whether our knowledge might be greater, had more attention been paid to Palaeolithic industries, even to the extent of studying 'collections' in museums or even in private hands, in the event that little else is available from a given area. For instance, the flint-bearing marine transgression zone of the Atlantic Sahara has been described as one vast knapping-floor (Mercer, 1976). Thousands of tools indubitably left the territory during European occupation: how many have been studied to date, regardless of the disadvantages of not being able to see the material *in situ*? How many sites studied by experts have previously been pillaged? Last but not least, how often have well-meaning amateurs been unable to find anyone willing, as well as competent, to study his or her Saharan finds, some of which may well be unique?

19.4. THE NEOLITHIC

In 1958 it was noted that one might expect to find oneself on firmer ground with the dawn of the Neolithic. Research on the period had often been neglected for the seductive charms of greater mystery clothing the more ancient past, plus the stronger appeal of a wealth of Classical sites, abounding in remains . . . and legible inscriptions. Can any Saharan specialist remain immune to this cry from the heart?

TABLE 19.2. Chronology of the Epi-Palaeolithic
(after Taute, 1978)

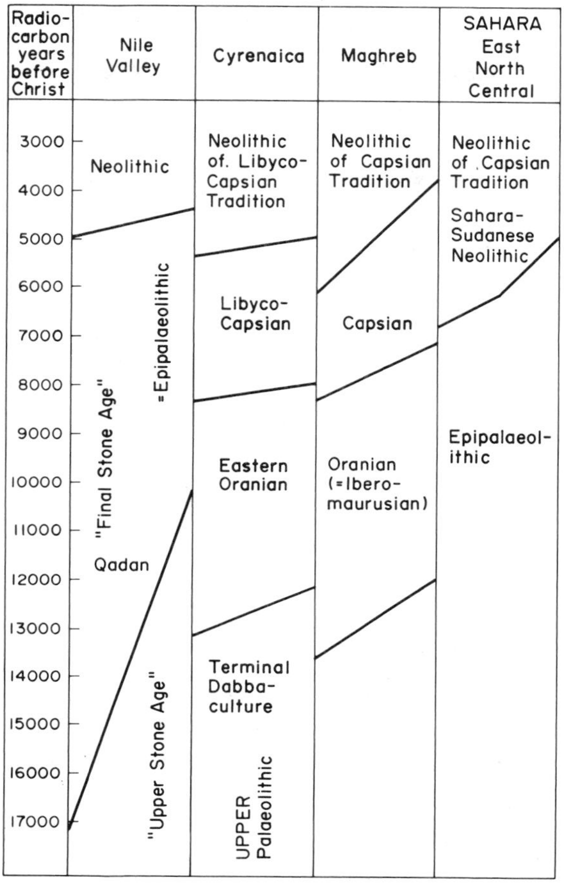

Radio-carbon years before Christ	Nile Valley	Cyrenaica	Maghreb	SAHARA East North Central
3000	Neolithic	Neolithic of. Libyco-Capsian Tradition	Neolithic of Capsian Tradition	Neolithic of Capsian Tradition
4000				
5000				Sahara-Sudanese Neolithic
6000		Libyco-Capsian		
7000			Capsian	
8000				
9000				Epipalaeol-ithic
10000		Eastern Oranian	Oranian (=Ibero-maurusian)	
11000				
12000	Qadan			
13000				
14000		Terminal Dabba-culture		
15000				
16000				
17000				

("Final Stone Age" / "Upper Stone Age" — Nile Valley column; "Epipalaeolithic" and "UPPER Palaeolithic" — Cyrenaica column)

Forde-Johnston was to write a year later of a regrettable gap in archaeological literature, a situation less evident by the time Whitehouse produced a world archaeological atlas in 1975, with African site maps almost bereft of Saharan sites. If the former feared that some of his conclusions were not always in agreement with those of some recognized authorities, a new situation has now arisen, in that various experts are clearly at loggerheads over subsequent finds and developments resulting. Meanwhile the amount of available literature has vastly increased. The point at issue is how to glean any coherent picture from the mass of differing opinions confronting us in 1982.

The process of 'neolithization' in a Saharan context is far from simple or even clear-cut, though a good many C14 dates are available. It must also be understood in a different way to that of Europe. If the term was first employed by Sir John Lubbock to imply polished stone tools and pottery, experts now consider domestication of animals and plants as most important too. In the Sahara, however, ceramics appeared long before the first sign of domestication of either, so that if we employ European criteria, then the process must be seen as covering some three millenia (Hays, 1975; A. Smith, 1980a).

It would be quite wrong to visualize groups of Epi-Palaeolithic hunter/gatherer/fishers suddenly acquiring all the above improvements to their mode *de vie*: indeed it is interesting to try to estimate how long in time some of them may have survived, albeit possessed of certain Neolithic traits, prior to being overcome by, or absorbed into, Neolithic societies. The presence of countless arrowheads attributed to the Neolithic could indicate a need for hunting and/or defence, a situation hardly compatible to a land of milk and honey where people potted, herded and perhaps reaped grain previously sown. Perhaps too the ability to cook in vessels was an innovation of huge significance.

At the same time there must have been some compelling reasons for men to abandon a life of hunting/gathering/fishing for more sedentary pastimes like herding and plant-production. Provisionally I see such societies as semi-sedentary, in view of the wide differences of opinion between those who see pastoralists as nomads and those who define them as transhumants. As to the still unproven plant cultivation, the system whereby 'noble' nomad classes left the tilling of their ground to specialized 'worker' sedentarists may well be a very ancient practise indeed.

Even at this late stage in prehistory, experts are not unanimous in their views of Neolithic cultures. At times there seems to be no clear differentiation between industry and facies (Camps, 1982; Hugot, 1981). All authors consulted, however, admit the existence of two major industries, the Saharan–Sudanese Neolithic (called by some the Neolithic of Sudanese Tradition) and the later Neolithic of Capsian Tradition. I propose to run quickly through those industries of only marginal interest, prior to tackling in some detail the Saharan–Sudanese Neolithic and two apparent facies, called the Bovidian and the Tenerian, both of which pose some fascinating problems. To summarize one opinion (Camps, 1982), the Neolithic is above all a change in behavioural pattern, non-uniform throughout the Sahara. The origin of the stimulus varied and had also to exert its influence on different industrial traditions.

19.4.1. The Mediterranean Neolithic

This need not concern us except in so far as it is found on the Atlantic coast. Impressed or stamped pottery decoration was followed by obsidian, while contacts with Iberia played an important role. The culture developed around the tell country of the Maghreb in the sixth millenium B.C.

19.4.2. The Neolithic of Capsian Tradition

This industry is deemed to extend to eastern Algeria, southern Tunisia, north Libya, the Saharan Atlas and the north part of the Sahara. Note that Balout (1981) and at least one other renowned specialist consider that there is no Neolithic of Capsian Tradition in the northern Sahara.

The Saharan facies is typified by countless arrowheads and scant pottery, while the Atlas facies is attributed an association with the rock art of the Saharan Atlas. Ostrich eggshell containers, some of them engraved, may account for sparse pottery. Indeed, some surface sites are aceramic. Bone industry was extremely rich, including even decorated potters' tools, plus the harpoons known further south. Animal engravings on small movable stone plaques are typical, as is the profusion of arrowheads at some sites, with elegant geometric microliths and bifacially retouched artefacts, including Egyptian-style knives. Nonetheless Epi-Palaeolithic behavioural patterns appear to have been followed. The population appears as partly Mechtoid and to some extent negroid, with a preponderance of Mediterraneans. In order to delineate the southern limit of this industry, a line would have to be drawn beyond which engraved ostrich eggshells, blade industries with geometric microliths and conical- (not round-) based pottery did not occur (Camps, 1974).

19.4.3. The Mauritanian Neolithic

This arose from various composing elements, the Mediterranean–Atlantic, Saharan–Sudanese and possibly Guinean, along with the Neolithic of Capsian Tradition: possibly this is the same culture mentioned by Hugot (1981, p. 599), little studied and typified by rough pottery with scant decoration, hearth-stones and occasional stone tools. A Neolithic of Mauritanian Tradition has recently been proposed by Petit-Maire (1981).

19.4.4. The Saharan–Sudanese Neolithic

From the seventh to the third millenium B.C. the south parts of Tibesti, Tassili-n-Ahaggar and Mauritania will have possessed wide areas covered by lakes, these being in their turn fed by non-permanent rivers like the Tafassasset and the Igharghar. In the central and southern regions one may reconstruct a lifestyle not dissimilar to that of modern peoples besides the Niger and Senegal rivers and Lake Chad. The lacustrine nature of some sites is attested, for example, by the remains left by fishermen of Oued Amded, nowadays over 50 km from water and lying in the western approaches to Ahaggar (Camps, 1982). Compare this with the 'acquatic' civilization—not all of it Saharan—mentioned by Sutton (1981).

The industry is suggested as having stretched from about the latitude of the Tropic of Cancer down to the fringes of the humid forests, developing pottery as early as the seventh millenium B.C. (to west of Ahaggar) and even earlier in the Fezzanese Acacus Mountains. The inhabitants are seen as negroid, on the one hand perhaps inventing agriculture without attributing much importance to it and, on the other, practising animal domestication only at a late stage, though cattle-raising was to assume great significance for the Bovidians and the Tenerians in the south. The Tenerian Neolithic, of which more anon, is said to be very closely related to the Egyptian Neolithic.

The expression 'of Sudanese tradition' is no longer held to imply influence from the lands along the Nile: the oldest dates for industries associated with pottery have indeed come not from the Nile but from central Sahara and the surrounding area. Wooden tools, matting fragments and cordage also make their appearance in the seventh millenium. The main features of this early period may be given as the negroid inhabitants of the central and southern areas, the importance of pottery (always round-based or nearly so), the frequent occurrence of celtis (nettle tree) seeds stored in receptacles and the fishing implements in stone, bone and ivory. There exists, however, a general mediocrity in the lithic tool-kit that contrasts strongly to the quality of the later Neolithic of Capsian Tradition and the Tenerian (Camps, 1982).

Quite another picture is that of polished stone axes, hoes and adzes playing a big role, with specialist drilling equipment for making beads, plus fine grinding material (Hugot, 1981).

A third opinion on this older phase is that these communities will have been specialist hunter/gatherer/ fishers. With game, water and plants remaining abundant up to c. 4000 B.C., the population probably became much more sedentary, with a more regular system of seasonal transhumance. If valid, this interpretation could mean that the amount of movement tied to rain will have diminished, allowing a closer relationship between game herds and hunters to build up. Perhaps an established form of taming was now prevalent, as seemingly shown by animals with collars and halters in rock art of Egypt and the Sahara (Clark, 1980).

The same work puts the pastoral Neolithic as lying between c. 4000–c. 2000 B.C., suggesting that much can be learned of the economy, customs and behaviour of the populations from systematic rock art studies. By 2000 B.C. it must have been problematic to maintain herds in the face of developing desiccation; large movements of nomads from the Libyan desert into the Nile Valley are known from then on. The C-group peoples of the northern Sudan are just one example. Settlements of Neolithic pastoralists are known in the Sahel, for example in the Tilemsi valley, by 2000 B.C., with the Kintampo culture appearing in Ghana by c. 1500 B.C., as evidence of these Saharan influences (compare Shaw, 1981).

19.4.5. The Bovidian pastoralist facies of the Saharan–Sudanese Neolithic

Without braving the complexities of speculation on the origins of domestic cattle (Camps, 1982; A.

Fig. 19.1. An unusual sculptured object, under 20 cm long, from northern Tenere Tafassasset. Although it will stand upright on two 'legs' today, wear at the ends of these may mean that it was not originally intended to do so. Attempts at positive identification have so far failed, though this surface find occurred adjacent to apparent Neolithic material.

Smith, 1980b), we should scrutinize the meaning of 'Bovidian'. Is it simply a style of rock art? Or a particular tradition within the present industry? Since Bovidian influence has been suggested right up to the Saharan Atlas, we should be wary: pottery and lithic industry are very different in that northerly zone.

The people who left paintings in the shelters of the Tassili-n-Ajjer and Tefedest were pastoralists, practising transhumance and spending long periods in the neighbouring valleys and plains where there now exist dunes (ergs). Their tools ought to be found in these ergs, as well as in the mountain rock shelters.

In fact this tool-kit from within the massifs, where people probably spent a comparatively short time, is badly known. Main elements are arrowheads, chipped and polished axes, pestles and upper stones and fragments of grindstones (querns). Fixed grinding/pounding hollows in flat rocks should also be mentioned. Importance may be attached to the presence of small plaques (slabs) or discs with retouched edges. In the ergs the tool assemblies are far richer: the elegant arrowheads and bifacial disc-knives shows influence of the Tenerian, not to be wondered at, since—in the case of the Tassili-n-Ajjer/Erg Admer zone—the Tenere Tafassasset lies not far to south.

Transhumance is attested by the presence in the mountains and on the plains of magnificent small stone sculptures, well executed and of a like standard to the paintings in the shelters; they include bovids, antelopes, sheep, rodents and anthropomorphic figures (Camps, 1982; Camps-Fabrer, 1978).

19.4.6. The Tenerian pastoralist facies

This is evidently contemporary with the Bovidian. Various attempts have been made to specify the composition of the tool-kit. Camps (1982) gives pestles, grindstones, rubbers, grooved axes (some of these appear to lack a cutting edge, I have noted): there are very fine arrowheads, tranchets and especially bifacial disc-knives, plus knives of Egyptian type as well as flat grindstones with notched edges to enable transport on ox-back (a pastime not shown in rock art to date).

Another attempt gives lotus-shaped projectile points, thick concave scrapers, elementary saws and

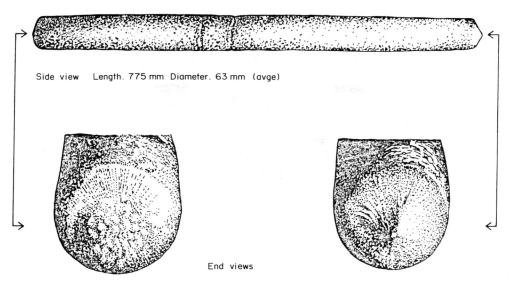

Side view   Length. 775 mm Diameter. 63 mm (avge)

End views

FIG. 19.2. A stone rod 77.5 cm long with two parallel grooves around its circumference. If the left-hand end looks as though it might have been used for pounding, the opposite end, shaped something like a blunt pencil, remains enigmatic, although a number of such ends are known. (Drawn by W. J. G. Godwin.)

hafted axes, with no clear Egyptian influence being found outside Tenere (Hugot, 1981). A third list mentions the Tenerian being found as far east as Borkou and to the west in the Aïr Mountains, with the type-tools bifacially flaked disc knives, flaked and polished stone axes and 'Shaheinab'-type adzes, plus numerous grindstones (Smith, 1980a; compare Krzyzaniak, 1978). Pottery decoration as well as adzes of Adrar Bous closely resemble those from Esh-Shaheinab, near Khartoum, and hunting large animals may have been as important as pastoralism. We must await publication of the long-expected main Adrar Bous report by Clark to obtain more details.

To all the above should perhaps be added the finely decorated ends of certain long and fairly symmetrical stone rods, whose function as pestles is not always clear, along the lines of the stone sculptures mentioned above. Were the carved end to be used as a pestle, the carved portion would suffer accordingly. Undecorated specimens can have varying ends, clearly indicating at least two different functions; when such a rod has two parallel grooves around its circumference, the imagination starts to boggle (Fig. 19.2).

Grindstones (saddle-querns) can also be decorated and indeed the whole question of the numerous types of upper stones, small, large and medium, in itself merits serious attention. At Kadero, Sudan, there appears to be only one type, for instance. Elsewhere I have raised the point as to whether a line of shallow cup-marks along one flank of a long stone rod should be seen as decorative or—as seems more likely—strictly functional. Stone mortars for use with such of the stone rods as may have been pestles—at the risk of certain breakage of one or other object—have not been found: they could have been of wood.

What I have termed static querns—large chunky stones with grinding-depressions caused by use, no other attempt to shape them, and massive upper stones weighting up to 5 kg or more—may have been intended to remain in one place. Saddle-querns, generally far smaller and lighter, evidently were not. Apart from their occurrence beside hearths (see below), in huge plains where no rock suitable for their extraction is present, modern nomads still use them, as well as the associated upper stones. Transport by camel is a simple matter.

We do not know how long Saharan populations continued to use these, or other, stone implements prior to metal being generally available. Thus a good many stone artefacts may occur outside a Neolithic context today.

19.4.7. Stone hearths

One aspect of cattle pastoralism hitherto accorded scant mention in English-language texts generally available are the hearths ('stone places') now studied for some years by Gabriel (1978: compare Gautier, 1980, 336). Seen as earth ovens where meals were cooked without pottery, they occur mostly on the great plains and appear to represent the temporary halting-places of cattle-herders. The life-style seems nomadic, with polished material and pottery being rare, though querns are quite common. No signs of permanent settlement are detectable adjacent.

Gabriel saw regular rings of stone, in each case apparently associated with large numbers of hearths, 170 km south-west of Kharga in Egypt as well as some 50 km south of Djanet, in the northern Tenere Tafassasset. If the rings remain enigmatic, their occurrence in two widely-varying zones is suggestive of long-range contact.

Fossil bones have been predominantly of game and there lack cattle bones in general. However, it seems probable that such herders nourished themselves with blood and milk of their animals, as well as by hunting and gathering. A 'stored' source of animal protein would significantly reduce the famine risk as well as providing wealth (Clark, 1980). Perhaps this could then be exchanged for other specialist products, such as pottery (?), abundant in mountain areas.

Gabriel further offers the idea that transhumance may not have been the rule, distance and the diversity of remains in mountain and plain habitat being the reasons. It is emphasized that the areas studied relate to his own travel routes. There remain zones of which little is known. Unpublished data currently indicates that the initial stage probably began in the seventh millenium (dates uncalibrated), with the main phase lasting from c. 5800–c. 5300 B.P. About 3800 B.P. the great plains had ceased to provide support and were abandoned, with people crowding into the mountain massifs or the southern margins of the Sahara, as well as into the present Sahel and Nile Valley, for about the next 1500 years.

One interesting point on cattle-domestication emerges elsewhere. A definite milking scene is known from the Fezzan, apparently carved in a style attributable to the earliest rock art. As usual, however, the representation has yet to be exactly dated (Jacquet, 1978).

19.4.8. Agriculture

Though a mass of persuasive argument has been advanced to suggest that Neolithic cultivation is indirectly proven, the evidence still appears inconclusive, except at the very end of the period (Dhar Tichitt, south Mauritania). Grain from wild cereal falls readily from the flower and grain-bearing structure, while the reverse is true for a fully domesticated plant. Farmers need to have a plant that will not scatter its grain on the ground in the days of approaching maturity. Nonetheless there is reasonable evidence in the form of grindstones and sickle sheen that Saharans harvested significant amounts of wild grain. We need carbonized remains of domesticated cereals (Stemler, 1982), the preservation of which on open sites is precarious.

It may well be asked whether animal domestication, with the resulting need for storage receptacles, aided the development of pottery in some areas. Notwithstanding a mass of products stored in pottery, however, the question of plant domestication must remain open, unless we can approach the problem indirectly. In the late 1960s, for instance, Beck and Huard were describing a good deal of 'ploughing material' from Chad. It is not easy to gain much idea of the reasoning behind such an assertion, except that various stone rods were mentioned. Although the question seems to have lapsed, I have since independently wondered about the use of some Niger rods, rough and ungainly, some of whose general shapes and dimensions recall those of certain ards of north-west Europe. Such a stone implement can be termed a scratch plough which stirs the soil but does not turn it: we need to prove, or disprove, its existence in the southern areas.

Before leaving the Neolithic, the abundance of pottery in some areas, plus its lack in others, makes one wonder about its transport in a Saharan context. Against the notion of nomads avoiding it, with its being made *in situ*, used *in situ* and smashed *in situ*, we do not know exactly who was nomadic and who was transhumant. Ceramic petrology has further shown that pottery and its contents were moved over great distances, also that specialist manufacture and marketing existed. The size of some pots, plus their weight and fragility when full, might imply scant movement, once installed in a site. Conversely—and ritual breakage (?) apart—might the bevy of smashed ceramics in many places indicate frequent transport and relative expendability, without forgetting receptacles of stone, skins or even wood?

In conclusion we should note that the presence of early and later Neolithic peoples in the Sahara was neither dependent on nor the result of diffusion from the Nile Valley. Evidence available further suggests the presence of Neolithic artefact assemblages prior to animal and plant domestication. It seems likely that invention of pottery, domestication of cattle and possibly that of millet may be the result of independent experimentation by autochtonous groups in the central Saharan massifs (Hays, 1975).

19.5. THE END OF THE NEOLITHIC AND THE START OF PROTO-HISTORY

It must be stressed that no hard-and-fast rule exists as to just when one era ended and the next began. Presumably there were variations between zones. We see some Neolithic groups able to survive within the encroaching desert by withdrawing on to the dwindling water resources. This was the case at Dhar Tichitt, where people lived by cattle-herding and grain cultivation and collecting, until Lybico-Berbers possessing metal weapons and horses put an end to Neolithic culture in the Sahara some time after 850 B.C. Possession of such equipment, in addition to chariots, is given as the reason why these northerners, who are stated to be nomads, were able to occupy large tracts of the central and southern Sahara at the expense of the late Neolithic negroids (Clark, 1978).

The question of metal—traditionally bound up with Proto-history—is fraught with unknowns. It may well be that those who possessed it early on, essentially foreigners/colonizers (?), took care not to let it get into the 'wrong' hands for as long as possible, The spread of metal may have been very gradual and irregular, with pockets of non-metal-owners managing as best they could. What happened, one wonders, if and when nomadic (?) northerners forging south came into contact with copper and iron-owners in northern Niger, these being known at least from the second millenium and late first millenium B.C. respectively? Once again, we lack adequate dates for both groups.

Much ink has flowed on supposed chariot routes across the Sahara, based on carved or painted likenesses of chariots of non-uniform type. Following much detailed experimentation by Spruytte (1977) it seems clear that those of the type called 'flying gallop' cannot have had more than limited use even for hunting, war or sheer prestige. Larger, different models could have been harnessed to bovids for local transport: animals are not known to have been shod in those days and Saharan terrain can be murderous. No chariot has yet been found and it is possible to construct one without using metal. From rock art it is not possible to say whether any pictures of armed men associated with chariots actually show metal weapons. This does appear likely in the case of later daggers, though again uncertain for spears, whose points could have been hardened by fire, as found in the Canary Islands by Spanish colonialists.

Associated with portrayals of heavily-armed horsemen of the central and southern Sahara it is common to find inscriptions of Libyco-Berber type, apparently going back to the last centuries before our era. Rather similar inscriptions occur in the Canary Islands and neither have yet been translated (Galand, 1979).

If these Equidians, Caballines or Horse Period people got into the Tassili-n-Ajjer around 1500 B.C. or even later, it is far from clear how they lived, once the late Neolithic populations were subdued. Maybe in the style of lordly aristocrats, with the negroids employed on menial tasks, perhaps a model for later Tuareg and their sedentary cultivators?

To these supposed chieftains, first charioteers and later intrepid horsemen, have been attributed the

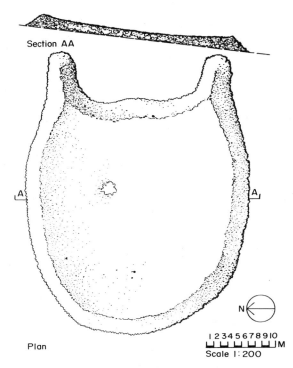

Section AA

Plan

Scale 1 : 200

FIG. 19.3. A heel-shaped cairn near Arlit, Niger. Built entirely of smallish stones, so far as can be seen, it contained an empty chamber in its upper surface: whether this held the initial burial is unknown. The curved façades of such Saharan cairns tend to face roughly east: those of Scotland—with which no connection has been demonstrated—favour a south-easterly or even south-south-easterly heading. May this reflect difference in latitude? (Drawn by W. J. G. Godwin.)

important dry-stone monuments that are such a striking though little-known feature of regions like Ahaggar, Ahnet and Tassili-n-Ajjer. Among the seemingly oldest, albeit not dated, are immense paved structures in crescentic or V-shape, plus the so-called keyhole monuments (Milburn, 1981). English-language texts relating to these and other types are minimal and some of it is unintelligible (Salama, 1981), probably due to indifferent description and translation.

Notwithstanding the usual attribution of these and many other structures to the 'pre-Islamic' or 'Garamantean' periods—whatever these mean—there are a few dates available in recent years. A pear-shaped flat paved structure with a central tumulus lies well into the Neolithic, as does another lone tumulus elsewhere, this latter being the commonest Saharan form. A third date, involving a drum-like 'choucha' of Adrar Bous, appears suspect. Of no small interest are C-group graves of Nubia, not dissimilar to the choucha. The earlier A-group graves, datable to c. 3000 B.C., consisted of the simple tumulus built over a pit, rather like some Islamic burials found in Ahnet.

During the Neolithic the usual practice appears to have been to inter the body in a simple trench made in the occupation level, whether this was a cave/shelter or in the open air. It was then covered with earth, with possible offerings and the use of red ochre (Forde-Johnston, 1959). Subsequently clear cases have occurred of overlap between Neolithic and proto-historic burial-practices, in that Neolithic remains exist with stone monuments raised above them.

Such graves have been consistently robbed and corrosion would have destroyed most of the metal, if any, placed within. High humidity and acid soils are not ideal for the preservation of bones: one may expect to find human remains of any great age in caves, rock shelters or artificial shell-mounds (Mauny, 1978). By their very nature, burials in stone structures in the open cannot conform to these requirements.

Broadly, the vast majority should coincide with the last 1500 years prior to Islam, with animistic forms, probably built by Tuareg and Tibu, continuing well into the Islamic era.

Another conundrum is that of typology: if by their very size a number of structures qualify as megaliths, the individual stones of which they are built generally do not. It is difficult on the whole to relate more than a few types to megalithic structures elsewhere. There do, however, exist heel-shaped cairns in the central and southern regions, while the general shape of the keyhole monuments has been compared to that of sacred wells in Sardinia.

Though some types can be pinned down to definite areas, others—like the simple tumulus, a mere stone pile—are universal. I have been surprised sometimes by seeing types apparently far away from their usual habitat, such as 'crescents in relief' as far north as eastern Ahnet or a bordered disc with two long arms forming a V way down below In Afellahlah (Tenere Tafassasset). Lacking dates for either, though suspecting their relative antiquity in a post-Neolithic context, their presence close to major water-courses indicates that people moved about at least at certain seasons in regions where sedentarist occupation had ceased to be feasible.

A book on Saharan megaliths and cairns is long overdue, though I am hampered by lack of C14 dates as well as difficulties of communication about areas to which I have not been in person. As that wise counsellor R. Mauny remarked some years ago, the Saharan populations are known to us today mostly through their rock art and their stone structures.

19.6. ROCK ART

This is found in the form of paintings and carvings, the latter seemingly more numerous than the former, although the careful observer will see that a myriad paintings have vanished under the onslaught of the elements. They are mostly present today in well-sheltered abris or caves, often containing ancient habitation sites. Carvings occur just about anywhere where suitable rock is present.

Approximate ages of the periods—subject to constant revision—are given in Table 19.1. There are now indications that subdivisions are necessary in some cases. All efforts to locate an English-language publication giving the main rock art periods AND adequate illustrations of each have proved abortive hitherto: perhaps this will be remedied by the forthcoming work by Willcox. Careful perusal of illustrations by Lhote (1978) is recommended. His classic 1959 work refers mainly to paintings, of which he distinguished twelve main styles, enough to deter all but the most determined student.

A most painstaking essay on the subject in general is that by Forde-Johnston (1959), subject to change necessitated by the passing of time. All illustrations are of carvings. Kuper (1978b) has asked, and partially answered, some useful questions embracing many aspects of the subject. A technical study devoted largely to dating methods and problems, periods, genres, techniques and styles is that by Ki-Zerbo (1981). Surprisingly enough, the Saharan horse and chariot is stated to have appeared about 6000 B.P., marking the emergence of what are presumed to be the earliest wheeled vehicles of all time.

Few studies hitherto have achieved much in solving a major problem, that of chronology established by scientific dating techniques, which are possible at least in the case of paintings. In this field Mori (1978) has established an impressive lead. Carvings are quite another matter, for the reasons I gave in 1978, though excavation of stone monuments apparently associated to carvings was suggested.

In 1958 an opinion was proffered to the effect that there is little to be learned from rock art that cannot be learned as easily, cheaper and with far greater certainty by systematic and thorough investigation of other types of evidence. It is just possible that this view crystallized after a period spent in close but involuntary contact with a group of pure rock-art theorists prone to suffer rises in blood-pressure when their own view is questioned.

FIG. 19.4. In this decadent Ahnet painting we encounter the frequent difficulty of identifying objects portrayed in rock art. Regardless of whether human and bovid are connected, it is not clear what the human is doing.

FIG. 19.5. Described by its finder, A. Simoneau, as a 'fairly megalithic design', this unusual carving of Imaoun, Moroccan Sahara, has no apparent truly megalithic structures nearby. (Photo A, Simoneau.)

Actually the value of rock-art studies as a source of history is becoming increasingly recognized by specialists in other fields (compare Striedter, 1978). As is to be expected, controlled and disciplined use of the imagination by such people results in a generally more valuable contribution than that possible by individuals whose sole interest is rock art. However, some of the latter have produced valuable pictorial records.

19.7. CONCLUSION

To illustrate our general lack of knowledge, I propose to end by presenting a situation to which I do not know the answers, in the hope of stimulating further research. It is further necessary to underline the unknown function(s) of certain artefacts, including those whose existence may never have been reported, coupled with the enigmas of comprehension—to the twentieth-century brain—posed by numerous objects portrayed in rock art.

In a south Saharan context, we find ourselves in a semi-upland zone of sand, liberally strewn with varying sandstone outcrops, often much eroded. Occasional lone hearths are in evidence, as are single static querns. The quantity of non-warlike lithic assemblage around the base of some outcrops leads one to surmise their association with the myriad carved bovids adjacent. Pottery is not often to be seen, except at one magnificent site high up and well above the desert floor.

This is a flat open space, flanked to south by a main shelter with carvings of bovids and humans, some painted in red in addition, while to west lies a high rock wall bearing decadent images of large wild animals. A fragment of an elegant stone receptacle lies at our feet; close by is a flattish stone, deeply embedded in the ground, with a number of probably functional cup-marks of varying dimensions in its horizontal upper surface.

A short way off to south is another rock pinnacle bearing faint carvings of Great Wild Animals on its vertical flanks, with later superimpositions; in the eastern wall is a cave with many grooves and cup-marks in the floor, plus Libyco-Berber/Tifinagh inscriptions and carved sandal-imprints. Scouting around the neighbouring outcrops we come upon strange 'pigeon-hole' niches. Whole fields of 'crescents in relief', facing both east and west, are evident in the intervening flattish ground: one measures around 250 paces from the tip of its southern extremity to the centre of the structure, while certain smaller specimens are actually built upon outcrops.

At intervals are numerous saddle-querns, upper stones of many shapes and dimensions, interspersed with other eroded items of tool-kit: the majority of these latter are comprised by flaked and polished axes, some of the former evidently re-sharpened and now seen as a usable blade on an otherwise near-fossil instrument, A broken stone rod has its sole surviving extremity very much in the shape of a polished axe blade. A grooved axe (or can it be a grooved pestle?) enters our field of vision. To complete this inventory, there is a strange ovaloid flattish blade, sharpened all round and notched on either side of its longest axis. It measures about 12 × 9 cm. If the notches assisted in affixing it to a wooden haft, then why does one need one continuous cutting-edge, some of which would be rendered unusable by the haft?

No warlike/hunting tools are seen here. But another surprise is in store. Surmounting a sandy ridge, we come to a mass of oval depressions, both vertical and horizontal, around the north face of yet another outcrop. It is at last evident that the strange depressions that have puzzled us for years can be interpreted as sites for the extraction of saddle-querns. One is all ready to be removed, though just how this may have been effected is a mystery. But is looks as though a huge amount of grinding went on long ago, so much so that the population could no longer (?) rely upon finding suitable lumps of sandstone lying around: seemingly unextracted querns could indicate abandonment of the area in a hurry.

The above phenomena are recounted solely in the hope of encouraging further research. Were we able to relate all this evidence of human activity to definite periods in time, our knowledge of Saharan Prehistory would have taken a huge step forward.

FIG. 19.6. A saddle-quern of the southern Sahara in process of manufacture. Beneath the hammer-handle is a larger embryo: the straight groove may well have served for tool-sharpening, though only once that particular embryo had been rejected. Many such sites occur on vertical rock walls, in spots where it would today be awkward to work without a ladder.

ACKNOWLEDGEMENTS

The author wishes to thank the following for allowing use of their material as illustrations: W. J. G. Godwin, Lagos; Dr R. Kuper, Cologne; Mme A. Simoneau, Nice; Prof Dr W. Taute, Cologne.

REFERENCES

BECK, P. and HUARD, P. (1969) *Tibesti. Carrefour de la préhistoire saharienne.* Arthaud, Paris.

BALOUT, L. (1981) The prehistory of North Africa. In *General History of Africa. I. Methodology and African Prehistory*, edited by J. KI-ZERBO, pp. 568–584. UNESCO and Heinemann, Paris and London.

CAMPS, G. (1974) *Les civilisations préhistoriques de l'Afrique du Nord et du Sahara.* Doin, Paris.

CAMPS, G. (1982) Beginnings of pastoralism and cultivation in north-west Africa and the Sahara: origins of the Berbers. In *The Cambridge History of Africa.* Volume I. *From the Earliest Times to c. 500 BC*, edited by J. DESMOND CLARK, pp. 548–623. Cambridge University Press, Cambridge.

CAMPS-FABRER, H. (1978) Steinplastiken aus der Wueste. In *Sahara. 10.000 Jahre zwischen Weide und Wueste*, edited by MUSEEN DER STADT KOELN, pp. 246–252. Museen der Stadt Koeln, Cologne.

CLARK, J. DESMOND (1978) The legacy of prehistory: an essay on the background to the individuality of African cultures. In *The Cambridge History of Africa*. Volume 2. *c. 500 B.C.–A.D. 1050*, edited by J. D. FAGE, pp. 11–86. Cambridge University Press, Cambridge.

CLARK, J. DESMOND (1980) Human populations and cultural adaptations in the Sahara and the Nile during prehistoric times. In *The Sahara and the Nile. Quaternary environments and prehistoric occupation in northern Africa*, edited by M. A. J. WILLIAMS and H. FAURE, pp. 527–582. A. A. Balkema, Rotterdam.

CLARK, J. DESMOND (1982) The cultures of the Middle Palaeolithic/Middle Stone Age. In *The Cambridge History of Africa*. Volume I. *From the Earliest Times to c. 500 BC*, edited by J. DESMOND CLARK, pp. 248–341. Cambridge University Press, Cambridge.

COPPENS, Y. and BALOUT, L. (1981). Hominization: general problems. In *General History of Africa*. I. *Methodology and African Prehistory*, edited by J. KI-ZERBO, pp. 400–436. UNESCO and Heinemann, Paris and London.

FORDE-JOHNSTON, J. (1959) *Neolithic Cultures of North Africa*. Liverpool University Press, Liverpool.

GABRIEL, B. (1978) Die Feuerstellen der neolithischen Rinderhirten. In *Sahara. 10.000 Jahre zwischen Weide und Wueste*, edited by MUSEEN DER STADT KOELN, pp. 214–219. Museen der Stadt Koeln, Cologne.

GALAND, L. (1979) *Langue et littérature berbères. Vingt cinq ans d'études*. C.N.R.S., Paris.

GAUTIER, A. (1980) Contributions to the Archaeozoology of Egypt. In *Prehistory of the Eastern Sahara*, edited by F. WENDORF and R. SCHILD, pp. 317–344. Academic Press, New York.

HAYS, T. R. (1975) Neolithic settlement of the Sahara as it relates to the Nile Valley. In *Problems in Prehistory: North Africa and the Levant*, edited by F. WENDORF and A. E. MARKS, pp. 193–204. Southern Methodist University Press, Dallas.

HUGOT, H. J. (1981) The prehistory of the Sahara. In *General History of Africa*. I. *Methodology and African Prehistory*, edited by J. KI-ZERBO, pp. 585–610. UNESCO and Heinemann, Paris and London.

JACQUET, G. (1978) Au coeur du Sahara libyen d'étranges gravures rupestres. *Archéologia*, **123**, 40–51.

KI-ZERBO, J. (1981) African prehistoric art. In *General History of Africa*. I. *Methodology and African History*, edited by J. KI-ZERBO, pp. 656–686. UNESCO and Heinemann, Paris and London.

KRZYANIAK, L. (1978) New light on early food-production in the central Sudan. *J. Afr. Hist.* XIX.2, 159–172.

KUPER, R. (1978a) Einfuehrung. In *Sahara. 10.000 Jahre zwischen Weide und Wueste*, edited by MUSEEN DER STADT KOELN, pp. 7–9, Museen der Stadt Koeln, Cologne.

KUPER, R. (1978b) Sieben Fragen zur Felsbildkunst. In *Sahara. 10.000 Jahre zwischen Weide und Wueste*, edited by MUSEEN DER STADT KOELN, pp. 98–103. Museen der Stadt Koeln, Cologne.

LHOTE, H. (1959) *The Search for the Tassili Frescoes*. London, Hutchinson & Co.

LHOTE, H. (1978) Die Felsbilder der Sahara. In *Sahara. 10.000 Jahre zwischen Weide und Wueste*, edited by MUSEEN DER STADT KOELN, pp. 70–80. Museen der Stadt Koeln, Cologne.

MAUNY, R. (1978) Trans-Saharan contacts and the Iron Age in West Africa. In *The Cambridge History of Africa*. Volume 2. *c. 500 B.C.–A.D. 1050*, edited by J. D. FAGE, pp. 272–341. Cambridge University Press, Cambridge.

MERCER, J. (1976) *The Spanish Sahara*. George Allen & Unwin Ltd., London.

MILBURN, M. (1978) Towards an absolute chronology of certain Saharan rock art. *Antiquity*, LII, 135–136.

MILBURN, M. (1981) Multi-arm stone tombs of central Sahara. *Antiquity*, LV, 210–214.

MONOD, Th. (1932) *L'Adrar Ahnet. Contribution à l'étude archéologique d'un district saharien*. Institut d'Ethnologie, Paris.

MORI, F. (1978) Zur Chronologie der Sahara-Felsbilder. In *Sahara. 10.000 Jahre zwischen Weide und Wueste*, edited by MUSEEN DER STADT KOELN, pp. 253–261. Museen der Stadt Koeln, Cologne.

PETIT-MAIRE, N. (1981) Aspects of human activity in the coastal occidental Sahara in the last 10,000 years. In *The Sahara. Ecological change and early economic history*, edited by J. A. ALLAN, pp. 81–91. Menas Press Ltd., Outwell.

SALAMA, P. (1981) The Sahara in classical antiquity. In *General History of Africa*. II. *Ancient Civilizations of Africa*, edited by G. MOKHTAR, pp. 513–532. UNESCO and Heinemann, Paris and London.

SHAW, C. T. (1981) The prehistory of West Africa. In *General History of Africa*. I. *Methodology and African Prehistory*, edited by J. KI-ZERBO, pp. 611–633. UNESCO and Heinemann, Paris and London.

SMITH, A. B. (1980a) The Neolithic tradition in the Sahara. In *The Sahara and the Nile. Quaternary environments and prehistoric occupation in northern Africa*, edited by M. A. J. WILLIAMS and H. FAURE, pp. 451–465. A. A. Balkema, Rotterdam.

SMITH, A. B. (1980b) Domesticated cattle in the Sahara and their introduction into West Africa. In *The Sahara and the Nile. Quaternary environments and prehistoric occupation in northern Africa*, edited by M. A. J. WILLIAMS and H. FAURE, pp. 489–501. A. A. Balkema, Rotterdam.

SMITH, P. E. L. (1982) The Late Palaeolithic and Epi-Palaeolithic of northern Africa. In *Cambridge History of Africa*. Volume I. *From the Earliest Times to c. 500 BC*, edited by J. DESMOND CLARK, pp. 342–409. Cambridge University Press, Cambridge.

SPRUYTTE, J. (1977) *Etudes expérimentales sur l'attelage*. Paris, Crépin-Leblond.

SPRUYTTE, J. (1983) *Early Harness Systems—Experimental Studies*. J. A. Allen, London.

STEMLER, A. B. (1980) Origins of plant domestication in the Sahara and the Nile Valley. In *The Sahara and the Nile. Quaternary environments and prehistoric occupations in northern Africa*, edited by M. A. J. WILLIAMS and H. FAURE, pp. 503–526. A. A. Balkema, Rotterdam.

STRIEDTER, K-H. (1978) Felsbilder als Geschichtsquelle. In *Sahara. 10.000 Jahre zwischen Weide und Wueste*, edited by MUSEEN DER STADT KOELN, pp. 262–271. Museen der Stadt Koeln, Cologne.

SUTTON, J. E. G. (1981) The prehistory of East Africa. In *General History of Africa*. I. *Methodology and African Prehistory*, edited by J. KI-ZERBO, pp. 452–486. UNESCO and Heinemann, Paris and London.

TAUTE, W. (1978) Das Ende der Altsteinzeit in Nordafrika. In *Sahara. 10.000 Jahre zwischen Weide und Wueste*, edited by MUSEEN DER STADT KOELN, pp. 48–59. Museen der Stadt Koeln, Cologne.
WHITEHOUSE, D. and WHITEHOUSE, R. (1975) *Archaeological Atlas of the World*. Thames and Hudson, London.
WILLCOX, A. R. (1983) *Rock Art of Africa*. Croom Helm Ltd., London (in press).
WYMER, J. (1982) *The Palaeolithic Age*. Croom Helm Ltd., London.
ZIEGERT, H. (1978) Die altsteinzeitlichen Kulturen in der Sahara. In *Sahara. 10.000 Jahre zwischen Weide und Wueste*, edited by MUSEEN DER STADT KOELN, pp. 35–47. Museen der Stadt Koeln, Cologne.

CHAPTER 20

Indigenous Peoples of the Sahara

H. T. NORRIS

Near and Middle East Dept., The School of Oriental and African Studies

CONTENTS

20.1. THE DESERT AND ITS PEOPLE

The Spanish poet, Luis de Camoens, described the Western Saharans as those,

> Who never enjoy fresh water
> And whom the grass of the fields inadequately supplies,
> A country you may say ill disposed to bear any fruit,
> Where the birds consume iron in their bellies,
> A land suffering extreme want of everything.

How true a description is this of the Great Desert? How true is it of all its 2 million people who live amidst its some 3,700,000 square miles of sand, rock and oasis date groves? The Sahara is indeed a region of climatic extremes. Rainfall varies from one year to the next. Often 2 years can pass without rain of note. Yet, sudden downpours can be a brutal blow to the nomads, precipitating floods in the *wādis*, destroying their tents but leaving a carpet of green herbage for their animals.

Parts of the Northern Sahara enjoy the seasonal rainfall of the Mediterranean lands, while the southern edge of the desert receives the heavy summer rains of the Sahelian *hivernage*. On the foggy Atlantic coast of the Rio de Oro there are autumn rains, conversely on the Gulf of Syrtes rain falls during winter and spring. Deep in the desert there are highlands which are often well watered; the Mauritanian Adrar, Guelta Zemmour in the Western Sahara and the Adrar of the Ifoghas in Mali where outlying Sudanic vegetation grows 200 miles to the north of the Sahelian border. Life with the herds is possible in the Central Sahara because of three high Massifs; Tibesti, the Ahaggar and Aïr (Asben), the latter enjoying 2 to 4 inches of rain annually. There are pasturages within these Massifs. There are also barrages, fields and date groves.

Here nomads may at times stay in one valley, sow their cereals or harvest wild grasses. On the edge of the Sahel, the Moors and the Tuareg will move north in summer with their cattle in order to browse on desert pasturage, while in the dry season their herds will move south into Mali and Hausaland in order to consume the dried straw or else drink at the water holes and lakes which retain water over long periods.

Mobility is characteristic of the Saharans. In their combat with drought they diversify these activities. They diversify their herds between camels, sheep and goats and cattle, they wander far or remain within certain valleys or around a water-hole. They store food or sell their animals or engage in fall-back activities such as hunting, or, in earlier days, they launched raids over great distances. Rain and pasturage has coloured the religious outlook of the Saharan peoples. Their holy men are also their rain-makers.

But the Saharans are not only herders and shepherds. They were, or are, great caravaneers. Their caravans once crossed the sand sea in order to collect the gold of Ghana and Sudanese slaves. They brought prized gifts and objects to North Africa; giraffes, ivory and ostrich feathers, hides, copper and alum. But it was the Saharan salt which provided the most important commodity in Saharan commerce. The salt mines were, and still are, key points on its ethnographical map. Ijjil in Mauritania, Taghaza and Taoudeni in Mali and Amadror in the Ahaggar are mines of rock salt worked by the Moors and the Tuareg and their former slaves.

A Mauritanian scholar has written in his *Kitāb al-Wasīt* regarding Ijjil,

> This salt is cut from the Sabkhat Ijjil. The latter is located some five or six days distance from the town of Shinqīt. It is owned by the Kunta. The people pay them a paltry sum in relation to that which they get in return. People travel from the Hodh and Tīshīt and the Rgība and Tagānit for the salt. The majority of them buy it from the town of Shinqīt, but sometimes some of them went to the Sabhka itself, and took it there. Then all of it is sold in the Sudan.

Kawar/Agram and Tagedda-n-Tasemt in Niger have major salt deposits obtained by evaporation. The saline plants which grow around the latter are the destination of the annual *cure salée* of the Tuareg and their camels. Adjacent Takedda was a key town of the medieval Sahara and nearby Azelik was a centre of copper production. Salt and copper ore combined under high pressure formed copper chloride which, when combined with coke, solidified into metallic copper. Ibn Battūta, the fourteenth-century traveller, has described how slaves made copper rods which were carried by the caravans and they purchased slaves, meat and wood, sorghum, butter and corn in the Sudan. When the nearby city of Agadez was established as the capital of a Tuareg Sultanate about 1430, the copper commerce of Takedda was destroyed.

The salt mines of Kawar and Bilma, which are located further to the east, belong by tradition to the Tuareg. However, the Kanuri people of the oases are the real owners and the workers. The salt is exchanged for millet in Damergu. The Kel Owi Tuareg are the master middlemen of the annual Kawar salt caravan (*taghalam* or *azelai*) which journeys south to Hausaland and which numbers thousands of camels. Kawar's population could not survive without the outlet possibilities these Tuareg furnish for their salt, yet without the Kawar salt trade the Tuareg themselves would suffer greviously since they would lose much of the economic surplus which has enabled them to become an important factor in the economic life of the Sudan. The Kawar salt production illustrates how different Saharan peoples have tried to use their particular assets in an inter-group or inter-regional relationship, by trading or raiding or by establishing political dominance.

Today, the principal Saharan peoples, from East to West, are the Tubu or Teda, the Tuareg and the Moors. Each has its own physiognomy, language and material culture, social organization and history. All, in certain respects, share common social systems and are the heirs to the prehistoric and early historical peoples of the interior of Libya and Numidia, among them the Garamantes of the Fezzan, who ruled there about 800 B.C. and whose horse-drawn chariots took them deep into the desert.

20.2. THE ARABS

Before describing these people individually, however, an initial introduction of the Saharan Arabs seems appropriate. It is the Arabs who, by the spread of Islam following their conquests in North Africa in the seventh century, have diffused their language throughout the desert and they have established small or large communities from the southern Atlas foothills to Timbuctoo and fom the Atlantic coast to the Nile. The Arabs may be likened to an ethnic 'cement' which bonds the other Saharan peoples together. They have marked the culture of these latter and have appended them to the greater Arab World. There are Arab peoples in Cyrenaica, the Fezzan, Chad, and southern Tunisia and Algeria. Some are bedouin, others traders, others the flotsam and jetsam of the waves of Islamic conquest which passed through the Sahara in its southward expansion into Black Africa.

The Saharan Arabs believe that the earliest Arab expeditions left descendants of the Arab commander, ʿOqba ibn Nāfiʿ, who died near Biskra in 683 in the Sahara. Some Kunta of Mauritania claim descent from Shākir who was a companion of ʿOqba. He was buried at Ribāt Shākir in Morocco. The historian, al-Bakrī (1068), mentioned wells which were attributed to the Arabs. The cameleers' well (bi'r al-jammālīn), on the route from Marrakech to Ghana, was allegedly dug by ʿAbd al-Rahmān ibn Habīb who was the Arab governor of Tunisia in 745. A previous Saharan expedition had occurred in 734. In the eleventh century during the push of the Saharan Almoravid rulers of Morocco towards Ghana and Gao many Arabs intermarried with the Berber Saharan tribes. At about this time the so-called Hilālī 'invasion' of North Africa took place. The eastern bedouin, the Banū Hilāl and Banū Sulaym, were in Tripoli in 1037/8. In 1052, at Haydarān, they defeated the Zīrid Sultan of Tunisia, Muʿizz ibn Bādis. This 'plague of locusts', as Ibn Khaldūn described them, had already settled in the Egyptian Sahara at a far earlier date. Ibn Hawqal (d. 988) described them as nomads who wandered with their flocks and herds but who returned in the summer to the Egyptian oases in order to harvest what they had sown. They intermarried with the Hawwāra Berbers. The Hilālīs bequeathed a rich tradition of poetry and folk epic. Although the Arabs of the Eastern Sahara, for example the Awlād Sulaymān of the deserts of Chad, arrived there far later from Libya, their way of life accords with the ancient Hilālī pattern.

Characteristic of the Arab bedouin of Libya, Tunisia and Algeria, more especially the Tunisian Souf, is semi-nomadism. The Arabs pasture their herds in the Chotts in spring and they live in tents or palm-branch huts in the oases where they have stores for dates and corn. The tribe (ʿarsh) is fundamentally an extension of the patrilineal family. Often a number of tribes share a common eponym whose tomb is known to them all. A sort of feudal system prevails whereby certain families are classed as mrābtīn or vassals within one tribe. For a tribute, which by hereditary obligation each mrābit family must pay, the weapon-bearing family to which it is attached is bound to defend and protect the mrābit family. The latter can herd their flocks in peace and offer religious and spiritual services to the group. As will be seen this sacerdotal class characterizes the Saharans, Arabs and Berbers alike.

Muhammad al-Marzuqī, himself of Tunisian bedouin origin, has described the life of the Saharan Arabs. Much has now changed through sedenterization. He remarks:

> The nomadic bedouins of the south have customs and traditions in regard to foraging and transhumance from one season to another and from one place to another. The vast majority of them used to divide the year into two parts, spending six months in the village and six months in the desert, moving between its pasturages and its distant water points. To the village belong the summer and the autumn. These are the seasons of cereals and crops, more especially the date season among the oasis dwellers. To the desert belong the winter and the spring. These are the seasons when the sheep and camels are born, the seasons of milk, butter, wool and the hunt. From all this it is apparent that the bedouin was good at planning in dividing his year thus, for he had no need to forsake the fruits of the village and the shade of its walls and the freshness of its waters in the summer and autumn. Likewise, he did not cease to watch over his camels and his sheep, and to derive benefit from what they had abundantly provided in milk and its products, and to enjoy the good things of the desert in the chase and the products of its trees and its grasses in winter and spring. . . .
>
> The bedouins of the Tunisian south used to live a simple life, depending, as they did, on resources which were limited to three things, tilling, commerce and handicrafts. These formed the basis of their livelihood since the turn of the century. In the last century, and before it, the tribes used to combine with these resources other activities; raiding and the forceful seizure of supplies. However, this profitable activity was confined to a few tribes which were known for the great number of their horsemen and the weakness of religious curb on their activities. In their proverbs they said, 'Provision is from tillage, inheritance or quarrelling'. In contrast to these tribes there were others the members of which were noted for their piety and the strength of the religious restraint. They showed no hostility against another's person although this did not prevent them from defending

themselves and protecting their livelihood, taking their revenge on him who had violated their sanctity. Though they eschewed the inflicting of injury on others they used to undertake the role of mediator between opponents in conflicting tribes in order to establish security and to return the aggressor to the straight path he should follow. (From *ma'a 'l-badū*, Tunis, 1980, pp. 105, 137.)

20.3. IBĀDĪ KHĀRIJITE BERBERS AND ARABS

Linked to the Arabs are the Ibādī towns and villages. Once, the whole of the Tunisian Djerid, the Jabal Nafūsa and Ghadames in Libya, Sijilmāsa in Morocco and the Drā were the domain of heretical Muslims known as the Khārijites, who revolted in the East in the seventh century and who found a refuge in Berber Africa. From their centre at Tahert in Algeria in the eighth and ninth centuries their merchants headed west and south towards Ghana to engage in commerce in gold. Khārijite architecture and austere practice were spread across the Sahara by these merchants. Today, the heart of the Ibādite Khārijite Sahara is situated in the Mzab region in the Algerian Sahara. Its urban Berbers are date-growers, merchants and traders who practise the strictest puritanism. Their ancestors first settled near Ouargla where they planted a date-palm grove, but in the eleventh century they moved to the Mzab for protection. Their government of Elders ('*azzāba*) was theocratic yet democratic. In time there evolved not a community of mystics, common in the Sahara, but a trading theocracy. The Mozabites excel as merchants and shopkeepers inside and outside the Sahara. Their womenfolk are secluded within these walled and gated, vault-covered complexes. Beni Izguen is possibly the best surviving town.

20.4. THE TUBU OR TEDA

The Tubu or 'Rock people' are the most easterly Saharans. They differ markedly from the rest. The heart of their realm is the Tibesti Massif which rises to 11,200 feet. It contains a population of some 12,000 Tubu. The latter are sometimes called the Teda. This is a linguistic term; it denotes on the one hand the Tubu of Tibesti, Kawar, Jado, formerly Kufra, the Fezzan, Borku and Ennedi, and, on the other, the Sahelian Teda Daza of Kanem, Wadai and Darfur in the Sudan Republic. Ancestors of the Teda Daza must at some time have moved south from the Eastern Sahara. Their historical fame lies in the fact that they were the ruling aristocracy of the kingdom of Kanem which dominated the Kanuri of the area around Lake Chad in the later Middle Ages.

In all, the Teda may possibly number some 200,000 people, though most of these are not Saharans. The majority live to the south of 18° latitude. The Teda blood-group is quite distinct from the Tuareg. The Teda language is related to Kanuri. Physically, the Teda are dark, though the Tubu of Tibesti are Eastern Hamites akin to the Beja and Somalis. Negro features are pronounced among the Daza near Lake Chad. The Tibesti Tubu are lean and agile. They live frugally on dates. It is said that a Tubu can live for 3 days on one date, the first on its skin, the second on its pulp and the third on its kernel. They collect edible grasses and they grind the fruit of the doum palm for flour. Their common domestic animals are camels, asses, sheep and goats. They live in huts made from doum palm branches or from acacia roots. Occasionally, their dwellings are built of stone and they use mats to make tents.

There is a sexual division of labour in Tibesti. The men herd camels, half of which are for riding, they hunt and engage in trade and they used to raid. They fabricate their own weapons; swords, daggers, throwing spears and knives, and utensils and equipment. The women tend the goats, they fetch wood and water, cook, tan leather and weave baskets and matting. Many Tubu are wholly nomadic and practise no agriculture. Others own date-groves and a large proportion of the present population of Tibesti is semi-nomadic. Some members of the family remain permanently in the gardens in Central Tibesti, especially Bardai and Zumri districts, while the others roam with their livestock from pasture to pasture.

The Tubu are made up of forty patrilineal clans. The clan is the foundation of their society and it is endogamous with in-law avoidance. Family ties are loose compared with other Saharan societies. Property is either in the clan, possession of a valley for example, or it is individual, for example possession of herds, tents and palm trees. They are suspicious of strangers and have earned a reputation for hostility to Europeans. This attitude is hardly inspired by religious fanaticism. The Islamization of the Tibesti region has been superficial save in Agram and Kawar and there are many pagan survivals including the adoration of trees.

Tubu society is divided into classes, the nobles—Tomagra, Gunda and Arna—whose chief, the Dardai, is elected by the council of the Tomagra, the commoners who are all the plebian clans of Tibesti and the Daza of Borku, the smiths (*aze*) who are a degraded caste, the serfs (*kemaje*) who till and tend the gardens and look after the palm-groves of their overlords, and finally domestic 'slaves' or servants. Caste structures are a feature of all Saharan societies but the Tubu are distinguished from the rest by the absence of a sacerdotal class of *mrābtīn*.

20.5. THE TUAREG

Among the Saharan peoples it is the Tuareg who have achieved the greatest fame. They inhabit the Sahara between 14° and 30° North latitude and 5° and 13° East longitude. Their name is hard to explain since they do not use it themselves. Tuareg is possibly derived from Targa, a district of the Fezzan which was the original homeland of many of the Tuareg groups. The Arabs dub them *Tawārik*, the 'abandoners of Allāh', the godless ones, and it is true that in general the Tuareg have been viewed as lax Muslims, though there are many pious Tuareg scholars in Mali and Niger. There are Christian survivals in some of their customs, the cross motif in their ornament (*teneghlit*) may be pagan, and there are Latin borrowings in their language. They speak Tamahaq, a dialect of Berber, and a few of them still use an ancient Libyco-Berber alphabet called Tifinagh.

The Tuareg call themselves Imohagh or Imajughen, 'noble ones'. They are the 'veiled people'. The mouth veil is donned by males when striplings and is sometimes worn loosely, sometimes covering the entire face save the eyes. The veil may have originated as a mere protective covering. Today there are many social and semi-religious traditions attached to the wearing of it. Tuareg women do not veil and are remarkably free for a Muslim people. The following is a description of a Targui by the Tunisian merchant Shaykh Muhammad ibn 'Uthmān al-Hashāshī who crossed their territory in 1896:

> The Tuareg wear white or black shirts, the sleeves of which, very long and floating, trail on the ground if they do not raise them. They wind a turban around the head. Sometimes the crown of the head is uncovered. They shave the head but keep a band of hairs of three fingers breadth or more which grows from the nape of the neck to the forehead. The head is covered by a veil, which they never remove, save sometimes if they need to eat. The men carry amulets around the neck, at least six, enclosed in a small skin pouch. I was held in great regard by them and I wrote more than one hundred and fifty of such amulets for their use. The chiefs wear caps some seven inches in height. They are made in Tunis. The Tuareg do not practise polygamy. (*Voyage du Pays des Senoussia*, Paris, 1912, page 181.)

The Tuareg are tall, long-headed Caucasians. They are independent, brave, impulsive, mendacious, chivalrous to women, fond of poetry and music, and, amongst the youth, delight in 'Courts of Love'. They are among the finest of all cameleers. In all they number upwards of 300,000, excluding slaves and ex-slaves who form or formed a large proportion of the community, one-third to one-sixth among the Ahaggar Tuareg of Algeria, and up to one-third in Aïr in Niger. One-half of all the Iwellemmeden Tuareg of Mali were slaves or ex-slaves.

It is probable that the Tuareg are the descendants of the people who introduced the camel into the Sahara before the Arab invasion of the seventh century. However, some of the noble Tuareg emigrated from Libya, traditionally from Awjila, after the Arabs had arrived and it is likely that in material culture, their

FIG. 20.1. A group of Tuareg of the Kel Aghlāl during afternoon prayer to the south of Agadez in the Niger Republic. This region has a number of scholars who are equally expert in Arabic and Tamacheq, the Tuareg language.

black skin tents with T-shaped posts for example, and in their class social structure, the Tuareg had come under Arab influence at an early date, though they have retained their language and remain distinct. The Algerian Tuareg of the Ahaggar and Ajjer are only 3% of the total. By far the greatest number live in the Saharan districts of the Sahelian states of Mali and Niger. From West to East they are the Kel Intasar, the Iwellemmeden, the Kel Adrar and Kel Air. Despite famine and drought they survive as tribal units or in major confederations, herding camels or plying as caravan men, northwards to the Tidikelt and southwards to Nigeria, with dates or rock salt which they exchange for millet or they sell the surplus of their livestock, their butter or their cheese, depending on the season.

The Tuareg are divided into 'drum-groups' (*ettebelen*). *Ettebel* also denotes 'chief' or the drum of his authority. The groups in the Ahaggar are the Kel Rela, Taitoq and Tegehe Mellet, all three descended matrilineally from a half-legendary ancestress, Tin Hinan, who is buried at Abalessa. A group which is descended matrilineally is called a *tawsit*. Until recently the Algerian Tuareg were characterized by a matrilineal system. Descent was through the female line, the 'mother's house' (*tegeze*). In the succession to chieftainship a sister's son was preferred, or else maternal cousins. This was the ancient Tuareg system in Takedda and it survived quite late in the Sultanate of Agadez. Elsewhere, under Islamic influence, especially among the Iwellemmeden, descent is through the male line and the brother's son. Tuareg groups show some compromise between the two systems in regard to succession or to rights to property and inheritance.

A class, even 'caste' structure is found among the Tuareg. The Imohagh nobles are camel-herders. They elect the chief, the Amenokal, who is sometimes invested as a nominal *Imām* or Sultan. Beneath them are the clerics, the Ineslemen, who correspond to the *mrābtīn* among the Saharan Arabs in Tunisia, and even more so the clerics among the Mauritanian Moors. Beneath these two come the vassals, the Imghad, who are principally goat-breeders. Next come the 'slaves', the Iklan, who, if they are manumitted, are called the Irawalan. The artisans, the Ineden, make swords and saddles. Pre-Arab Saharan Berber society is most faithfully represented among the Tuareg, whom the medieval Arabs called the Sanhāja.

The Ineslemen class of the Tuareg, especially the Kel es-Sūq of the Adrar of the Ifoghas, represent the greatest attainments which the Saharans have achieved in spiritual profundity, in Sūfī mysticism, legal expertise and mastery of Classical Arabic. Here is a typical poem of theirs by Shaykh Muhammad Hāmma al-Sūqī, who lived during the last century. Allusions to rain and harvest, to caves and mountains are to be expected in the Malian Adrar which is both Saharan and Sahelian in its climate and vegetation.

Art Thou other than a Lord whose grace embraces all things?
That sinner whom Thou banisheth afar, lo how he is shunned,
Whilst him whom Thou brought close is by our side.
If a matter be one the Lord deems faultless and correct,
We deem Thee to be faulty in that matter, it is we who err.
Thou art the succour of the worlds, their helper
and a shower of favours, whereby the water and the yield
so longed for is stored in rock pools or is harvested.
Thou art a cave wherein we hide from ills when they
o'ershadow us and
a place where sickness which afflicts us is made good.
Were it not for the mountain summits firmly fixed,
Then all the hollows of the earth would shaken be,
And changed entirely in their form and hue.
Control or change the marvels of the age and govern them.
If not, then who will be the stripped, who the enriched?
Thou art among the shepherds of the creation, secretly or seen.
If Thou sleepest and ignore us for an hour or for a day,
Then we are those who in a desert lose their way.

20.6. THE MOORS

It is tempting to include the Moors among the Saharan Arabs. In fact they are a distinct and markedly individual people. They call themselves 'the Whites', *al-Bīdān*. They number about a million in the Western Sahara, extending from the Wād Nūn in Morocco to the Senegal River and from the Atlantic coast to the Adrar of the Ifoghas in Mali. As they are the dominant ethnic group in the Islamic Republic of Mauritania, among them half a million nomads, it may be said with justice that they are the one and only Saharan people to have created an independent state in the Sahara in the twentieth century.

The Moors are descended from the Berber Sanhāja who founded the Almoravid movement and empire in the eleventh century. They are cousins of the Tuareg. However, when sub-groups of the Arab Banū Maʿqil, known as the Banū Hassān, entered the Western Sahara in the fifteenth century, they dominated and intermarried with the Sanhāja and imposed their language upon them. The Moors are a mixture of Berbers and Arabs of South Arabian stock. Though their dress, indigo-dyed, somewhat resembles that of the Tuareg, the Moorish Saharan face muffler is looser and less formalized than that of their cousins. The Moors are shorter and more heavily bearded than the Tuareg.

Much of Mauritania is situated in the Sahel. The south of the country around the capital, Nouakchott, consists of 'dead dunes' which are covered with gum trees and acacia shrubs. There is pasturage for cattle and sheep. It is there that a handful of Zenāga Berber-speaking Moors live in goat- or camel-hair tents and teach the Koran or Arabic grammar to pupils who write on wooden boards instead of in exercise books. The true Sahara in the north of the country is made up of 'live dunes' and this is the territory of the great camel herds of the Rgaybāt and Tājakānt. The Saharan North is healthy and bracing. It is free of the malaria which afflicts the South. The Mauritanian Adrar has 400,000 date-palms and Tagānit is also rich in dates. In July those Moors who own garden plots in these districts go there for the *gaytna*, the date-picking festival. Wheat and barley are grown in the fields and plots, irrigated by irrigation canals (*sawāqī*) and well rope pulleys. The coloured tributaries, the *Harātīn*, who are ex-slaves, collect the dates in July and August. Then in September they sow the fields with melons and millet. In November more millet or beans are sown and the millet is harvested the following April or May. The Adrar region is famous for its medieval towns (*qsūr*) such as Shinqītī and Wādān which were once caravan centres. More famous still is Walāta which is situated in the province of the Hodh in the south-east near the Mali border. This town has artistically painted dwellings and it is built at no great distance from the site of ancient Ghana.

FIG. 20.2. A general view of Chinguetti in Mauritania. This town, which is located in the Mauritanian Adrar, gave its name to the whole area of Mauritania prior to the French occupation. The Arabs called it *Turāb Shinqīt*, 'the land of Chinguetti'.

FIG. 20.3. A view of the mosque of Chinguetti. Its minaret is capped with ostrich eggs. The architecture is very characteristic of the Mauritanian Adrar.

FIG. 20.4. Two Mauritanian smiths among the Awlād al-Nāsir. As among the Tuareg, the smiths (*haddādīn*) are a low social class in Saharan societies, although they are often poets, like the other artisans, and are feared for their magic powers.

FIG. 20.5. A young Moorish scholar at Walāta in the Mauritanian Hodh. In the later Middle Ages Walāta was an important caravan town on the southern edge of the Western Sahara.

The Moors are a tribal society. Each tribe (*qabīla*) is divided into a number of clans (*fakhdh*). Descent is patrilineal, although there is survival of matrilineal customs inherited from the Sanhāja Berbers. The Moorish woman has authority in the tent, a portion of which is hers where she keeps her utensils. She grinds the corn, milks the ewes, cards and spins wool and makes and decorates skin cushions and bags.

Moorish society is stratified into classes which match those of the Tuareg. The 'Arabs' (the *Maghāfira*) claim to be descendants of the Banū Ma'qil aristocracy. The Banū Hassān include all those who claim to have Arab blood, aristocracy or plebian, who speak the Mauritanian dialect of Arabic called Hassāniyya. Beneath the 'Arabs' are the Zawāyā, who are *mrābtīn*. They claim descent from the Sanhāja Almoravids or they are descendants of the Prophet or have an Arab pedigree. In former days they did not pay tribute (*hurma*), as vassal tribes, but engaged in teaching, well digging and caravan trade. They were protected by the Hassānī tribes which carried arms. The assembly of the tribe (*jamā'a*) is a council of chiefs who represent each fraction or sub-division of the tribe.

Beneath these 'free men' are a number of tributaries and 'slaves'. The tributaries—'white' (*Znāgiyya*) and 'coloured' (*Harātīn*)—are manumitted slaves. Officially slavery is now abolished in Mauritania. On the fringes of the whole social system are degraded classes, such as the dog-owning Nmadi of the Hodh who hunt antelope, addax, oryx, bustard, ostrich and eagle, the Imragen fishermen on the coast and the smiths and artisans all of whom once paid tribute and who are either nomadic or sedentary.

The Moors live in small camps. A *frīq* comprises fifteen tents, a *mahsar* up to fifty. The major camp of a Shaykh or Amīr is called a *halla*. In periods of danger in olden times each section of a tribe would send a contingent of warriors to a *mahsar*. The latter is a veritable nomad village and will often have a chiefly tent, tents for the artisans and a little mosque marked out in stones or by thorn branches in the sand.

Like the Tuareg, the Moors migrate far beyond their Saharan borders. Some go to Southern Morocco, others, particularly Zawāyā traders, settle in Senegal, Gambia and elsewhere in West Africa. By contrast there are many Wolof and Toucouleur in Mauritania, not only farmers on the banks of the Senegal but also as manual workers and technicians in Nouakchott and as far north as Zouérat and Fdérik where the iron ore workings are located.

20.6.1. Moors of the Rio de Oro

The Mauritanian historian, al-Mukhtār ibn Hāmidun, has provided the following sketch of Moorish relatives of the Mauritanians who inhabit the Rio de Oro. The territory is now disputed between the Moroccans and the Western Saharans. The nomads concerned are the Ahl Bārikalla:

In Mauritania, history is conceived as a chronicle of noteworthy events such as wars, droughts and the death of famous personalities. Such and such a year, for example, was the year of the death of Muhammad al Habīb by treachery, or the year of the meeting of the delegates which had assembled to conclude peace in such and such a dispute, or the year of the outbreak of war between the tribe of such and such, or the year they made peace, or the year of the eclipse of the sun, or the year of noted fertility, or the year of 'Dhrayr.

'Dhrayr was a famous year. It is said that a long time ago rains fell repeatedly in Tīris in the Rio de Oro. This lasted for some years. The whole district had abundant pasturage for the camels. In this situation miracles occur. For example, the little weaned camel is ridden before it is a year old and a she camel will give birth every year. It is said that the bulk of the wells were dug by the Ahl Bārikalla, the largest Moorish tribe in Tīris, at their own expense. They dwelt forty years in Tīris without seeing a funeral. Then Allāh decreed that some of the tribes should dominate others. They fought amongst themselves. Some were envious of others. Drought befell and the people fled to the south-west. It is said that the people of Tīris began to leave their offspring behind.

One of the events of that time concerned a man called Ibn al-Dīk. Another called al-Bukhārī fell in love with his wife. Al-Bukhārī 'ransomed' her from Ibn al-Dīk by a sum of money and he divorced her. While al-Bukhārī waited for her *'idda* waiting period of three months to pass before he married her, raiding began in Tīris. That woman departed among the women who escaped on foot from the battle. They were very hungry and they found 'cattle'* belonging to her lover. She slaughtered a 'calf'.* Her lover came to her and spoke roughly to her. He threatened to beat her and asked her to reimburse him for his 'calf'. One of the lettered said considering their circumstance:

> Behold the beloved of al-Bukhārī,
> See the familial tenderness for the young.
> She was a coy and timid spouse for Ibn al-Dīk.
> He sought to change her for another wife.
> By his ransom, al-Bukhārī split the two,
> To soothe a burning love and, for a while,
> to sip the goblet of his heart's desire.
> Her story was the talk of ladies' tents,
> The whole tribe took a pleasure in her tale.
> Men chatted through the night about divorce,
> While camels spread the news of her betrothal.
> Her skin was deeply dyed by costly clothes
> and sumptuous stuffs from India and Damas.
> For her, pure camel's milk was poured in bowls
> in plenty. Fate takes many turns.
> Her stars of fortune shortly set in murky heavens.
> Her winds changed their direction to the north.
> The rainless clouds massed in her wilderness.
> The ease of yesterday became a memory.
> Her slender, tender neck felt cruel hands—
> Only my Lord is Everlasting—thus she fared.
> We beg for Allāh's refuge in adversities.
> Until she sacrificed a female 'calf', born
> of a 'cow' abundant in its milk,
> belonging to the lover who enjoyed her fame.
> He summoned her to pay for it forthwith,
> After harsh words, with buffetings,
> He dared to threaten her with blows
> upon those cheeks he once in love caressed.
>
> O spirit wise, be warned, consider well,
> Renew your shedding of a tear, behold
> the gulf between the tenderness of love's tutor,
> and the cuffs of him who presses his demand,
> Between a kiss bestowed by sensual suitor
> and painful bruisings from his furious hand.

20.7. THE FUTURE OF THE SAHARANS

Much of the Saharan life which I have described has been changed out of all recognition by events of this century. The exploitation of iron and copper ore in Mauritania, uranium at Arlit in Niger and oil in Algeria and Libya has offered, it is true, new opportunities and employment to the Saharans, but it has sounded the knell on old tribal systems and customs and the liberation of the slaves has left no camp unchanged. Severe drought has driven a large number of the Moors and the Tuareg to starvation and total destitution. In the phosphate-rich Western Sahara, the Polisario movement is engaged in a struggle to reverse Morocco's annexation of former Spanish territories. In Algeria the Ahaggar Tuareg are being changed into good socialists and in other states the Tuareg are kept down by the military of the Sahelian states. In

---

* I have translated *baqar* as 'cattle' and *'ijl* or *'ijla* as 'calf', or heifer, as it appears thus in the Arabic original. A reference to bovids in the Rio de Oro is truly extraordinary! What appears to have happened is that the 'calf' of al-Bukhārī was in a herd of his in 'the southwest', the Trarza, the area to the north of the Senegal River to which his wife to be had fled. The whole story contrasts the Moors of the true Sahara with those who are in part Sahelian in their habitat. It shows the value the Sahelians place on cattle and the ignorance of their worth among the Moors of the Rio de Oro and the Western Sahara in general.

FIG. 20.6. A view of the mosque of Ghardaia in the Algerian Mzab. This city is one of a group of Ibādī Berber towns. The Mozabites who dwell in it enjoy a great reputation for piety and commerce throughout the Central Sahara. They are to be found in Kano and Agadez as well as in Algeria.

FIG. 20.7. The palace of Shaykh Mā' al-ʿAynayn, scholar and religious leader, at Smāra in the Western Sahara, a region of dispute between Morocco and the Polisario movement which seeks independence for this Saharan region. The *Qasba*, which contained a mosque, a college, a library and private quarters for the Shaykh and his family, was built at the turn of the century. It was designed by Moroccan architects in Saharan style.

Tibesti and Chad, the Tubu have joined the Front de Libération National and have died in the fratricide struggle in Chad. There is talk of an Arab Empire stretching across the Sahara from the Atlantic to the Libyan coast. Amidst Saharan turmoil and social distress it looks increasingly likely that Mauritania, however shakily, will remain the sole independent Saharan community and that the other Saharans will be absorbed by the Arab states to the north and the black Sahelian states to the south. More and more the Saharans will seek employment in Nigeria, Libya and Morocco. Yet, traditional Saharan life is unlikely to disappear entirely. It has an ability to adapt.

The Saharans have preserved a stability and continuity in the ideals and spiritual values of their societies, unruffled by the dust-storms of change which have blown across the desert. This sentiment has been expressed time and time again by Saharan poets. One Moor has left his testament:

> The day will surely come when I shall be a memory.
> When I shall only be a poem and an anecdote. My days will be over.
> Yes, you will question those old gossiping comrades who knew me so well.
> I know my death will not stop time. Nothing ever changes.
> Nawashid, Ekvis and Jouinni and the water pool
> Will see once more the encampments and the tents.
> Time is eternal;
> For it destroys the stones of the builders, it spares neither kings nor slaves.
> I also must disappear. Allāh, cover me with Thy grace when alone I stand before Thee.
> On Thou, the Immutable, the Merciful, the Most High, deceive not my hope.

SELECTED BIBLIOGRAPHY OF THE SAHARANS

General
BOVILL, E. W. (1970) *The Golden Trade of the Moors*. University Press, Oxford.
BRIGGS, L. C. (1960) *Tribes of the Sahara*. Harvard University Press, Cambridge, Mass.
LEWICKI, T. (1974) *West African Food in the Middle Ages*. Cambridge University Press, Cambridge.
UNESCO (1963) *Nomades et Nomadisme au Sahara*. Recherches sur la zone aride, XIX.
VIKØR, K. S. (1979) *The Oasis of Salt, the history of Kawar, a Saharan centre of salt production*. University of Bergen, Bergen.

Arabs
EVANS-PRITCHARD, E. (1949) *The Sanusi of Cyrenaica*. University Press, Oxford.
GELLNER, E. and MICAUD, C. (1972) *Arabs and Berbers*. Duckworth, London.

Tubu and Teda
CHAPELLE, J. (1957) *Nomades noirs du Sahara*. Plon, Paris.
CLINE, W. (1950) *The Teda of Tibesti, Borku and Kawar in the Eastern Sahara*, General Series in Anthropology, Number 12, George Banta Publishing Company Agent, Menasha, Wisconsin.
NACHTIGALL, G. (1980) *Sahara and Sudan*, Vol. 11, Kawar, Bornu, Kanem, Borku, Ennedi, translated by Allan and Humphrey J. Fisher, Hurst & Co.
TUBIANA, MARIE-JOSÉ and JOSEPH (1977) *The Zaghawa from an Ecological Perspective*. A. A. Balkema, Rotterdam.

Tuareg
ADAMOU ABOUBACAR (1979) *Agadez et sa Région*, Etudes Nigériennes, No. 44, Centre Nigérien de Recherches en Sciences Humaines, Niamy and Paris.
BERNUS, E. and S. (1972) *Du Sel et des Dattes*, Etudes Nigériennes, No. 31, Centre Nigérien de Recherches en Sciences Humaines, Niamey.
KEENAN, J. (1977) *The Tuareg, People of Ahaggar*. Allan Lane, London.
LEUPEN, A. H. A. (1978) *Bibliographie des Populations Touarègues*. Africa-Studiecentrum, Leyden.
NICOLAISEN, J. (1963) *Ecology and Culture of the Pastoral Tuareg*. The National Museum, Copenhagen.
NORRIS, H. T. (1975) *The Tuaregs, their Islamic Legacy and its Diffusion in the Sahel*. Aris and Phillips, Warminster.
RENNELL RODD, F. (1926) *People of the Veil*. Macmillan, London.
ZAKARA, MOHAMED AGHALI and DROUIN JEANNINE (1979) *Traditions Touarègues Nigériennes*. Editions l'Harmattan, Paris.

Moors

GERTEINY, G. (1967) *Mauritania*. Praeger, New York.

NORRIS, H. T. (1968) *Saharan Myth and Saga* (Oxford Library of African Literature). Clarendon Press, Oxford.

STEWART, C. C. and E. K. (1973) *Islam and Social Order in Mauritania*. Oxford Studies in African Affairs, Clarendon Press, Oxford.

Introduction à la Mauritanie (1979) Centre de Recherches et d'Etudes sur les Sociétés Miditérranéennes, Centre d'Etudes l'Afrique Noire, Editions du Centre National de la Recherche Scientifique, Paris.

Rio de Oro

BAROJA JULIO CARO (1955) *Estudios Saharianos*. Instituto de Estudios Africanos, Madrid.

MERCER, J. (1976) *Spanish Sahara*. Allen & Unwin, London.

TROUT, F. E. (1969) *Morocco's Saharan Frontiers*. Geneva.

CHAPTER 21

Oases

J. A. ALLAN

School of Oriental and African Studies, London

CONTENTS

What indeed is an oasis without palm trees? An uninhabitable pasture-ground with some paltry vegetation upon which, in the absence of the refreshing shade of its protector, an untimely death would, after a brief existence, descend . . . it is the comfort of the poor, a helper and a saviour for all (Nachtigal, 1879, p. 113).

21.1. INTRODUCTION

The oases of the Sahara have performed a key role in the history of the region. The oasis environment has always been a very special one mainly because of the significance of the human settlements located there but also because the oasis has water and agriculture which sustain them. It is water upon which the whole system depends and so there is some point in discussing the water resources of the Sahara and their potential for development in the light of the technology which has been used in the past and the new technology which has since the 1960s been deployed to support crops, livestock and human settlements. The chapter will conclude by examining current 'oasis' development which involves extensive tracts of irrigated farming made possible by modern technology for lifting and distributing irrigation water (Allan, 1979).

21.2. THE WATER RESOURCES OF THE SAHARA

Research completed in the past three decades has done a great deal to transform the awareness of the water resources of the Sahara. It has become clear that the minute and scattered oases reflect only a tiny proportion of the land which could be irrigated from the fossil and/or recharging waters contained in the Nubian Sandstone structures of the Sarir region, the Kufrah Basin, the Murzuq Basin, the Continental

325

Intercalaire of Algeria and of the Chad Basin (Pallas, 1980, p. 574; Burdon, 1980, p. 597; Unesco, 1972 a, b). The comprehensive exploratory drilling conducted since the 1950s together with the above scientific study has shown there to be aquifers capable of supplying water at the rate of the flow of the Nile at Aswan (55 cubic kilometres per year) for a number of years. Moreover, the water is of good quality and over substantial tracts of the northern Sahara the water is close to the surface.

The traditional oases were located at points where water flowed naturally, such as at al Kharga in south-eastern Egypt, at Braq in central Libya, or was clearly very close to the surface as evidenced by surface lakes such as those at al Kufrah. Other oases were sited in the dry valleys or wadis, for example Sabha. In any Saharan wadi (dry valley) water was likely to lie within a metre or two of the surface, and with the small demands made by traditional water-lifting systems the groundwater was safe from overuse. Water can be raised by animal or human power many tens of metres, but only in small quantities from depths over 10 metres. At shallower depths than 10 metres the skin bucket (dalu) system could irrigate up to a hectare of grain or vegetables, and with the persian wheel system larger areas could be commanded. When the water table was close to the surface or at up to 10 metres the palm trees and the oasis micro-climate could be sustained.

21.3. THE FUNCTION OF THE OASES AND THEIR LOCAL RELATIONSHIPS

Human settlements owe their origin to a variety of stimuli. Renewable and non-renewable natural resources attract settlement for the purpose of their development or extraction; traders require points at which their means of transportation and themselves can be fed or refuelled. In the Sahara to these two oases' functions can be added the role of sanctuary and that of a place of religious witness and related scholarship. In the eastern Sahara, the oasis of Siwa diverted the great Alexander from his incredible conquests in the fourth century B.C. He visited the oasis, now in western Egypt, to consult the priestess of Amon. More recently Jaghboub in eastern Libya and Kufra in southern Libya had reputations for

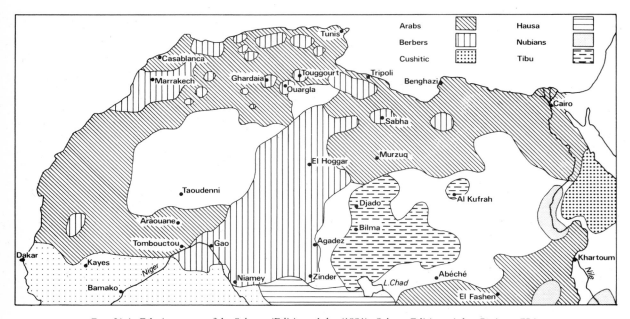

FIG. 21.1. Ethnic groups of the Sahara. (Editions Atlas (1981), *Sahara*, Editions Atlas, Paris, p. 75.)

learning, and were places sacred to the Sanusiyah sect, throughout the second half of the nineteenth century and even until the Libyan revolution of 1969. The Coptic monastries of the Wadi al-Natrun oases in northern Egypt are current examples of the religious role of oasis settlement.

The renewable and non-renewable resource base of the Sahara was poor until the discovery of oil in the 1950s in Algeria and Libya. Related to these developments was the discovery of groundwater resources which were clearly exploitable by modern well and irrigation technology. The past two decades have, as a result, witnessed dramatic changes in the perception of the potential of the Sahara in terms of mineral resources and agricultural productivity. Such attitudes contrast strongly with those of the past when oasis settlements were viewed in the 1860s, for example, as places capable of sustaining local subsistence but little more (Nachtigal, 1879, p. 119).

The only major resource of economic value located within the Sahara itself before the discovery of oil was salt. Oasis settlements grew up close to salt deposits at Idjil in Mauritania, at Taoudenni in Mali, near Tamanrasset in southern Algeria, south of Djado in Niger as well as in northern Chad. Salt moved mainly south, but it also transported to the north along with merchandise from the sahel and more humid lands of West Africa.

The movement of valuable merchandise had political consequences. Trade requires secure routes for transportation and oases have performed a major role in ensuring the safe passage of caravans. The oases settlers were sometimes clients of the local tribal group or confederation as in the case of the Tuareg in Algeria until the coming of the French in the late nineteenth century (Baier, 1978, p. 17). The arid environment encouraged the evolution of decentralized political systems amongst nomadic peoples explained by the general need for widespread movement to optimize the utilization of pasture and the occasional but unpredictable need to act in a unified way against an external threat. Individuals were organized in a series of widening and overlapping lineages descended from a common ancestor. The system of relationships enabled the resolution of conflict at local level as well as affording a ready identity for broader alliances when necessary (Baier, 1979, p. 6 and Peters, 1982, p. 106).

The nomads of the Sahara generally controlled the tracts between the isolated oases and could extract

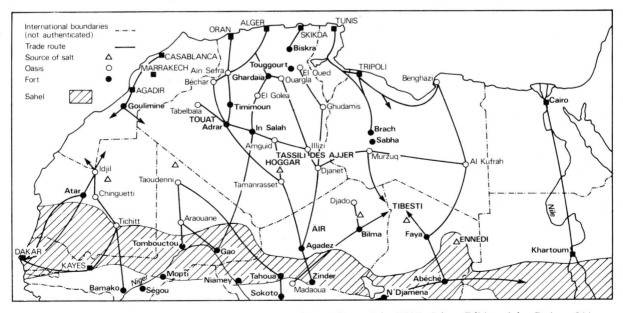

Fig. 21.2. Oases and trade routes in northern and western Africa. (Editions Atlas (1981), *Sahara*, Editions Atlas, Paris, p. 96.)

tribute from the traders who passed through. The groups of Sahara nomads had a major stake in the trade which passed through their territory. Thus the Regibat organized trade in the route in western Morocco, the Ait Khabbash (a segment of the Ait Atta) that on the Tafilalt–Tuat road, the Chaamba on the routes north of Tuat, the Tuareg on the Tripoli–Kano route, the Tibu on the Bornu–Fezzan road, the Mujabra and the Zuwaya on the Benghazi–Wadai, and the Khabbabish on the Darfur road to the Nile. The simple arrangement by which tribute was extracted from the transiting traders was complicated by the need for grain and other necessities which came from agricultural settlements on the margins of the desert or via the trading links of the oasis peoples. At the same time the military significance of the nomads was limited to the remote areas where their mobility gave a tactical edge but one not of much value in the better watered and consequently more populous margins of the desert.

Saharan traders and trading nations involved with them accommodated to the security realities of their time or intervened to control the trade routes or at least the oasis foci which marked them. The Romans were amongst the first recorded as moving into the desert. In the first century A.D. a Roman force advanced to Fezzan in what is now southern Libya and defeated the Garamantes. Roman trade as a result was secure thence and further to the south. Ottoman influence was exerted at various times and notably in the early nineteenth century when the Turkish authorities revealed themselves adept in manipulating segmentary lineage systems, as they were familiar with such socio-political forms. Not so the French who occupied Algeria, who in doing so caused the trade via Tamanrasset to decline dramatically in the second half of the nineteenth century. Meanwhile the main trading nation of Europe at that time, Britain, sought knowledge about the region by commissioning scientific exploration by Lyons, Oudney, Clapperton, Denham and Laing, as well as encouraging other Europeans such as Barth, to travel in Saharan Africa. Britain had located a vice-consulate at Murzuq in the mid-nineteenth century on the then major trade route linking Kano and the southern Sahara with Tripoli, although Barth remarked in 1850 that Murzuq 'is rather the thoroughfare than the seat of considerable commerce' (Barth, 1890, p. 77) and Nachtigal said that by the 1860s Murzuq 'had long since lost its importance as a trading centre' (Nachtigal, 1879, p. 87). There was a resurgence in the Kano–Tripoli trade in the 1870s but the negative influence of European nations on Saharan trade did prevail. Their alternative sea routes superseded the desert trade with West Africa and in addition their pressure on the Ottoman rulers of the Mediterranean ports brought about the termination of the Saharan slave trade. This pattern of manipulation by the northern coastal power of the authority and trade in the central Sahara was not echoed on the southern margin of the desert. There was a greater tendency to involve nomads in the economy of the fertile areas, with participation in commerce and agriculture (Baier, 1978, p. 8).

The changing political context in the Sahara over the past three millenia is a complex and fascinating story and one which substantially explains the manner in which the environment was managed. Since the 1880s, the progressive domination of the Sahara by European powers and subsequently by the successor independent governments has transformed the relationship of the peoples with each other and with their environment. Dunn (1977, p. 225) argues that the French advance posed the desert peoples with a new set of economic uncertainties. Compromises with the French necessitated unpleasant choices with respect to tribal allegiances, and on occasions led to more intensive resource use practices necessary to avoid submission to the invader. Such was the case in Morocco, but the Tuareg in Algeria also resorted to new land use strategies. They transformed the basis of their economy by extending cultivation with the labour of dependent agriculturalists. In due course as the French occupied the oases of the central Sahara the Ahaggar responded by using pastures in north-western Niger and by engaging in caravan trade moving salt to countries on the northern edge of sedentary agriculture near Damergu. These examples show the strength of the desert confederations, and their flexibility in responding to human, as well as the eternal environmental, challenges.

The history of these years also revealed the process by which the central Sahara was dominated. European forces armed with a military technology created in the factories of Europe captured the oases,

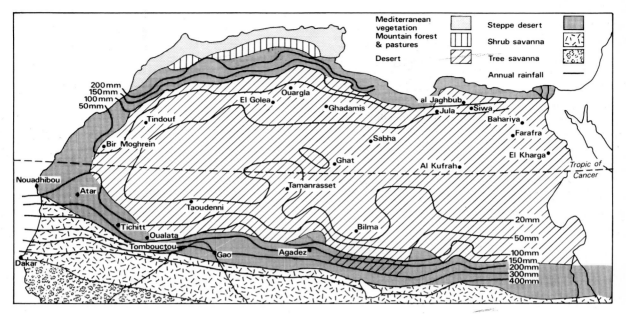

Fig. 21.3. Oases in relation to rainfall and vegetation. (Editions Atlas (1981), *Sahara*, Editions Atlas, Paris, p. 43.)

the French in the 1880s in Algeria and in what is now Mali and later in Niger and Chad, the Italians in southern Libya in the 1920s. By taking the oases the desert was effectively controlled militarily; by reducing the trading function the economic significance of the oases communities was eroded. Despite these assaults the traditional networks have often been sustained. One remarkable example of the durability of the links between a nomadic group and the oases within its sphere is that of the Zuwaya of eastern Libya. The tribe controlled the camel-based trade of the region throughout the nineteenth century and when the transport technology changed it adapted, and continued to finance and organize the truck traffic from Benghazi, through Ajdabiyah, Jalu, al Kufrah, and onto the Sudan and Chad (Peters, 1982, p. 112). Even more skilful was the political nimbleness of the group, in that it achieved a place in the highest councils of both the Idris regime and the revolutionary government since 1969. An even more extensive and significant network was that of the Sanusiyah, a religious community which from the middle of the nineteenth century used the oases of the central Sahara as a means of diffusing their fundamentalist Islamic ethic through the foundation of lodges (*zawiyah*) which had initially a religious and educational function but which eventually assumed a commercial role and finally became the dominant political influence in the eastern and north-eastern Sahara. This was *par excellence* an oasis network and points up the deterministic nature of the oases and their linkages on economic and political affairs of the desert.

21.4. AGRICULTURE IN THE OASES

One of the essential activities in the Saharan oases was the raising of crops. Studies of prehistory indicate that the Sahara was a different environment before 10000 B.P. and that there was some amelioration in the arid conditions *c.* 3500 B.P. Hunter gathering, livestock rearing and crop raising have all been proven as a means of livelihood in the north-eastern, central and southern Sahara (Shaw, 1981, pp. 102–104, 127 and 129). Crops, other than vegetables, have always had to accommodate to the differing climatic regimes of

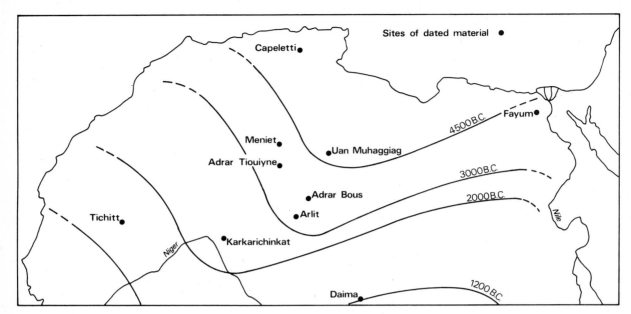

FIG. 21.4. The spread of cattle in northern and western Africa. (Shaw, T. (1980), The Late Stone Age in West Africa, in J. A. Allan, *The Sahara: Ecological change and early economic history*, Menas, London, p. 127.)

the northern and southern parts of the desert, and the choice of crops grown is still determined by the temperatures of winter and summer, with temperate grains possible in the winter and African grains in the summer. The relatively high temperatures throughout the year enable the raising of vegetables throughout the year and where water could be economically raised there has always been a sound rationale in taking advantage of the constant availability, throughout the year, of solar energy for agricultural production.

The fragmentary evidence for crop production from the prehistorian was not significantly updated until the nineteenth-century travellers published their chronicles. In Bornu Barth talked of rice growing wild, and of millet (*Pennisetum typhoideum*) and sorghum being plentiful. Wheat had been introduced a hundred years before. Onions were a favourite food of the Arabs and groundnuts were a very important article of food and the most common vegetables were beans of various sorts (Barth, 1890, pp. 383–4). In the Fezzan at Murzuq, Nachtigal waxed lyrical about the palm tree where he remarked that it seemed everywhere to reach the underground water level and no irrigation was needed to make it flourish luxuriantly. The fruit of the palm was regarded by many more highly than cereals. And the other parts of the tree were also of inestimable value; the trunk provided the beams for houses, the posts of doors, the apparatus for lifting water at the wells; the leaves were used for building huts and for fencing plots of land; the ribs of the leaves served as walking sticks, their pinnacles were woven into sandals and baskets, and the broad base was used as fuel and for charcoal. The fibrous tissue which enclosed the trunk of the palm tree was worked into durable ropes and the upper tip of the trunk was fermented into 'legbi', a strong sweet wine (Nachtigal, 1879, p. 113).

In the gardens of Murzuq in the 1860s Nachtigal saw the 'northern types of grain', wheat (*Triticum*), barley (*Hordeum*) 'join hands with the Negro cereals', durra (sorghum) and dukha (*Penicillania*). The former ripened in winter and the latter in summer, a pattern which could be observed in the vast 100-hectare centre pivot circles of the irrigated agricultural projects of Kufrah in southern Libya a century later. The following vegetables were also observed by Nachtigal (1879, p. 116):

broad beans (*Faba vulgaris*) purslane (*Portulaca*)
lubia bean (*Dolichos lubia*) aubergine (*Solanum melongena*)
peas (*Pisum sativum*) tomatoes (*Lycopersicum esculentum*)
carrots (*Daucus carota*)
melukhia (*Corchorus olitorius*) radishes (*Raphanus sativus*)
bamia (*Hibiscus esculentus*) beetroot (*Beta vulgaris*)
swedes (*Brassica oleracea* onions (*Allium cepa*)
cucumbers (*Cucumis sativus*) garlic (*Allium sativum*)
melons and water melons chillies (*Capsicum annuum*)
pumpkins (*Cucurbita pepo*)

Also cultivated were cummin (*Cuminum cyminum*), coriander (*Coriandrum sativum*), mallow (*Malva parviflora*) and Sudan chillies (*Capsicum conicum* variet. orient.). Tree crops in addition to the date palm included:

fig (*Ficus carica*) orange (*Citrus aurantium*)
lemon (*Citrus limonium*) vine

There were some few specimens of the quince, apple, almond, peach, apricot, and the olive. Nachtigal remarked that the fruit from those trees other than date was less important than fruit obtained from uncultivated plants and trees, for example the colocynth, and the berries known as *nabaq* (*Zizyphus spinachristi*), sidr (*Zizyphus lotus*) and the dum palm (*Hyphaene thebica*). Truffles were found frequently and when times were hard resort was made to the ghardeq (*Nitraria tridentata*) and sebat (*Aristida pungens*). The people of Fezzan 'also coaxed from their gardens a few cotton bushes (*Gossypium herbaceum*) indigo (*Indigofera argentea*)'. In all the larger gardens lucerne (*Medicago sativa*) and clover locally known as safsafa were found providing essential fodder and human food in emergencies. Tobacco was 'quite indispensable as a stimulant'. The agriculture of a central Saharan oasis was therefore extremely varied but sufficed to provide 'no more than a bare living'. Cattle rearing in addition assured subsistence, but agriculture did not support the community. Like many oases, Murzuq depended on trade, and even if trade was somewhat depressed in the 1860s at the time of Nachtigal's visit, 'traditionally, every tolerably well-to-do man was a merchant' (Nachtigal, 1879, p. 118). It must be remembered that the variety of cropping was achieved until the 1960s in many of the Sahara oases by technologies virtually unchanged for millenia. The intervention of colonial powers had some effect in Algeria, but elsewhere it was not until the search for oil began in the 1950s that new methods were deployed to irrigate, and not until the 1970s that modern oases based on completely new technologies began to appear.

21.5. THE OASES OF THE 1970S

The old oases for the most part remain. The oasis at Siwa in Egypt, sacred because of its ancient temples, still functions as an isolated agricultural settlement. The other oases of the Western Desert have been connected by roads and iron-ore extraction has affected Bahariya; the monastries at Wadi al Natrun are still Coptic sanctuaries. Al Kharga was designated as ripe for agricultural development and the groundwater flowing with a natural head has been utilized since the 1960s. The ambitious scheme has not been economically successful, however, and the artesian water no longer flows and water has to be pumped.

But Egypt has not been as well placed as other Saharan states to discover whether capital could be effectively combined with groundwater, indifferent desert soils and abundant solar energy in economically viable agricultural systems. The oil-rich countries such as Libya and Algeria have understandably

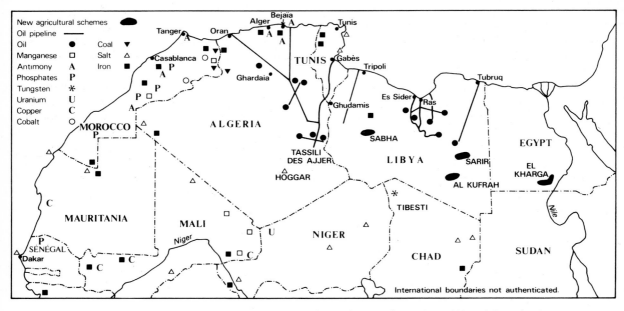

FIG. 21.5. The new oases. (Editions Atlas (1981), *Sahara*, Editions Atlas, Paris, p. 113, and the author.)

deployed more capital to test the idea that the desert can produce specialized agricultural products. In Libya two projects were set alongside the old oasis, at al Kufrah, formerly famous as a stopping point on the Wadai to Benghazi trade route and as a Sanusiyah centre. The production projection of some 10,000 hectares was initially a wheat project in 1971 then became in 1973 an intensive sheep-rearing scheme based on alfalfa, and reverted to a wheat scheme by 1975. The changing strategy reflected the uncertainties experienced by the project management especially those connected with water management. Water levels declined unexpectedly quickly in the early seventies and after two years of operation had declined almost 30 metres, a substantial proportion of the 40 metres which it was predicted they would fall in 20 years (Tipton and Kalambach, 1971). The second scheme at al Kufrah was a settlement scheme. The proposal was that desert people from al Kufrah and neighbouring oases would settle on schemes engineered to optimize the use of water by sprinkler and trickle methods and provide comfortable family accommodation in small hamlets (Allan, 1976). These schemes have not to date been agriculturally successful in economic terms and the effect on the local oasis has been negative. While the fall in water levels has not been unacceptable with respect to the continued working of the new schemes, the drawdown of the water upon which the oasis date palms has depended has been very serious. Trees have died and the standing water for which the oasis was famous has been reduced in volume.

Further north, at Sarir, a completely new oasis was created after 1975 when centre-pivot irrigation circles were set up to water upwards of 20,000 hectares. Similar to the Kufrah Production Project the purpose was to produce grain; wheat in winter and sorghum in summer. Again the project encountered difficulties and the commitment of the Libyan government to such ventures has faltered. In future instead of creating twentieth-century oases in mid-Sahara by developing water *in situ*, the next phase of agricultural development in Libya will be the utilization of Saharan water at the coast where it is hoped that effective management will be more easily arranged and Libyan labour more readily motivated to participate. Lessons learned in the Sahara to date have been that institutional impediments are quite as important as resource difficulties in hindering viable agriculture in the new schemes, and that Libyan staff and workers do not welcome working in remote areas. Hence the new strategy will be to move the

Saharan water at the rate of two cubic kilometres per year to the coast near Ajdabiyah from Sarir and at the rate of one cubic kilometre per day from the Braq area, and later from the Murzuq Basin, to provide water for the Gefara Plain near Tripoli where groundwater mismanagement has brought about an ecological disaster.

Interestingly the crops grown in the remaining oases in the private farms resemble very closely the list recorded by Nachtigal over a century ago. A study in the later 1960s (Allan *et al.*, 1973) showed that the farms of Fezzan and in the Jufrah oases were producing exactly the same range of tree, field and fodder crops as a century ago. However, in a number of places new schemes have caused declines in groundwater levels which have impaired the date gardens, for example at Sawknah. By the end of the 1970s, the traditional oasis was in decline in the oil-rich countries, although it continued as the only viable livelihood in less-well-endowed countries. The massive public sector developments in the oil-rich countries, such as those at al Kufrah and Sarir in Libya, do hark back to the past in one way in that they draw their inspiration from crops native both to the Mediterranean and Africa. Wheat and barley are raised in the winter and millet and sorghum in the summer, just as in Nachtigal's description.

The other major settlements of the modern Sahara owe nothing to the traditional oasis. They may be associated with transportation as were the traditional oases, but the transportation involves modern technology and often massive airstrips (4 kilometres long at al Kufrah). Others owe their existence to the oil industry. None of the new centres of settlement have, however, a trading function. The settlements in Libya and Algeria based on petroleum extraction grew rapidly in the 1950s and 1960s. Meanwhile urbanization has been rapid in all Saharan countries and especially in the oil-rich states reflecting the restless and rapidly changing state of society in Saharan Africa. But the developments do differ considerably in one respect from those of the past in that with very few exceptions, the changes have been effected in circumstances of relative political stability and the insecurity which marked earlier centuries has not featured as a dilemma for those responsible for devising development strategies.

REFERENCES

ALLAN, J. A., MCLACHLAN, K. S. and PENROSE, E. T. (1973) *Libya: Agriculture and economic development*. Cass, London.
ALLAN, J. A. (1976) The hexagon is alive and reproducing in the Sahara. *Area*, VII, 4, 275–278.
ALLAN, J. A. (1979) Managing agricultural resources in Libya: recent experience. *Libyan Studies*, X, 17–28.
BURDON, D. J. (1980) Infiltration conditions of a major sandstone aquifer around Ghat. In SALEM, M. J. and BUSREWIL, M. T., *The Geology of Libya*, Vol. II, Academic Press, London.
BAIER, S. (1978) A history of the Sahara in the nineteenth century, Working Paper No 4, African Studies Center, Boston University.
BARTH, H. (1980) *Travels and Discoveries in North and Central Africa*, Ward Lock, London, being a reprint of the original 1857 edition.
DUNN, R. E. (1977) *Resistance in the Desert: Moroccan responses to French imperialism, 1881–1912*. Croom Helm, London.
KEENAN, J. (1977) *The Tuareg: People of Ahaggar*. Lane, London.
NACHTIGAL, G. (1879) *Sahara and Sudan*, Vol. 1, *Fezzan and Tibesti*, translated by FISHER, A. G. B. and FISHER, H. J. and published in 1974 by Hurst, London.
PALLAS, P. (1980) Water resources of Libya (SPLAJ). In SALEM, M. J. and BUSREWIL, M. T., *The Geology of Libya*, Vol. II. Academic Press, London.
PETERS, E. (1982) Cultural and social diversity in Libya. In ALLAN, J. A., *Libya Since Independence*, pp. 103–120. Croom Helm, London.
SHAW, T. (1981) The late stone age in West Africa. In ALLAN, J. A., *The Sahara: Ecological change and early economic history*. Menas Press, London.
TIPTON and KALMBACH (1971) Report on the Kufrah Project, Tipton and Kalmbach, Tripoli.
UNESCO (1972a) *Etude des resources en eau du Sahara septentrional*. Unesco/UNDP Proj. Reg. 100.
UNESCO (1972b) *Study of the water resources of the Chad Basin*. UNDP/Unesco, Tech. Report Reg. 71.

CHAPTER 22

The Problems of Desertification

A. WARREN

University College, University of London

'DESERTIFICATION' is a difficult word to define. The ready image is of parched earth, dead trees, and the skeletons of cattle; but a precise and practical meaning is much more problematical. What can be said from the start is that it is an ecological change (Warren and Maizels, 1977) which takes place only on the desert's edges. Within true desert, there is very little to change. In getting nearer to a meaning, it would be helpful to be more precise about the idea of the desert itself. There have been many attempts at this. Some have been graphic: Capot-Rey's (1953) vivid definitions involved things like the distribution of bed-bugs; the line round the productive palmeries on the north; or the point at which on demounting from a car or a camel one's clothes become covered with the burs of the cram-cram grass. Another graphic image is of moving sand dunes. Other definitions are more statistical and, of these, the most widely employed today is Peveril Meig's (1953) climatic index which uses a moisture index developed by Thornthwaite (relating precipitation and evaporation). Other classifications depend on abstract ideas such as 'vegetational formations' or 'soil systems'. If we speak about the spread of deserts we must mean the movement of a boundary defined in one of these ways.

Bed-bugs, cram-cram, and moving dunes have patchy, irregular distributions, and are very difficult to delineate accurately except on very detailed maps. Their distribution, too, may have little practical meaning to many of the people who inhabit the desert. The nomads of the desert edge move back and forth across these boundaries with the seasons. In the Sudan, for example, the Kababish make use of the Gizu grazings by the Wadi Howar in wet seasons, and the northern edge of the savanna on the Qoz sands in the dry season. To them, nice distinctions about whether or not they are in or out of the desert have little real meaning. These difficulties also apply to more abstract definitions but, in addition, there are more serious problems associated with them. They are based, inevitably, on long-term averages, and in this way compel the inclusion of a discussion of the notion of the dynamic movement of the edge of the desert. We can exclude the seemingly ridiculous idea that the desert moves back and forth every year, but note that this exclusion is one of degree not of kind, for if anything is certain at the desert edge it is uncertainty: rainfall, which is obviously the prime limiting factor, is extremely unreliable. It is fickle at all scales: within a season, from year to year, from decade to decade, and so on. A striking example of fluctuation on the Saharan edge is from Agadez in Niger (Fig. 22.1). The apparently ominous decline in rainfall there between 1953 and 1973, brings us to the next conundrum about semi-arid climates for, as one moves along the spectrum of scales of climatic variability, dry seasons become droughts and droughts at some point become desertification.

Attempts to classify the idea of the desert using climatic indices must use averages; but the problem is to choose a suitable period for averaging. For instance, should one use 1945 to 1975, 1957 to 1973 or 1945

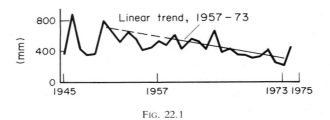

FIG. 22.1

to 1957 as the average at Agadez? Those who base their classification on vegetation usually stick to perennial shrubs and ignore ephemerals. In doing so, they make assumptions (not always explicitly) about the survival of shrubs and the age-class structure of the plant community. With soils there are even more dubious assumptions regarding the speed of adaptation of soils to new conditions.

At one end of the scale of climatic variation there are questions of climatic change, i.e. desertification due to some external cause. On the alarmist side of the debate are people like Reid Bryson (1974) and Bunting *et al.* (1976) who believe that the late sixties and early seventies of this century saw the start of a climatic change towards aridity on the south side of the Sahara (i.e. a secular, or long-term change, not defined more precisely than that). These prophesisers of doom used as part of their argument the no longer disputable fact that the Sahara did indeed expand during the Pleistocene and even the Recent (the last two million years or so). Part of the evidence is the widespread occurrence of sands blown by the wind in earlier times, into dune forms that are now anchored by trees, shrubs and grasses. In the Sudan, for example, these occur as far south as 8°N where the mean annual rainfall is over 1000 mm today (Warren, 1970). Desert sands occur widely in West Africa, notably in the Kano area, south of the Niger bend in Mali, and south of the Sénégal River in the Ferlo area of the Sénégal Republic (Grove and Warren, 1968). Similar sands even occur in some of the wettest parts of the Zaire basin. In North Africa, the signs of ancient deserts are less obvious, perhaps because of the barriers provided by the Atlas Mountains and the Mediterranean Sea, but fixed sands are found in north-eastern Sinai.

There is evidence, too, that Pleistocene and Recent changes have not all been towards more aridity but that there have been many fluctuations. Dry diatomaceous lake deposits are found in the northern Ténéré Desert in pockets between dunes, often with an ancient shore-line fringed with neolithic harpoons and fish vertebrae (Hall *et al.*, 1971). Lake levels across Africa seem to point to alternating wet and dry periods, continuing until very recent times (Street and Grove, 1979). The fascinating isolated fauna on mountain 'islands' in the Sahara—birds, baboons and crocodiles—are remnants of a wetter past.

Some climatologists point to rainfall figures for the last few decades and claim to detect minor cycles there too: drought in the latter part of the first decade of the century; dry again in the early thirties; and dry in the late sixties and seventies. The flood levels of the Sénégal River, for example, seem to have followed this sequence (see Rapp, 1974). Could there be a 20-year cycle of dry and wet years? When we come to the present, however, short-term climatic unreliability confounds the arguments of those who believe in long or short cycles. The past leaves convenient time-averaged deposits—lake levels and dune fields—but we are too close to the evidence of the present to know if what we experience today (a transient flooded hollow, or a small reactivated dune) is likely to last. Moreover, human activities now complicate the picture. Cautious meteorologists, applying time-series analysis to the rainfall figures over the last few decades, confine their predictions to saying that wet years are most likely to follow wet ones and dry years dry ones (Hare, 1977). Moreover, a careful look at the spatial pattern of the drought of the late sixties shows a very variable picture (Bradley, 1977).

Whether they recur regularly or at random, there is little doubt that there have been serious droughts, notably in the late sixties and early seventies of this century. But drought should and can be distinguished from desertification. This can be done for instance in the case of the results of a French team who happened to be studying an ecosystem in central Sénégal as part of the International Biological Programme (Bille,

1974). They departed before the height of the drought, but returned with the rains. Thus they found that annual and ephemeral plants had disappeared in the drought, but grew from well-protected seed and other propagules when the rains came again. Perennials did not flower or leafed less prolifically in the dry years. Animal populations contracted to a few favoured sites or emigrated, but proliferated again or returned after the drought. In other words, semi-arid ecosystems are adjusted to cope with drought that is a 'normal' recurring part of their environment. Desertification is something different: the term can only be useful if it is distinguished from drought and is applied to changes of longer amplitude that somehow damage the ability of the ecosystem to recover from drought.

Nonetheless, desertification and drought could be, and probably are, linked. Desertification may either be made apparent or may be triggered by a drought. The slow sapping of the ability of the ecosystems to recover, caused by the cutting of the key shrubs that create small oases under which other species survive, by reducing seed production, or by soil erosion, may not be apparent during a run of good years, but will suddenly become obvious when the drought comes. A run of good years too may encourage the growth of herds which will be left with very poor pasture when the dry years begin. They then resort to eating almost anything, and are compelled to devour the few remaining plants, thus damaging the resilience of the system as a whole: desertification would then have been initiated by the drought.

If meteorologists disagree about climatic changes in recent decades, they are a little more confident about another climatic effect sometimes known as the 'Charney' effect after one of the meteorologists who investigated it in the context of the Sahara (Charney, 1975). According to Charney, when the vegetation is removed from an area—by over-grazing or cutting trees for firewood—the 'albedo' (the amount of the sun's energy reflected by the earth's surface) increases. In consequence, the temperature in the column of air above the desert surface decreases, so that there is a tendency for air subsidence. This, in turn, decreases the likelihood of rainfall. The effect has nowhere been demonstrated conclusively, but the results of experiments on large computer numerical models of the atmosphere (notably at Boulder, Colorado) are suggestive. Fig. 22.2 shows how the effect might work in West Africa. If it actually occurs, such an effect would clearly damage the ecosystem's powers of recovery from a drought—i.e. would be desertification.

There are, of course, other ways in which human activity might encourage desertification. The

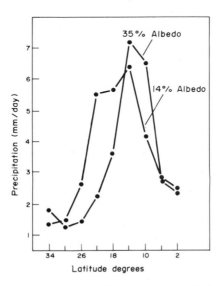

FIG. 22.2. The possible effect of removing vegetation from the Sahel: the increased albedo (14 to 38%) discourages rainfall in the critical 14° to 26°N zone—i.e. the Sahel (after Charney, 1975).

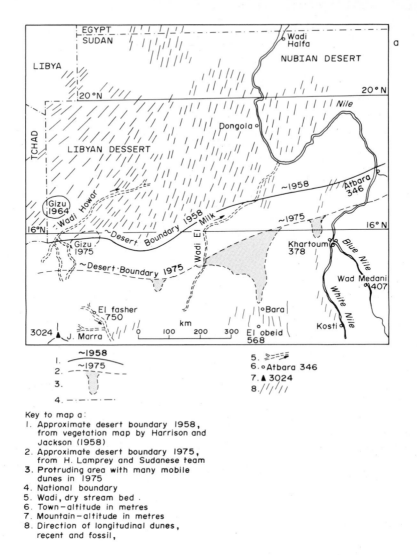

Key to map a:
1. Approximate desert boundary 1958, from vegetation map by Harrison and Jackson (1958)
2. Approximate desert boundary 1975, from H. Lamprey and Sudanese team
3. Protruding area with many mobile dunes in 1975
4. National boundary
5. Wadi, dry stream bed.
6. Town—altitude in metres
7. Mountain—altitude in metres
8. Direction of longitudinal dunes, recent and fossil,

FIG. 22.3. Changes in the desert boundary in the Sudan (after Lamprey, 1976).

hypotheses are easily understood; the problem is to decide whether the processes suggested have taken place in reality. Theoretically, and in controlled experiments, the removal of plant cover by grazing, trampling or cutting for firewood undoubtedly damages the resilience of ecosystems. Plants intercept rain and channel it down their stems to the roots. In the Sudan, there is a sub-species of *Acacia melifera* which does this. Plants also roughen soil surfaces, slowing and deepening the overland flow, so that there is more time (and more 'pressure head') for infiltration. Their litter and the community of decomposer organisms that depend on it, break up the surface soil, and encourage faster inflow than in purely mineral soil. Where there are no plants, the water runs off quickly and a crust forms on the surface which further discourages infiltration. Trampling exacerbates the problem by beating down the surface. Erosion is particularly damaging in deserts where most of the soil nutrients are held within a few centimetres of the surface (Charney and Cowling, 1968). Desertification, if it takes place, could occur at various degrees of severity:

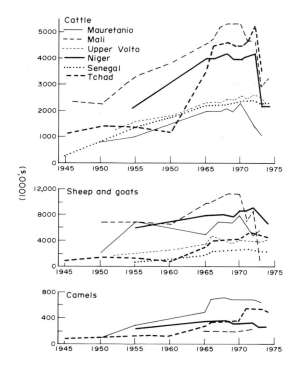

FIG. 22.4. Reported trends in domestic animal numbers in the Sahel (after SEDES 1975).

it could be drastic on a hard clayey soil on slopes that allow runoff; but it could be a phenomenon little more lasting than drought on a nearby sandy *Qoz*.

The colour magazines in the early seventies showed that it was easy enough to find examples of frightening landscapes. But neither they, nor many of the scientists who examined the Sahel, were concerned with the fine but vital distinction between drought and desertification. A survey of the central Sudan in which the vegetation in 1975 was compared with that of an earlier survey in 1958 showed a movement southward of vegetation zones of 150 km classified using mainly perennial species (Lamprey, 1976) (Fig. 22.3). The Niger River was becoming much more erratic in its flow, a sign of fast run-off and little storage, probably associated with devegetation (Delwaulle, 1973). Off the West African coast dust clouds were thicker in the early seventies than before (Rapp, 1974), and there were reports of hardship, and of the death of many cattle (SEDES, 1975) (Fig. 22.4). Many experienced scientists believed passionately that the Sahara was surrounded by a zone of desertification (see Boudet, 1972; Le Houérou, 1974; Floret and Le Floc'h, 1973).

To observe some caution over this alarm is not to deny that there has been some long-term degradation, but care is needed. Similar cries arose in the 1930s and 1940s when the chief prophet of doom was Stebbing (e.g. 1953)—but what damage he found was subsequently seen to be more the result of drought than longer-term damage. Human populations were lower then, however, and the cries not as shrill. More important are the nearly intractable problems of arriving at any certain assessment that there has been long-term damage. In such a temporally and spatially variable environment, where plant production alone can vary by many hundreds of percent from place to place, season to season, and year to year, due to topographic position, rainfall variability and the vagaries of local grazing strategies (Bourlière and Hadley, 1970), one would need several hundred square kilometres, fenced and managed in a complex experiment on grazing, and taken over a lean and a drought cycle (?30 years) to find a certain answer.

The results of the less ambitious experiments that have been undertaken merely suggest clues to what is happening, and they also raise additional questions. There have been a number of 'exclosure' experiments on both sides of the Sahara; some in the Sahel and the Sudan have been reviewed by Bradley (1977). What happened in Halwagy's (1962) experiment was fairly typical. One exclosure of 31 ha was fenced off in 1955; by 1962 the perennials were flourishing, but there were few new plants, simply larger ones. Annuals recovered rapidly but then remained at a constant level.

How are these results to be interpreted? If recovery is possible within such a short period, does it not show that the damage was not severe i.e. not desertification? Of course, it may be that recovery was more apparent than real—i.e. that the vegetation had been damaged so severely by long-continued grazing that the propagules of some plants had been completely lost, so that recovery was only partial. In New South Wales, where there is more historical data from the period before the introduction of domestic grazers and browsers, Williams (1969) believed this to be the case. The long history of domestic animals in Africa means that we may never know if this has been the case on the edges of the Sahara.

This may seem an academic quibble but there are practical questions as well. Above all—is the 'recovery' desirable? At first sight and, even with some theory, one might say 'of course'. Do not domestic animals damage first those plants they find most palatable? This theory would have it that grazing makes a pasture less attractive because the preferred species are eliminated, and that the 'recovery' after grazing includes a resurgence of the palatable species. This may be true of some pastures some of the time, as Boudet et al. (1969) showed for the pastures of the Dallol Mauri in Niger. But such a conclusion needs to be tempered with caution.

First, many abandoned pastures become infested with large perennial trees and shrubs that would otherwise have been kept low by browsing animals. They grow out of the reach of most domestic browsers and shade many of the lower, palatable species. The degeneration of pastures into scrub is an especially serious problem in East Africa (see Pratt and Gwynn, 1974).

Second, is the now well-known process whereby moderate grazing and browsing can actually increase the productivity (rate of growth) of some plants while keeping down the total biomass (or production). Pruning stimulates growth, the removal of dead growth allows light to the new shoots and animal faeces provide a source of nitrogen (e.g. Mott, 1960). Unbrowsed shrubs tend to be more woody and have a smaller proportion of fresh green growth than grazed ones. Many pastoralists use fire as a tool to keep productivity high.

In other words, grazing and the deliberate suppression of some kinds of growth can actually promote the useful productivity of pastures. The question is now—what is the optimal amount and best type of grazing? There can be no unequivocable answer to this question. It will depend among many other things on what period and in what season the grazing takes place, the nature of each particular season (good, bad or indifferent), and the nature of the grazing animal favoured by the local pastoralists. If pastoralists graze goats—as do the Kel Ulli caste ('They of the goats') among the Taureg—they want low shrubs. Their aristocratic erstwhile over-lords require taller trees for their camels. Their neighbours, the cattle-keeping Fulani, with whom they share some pastures, want grass not shrubs and therefore favour pastures that have fewer woody perennials. Each group is engaged in a complex game with the environment, with their neighbours, and with their own cultural norm, that produces different demands on the pasture. The innocent ecologist from Europe or America, let alone the paparazzi of the colour magazines, can all too readily jump to facile conclusions about such involved systems of land use.

The caution expressed above about desertification must not be interpreted as a dismissal of the problem. Even without concrete evidence, the theory and the judgement of so many experts cannot be discounted. The reasons for the problem are an important part of the diagnosis, but they are even harder to establish.

Nomadic pastoralists are the only users of the greatest part of the desert edge. In addition, cultivators crept north into the Sahel of West Africa and the Sudan during the goods years of the fifties, and there have always been adventurous dry farmers on the Mediterranean edge of the Sahara. Such people have undoubtedly contributed to the destabilization of soils and, in places, to soil erosion by wind and rain; but their total contribution is much less than that of the pastoralists.

A charitable view of the pastoralists, common among commentators today, sees them as 'natural ecologists' (e.g. Toupet, 1975). They are said to be able to balance the vagaries of the climate and the requirements of their stock to achieve a long-term equilibrium of men, animals and pastures. One supporting piece of evidence for this view is that their numbers seem to change remarkably little; another is that since the early domestication of cattle and the later taming of camels, there have always been pastoralists in and around the deserts of the Old World who have not, so far, degraded their habitat. But it is also true that these people have always depended on more than just the desert pastures: on trading, caravaneering, salt, sorcery, protection, warfare, social dominance and so on. In Theodore Monod's words, they live 'in the desert but not on the desert' (Monod, 1975). Most of these additional sources of support have now been removed.

The uncharitable view, common among frustrated colonial administrators in the fifties and sixties, was that nomads were deeply irrational, building up herds regardless of the consequences, burning the range for the mere enjoyment of it, hide-bound by antiquated custom, victims of 'cowdolatory' (see a review by Baker, 1975). A third view, equally sweeping, sees these societies as the victims of colonial dominance, deprived of responsibility for their own fate, destabilized by ill-informed edicts emanating from distant imperial capitals, and robbed of part of their production by colonial trade (Raynaut, 1977).

If a simple truth is to be found it seems to be that, if their own members did not rise, then the nomads' herds did grow during the fifties and sixties. Statistics are remarkably unreliable as Sandford (1978) has pointed out for, to count the cattle of a suspicious, mobile and wily group like Fulani, Taureg or Bedouin is wellnigh impossible! Nevertheless, what evidence there was pointed to increasing numbers on diminishing pasture. This was, in turn, most probably the result of eight forces. First among these was the peace and control of colonial administration: fewer raids, less stealing, more recourse to litigation (see Horowitz, 1975). Second was the veterinary control of epizootic diseases such as rinderpest, which decimated herds in the past. Third was the opening of lightly grazed pastures by the development of bore-holes for water with new deep-drilling equipment financed by central governments (as in Niger, see UNCOD, 1977a). Fourth was the run of particularly good years (before about 1965 in the Sahel) that allowed cattle numbers to grow. Fifth was the burgeoning market for meat both in the central West African cities and in the cities and new oil settlements of North Africa and Arabia. Sixth was the influx of capital to the expanding industries of the local nations (as in Tunisia, see UNCOD, 1977b); pastoralists saw the most secure investment for their new-found cash to be more cattle. Seventh has been the widespread meddling with pastoral strategies by central government officers (Baker, 1977). And eighth has been the appropriation of some of the pastoralists' best pasture by agriculturalists (Sandford, 1978).

Finally, what should be done to decelerate desertification? The UN Desertification conference at Nairobi suggested an outline programme for each country: surveys, central directorates, etc. It also encouraged some co-operative programmes: a survey of groundwater in the 'continental intercalaire' formation in the Sahara: a programme for marketing cattle more effectively and with more benefit to the producer, and an organization based in Ouagadougou to encourage West African co-operation.

One of the more charismatic schemes was the idea of the planting of Green Belts along the desert edges, north and south. The southern scheme has been transformed, mercifully, into the encouragement of fuel-wood plantations, for an unbroken belt of trees would be impossible to co-ordinate. In the north, there is more movement towards an actual '*cienture verte*'. There is enough money, particularly in Algeria, to channel some into planting. As a measure against desertification, however, such a scheme is ludicrous. Herds will be grazed behind the barrier, and this will continue to destroy the ecosystem there. Desertification is not some contagion that spreads without human agency and whose advance can be stopped by a few lines of trees. It is not like a tidal wave spreading in from the desert. It *could* result from a change in climate and, if so, nothing will stop it. It is, however, more likely to be the result of over-grazing by domestic herds, in which case its cure lies in some kind of social change.

REFERENCES

BAKER, R. (1975) 'Development' and the pastoral peoples of Karamoja, North-East Uganda: an example of the treatment of symptoms. In *Pastoralism in Tropical Africa*, edited by T. MONOD, pp. 187–205. Oxford Univ. Press, London.

BILLE, J. C. (1974) Recherches écologiques sur un savane sahélienne du Ferlo septentrional, Sénégal: 1972, Année sèche au Sahel. *Terre et Vie*, **28**, 5.20.

BOUDET, G. (1972) Désertification de l'Afrique tropicale sèche. *Adansonia*, sér. 2, **12**, 505–524.

BOUDET, G., LAMARQUE, G., LEBRUN, J. P. and RIVIÈRE, R. (1969) Pâturages naturels du Dallol Mauri (République du Niger). *Étude Agrostologique*, **26**, IEMVT Maisons Alfort, France.

BOURLIÈRE, F. and HADLEY, M. (1970) The ecology of tropical savannas. *Ann. Rev. Ecol. & Systematics*, **1**, 125–152.

BRADLEY, P. N. (1977) Vegetation and environmental change in the West African Sahel. In *Land Use and Development*, edited by P. O'KEEFE and B. WISNER, pp. 34–54. Internat. Afr. Inst. London.

BRYSON, R. A. (1974) A perspective on climatic change. *Science*, **184**, 753–760.

BUNTING, A. H., DENNETT, M. D., ELSTON, J. and MILFORD, J. R. (1976) Rainfall trends in the West African Sahel. *Quart. Jnl. Royal Meteor. Soc.* **102**, 59–64.

CAPOT-REY, R. (1953) *Le Sahara français*. Presses Universitaires de France, Paris, 564 pp.

CHARLEY, J. L. and COWLING, S. L. (1968) Changes in soil nutrient status resulting from overgrazing and their consequences in plant communities of semi-arid areas. *Proc. ecol. Soc. Anstr.* **3**, 28–38.

CHARNEY, J. (1975) Dynamics of deserts and drought in the Sahel. *Quart. Jnl. Roy. Meteorol. Soc.* **101**, 193–202.

DELWAULLE, J. C. (1973) Désertification de L'Afrique au Sud du Sahara. *Bois et Forêts Tropiques*, **149**, 3–20.

FLORET, C. and LE FLOC'H, E. (1973) Production, sensibilité et évolution de la végétation et du milieu en Tunisie présaharienne. *CNRS, CEPE,* Montpellier, France, No. 71, 45 pp.

GROVE, A. T. and WARREN, A. (1968) Quaternary landforms and climate on the south side of the Sahara. *Geogr. J.* **134**, 194–208.

HALL, D. N., WILLIAMS, M. A. J., CLARK, J. D., WARREN, A., BRADLEY, P. and BEIGHTON, P. (1971) The British Expedition to the Aïr Mountains. *Geogr. J.* **137**, 445–467.

HALWAGY, R. (1962) The impact of man on semi-desert vegetation in the Sudan. *J. Ecol.* **50**, 263–273.

HARE, F. K. (1977) Climate and desertification report to the UN conference on Desertification. Also in *Desertification: its causes and consequences*, ed. U.N. Conf. on Desertification, pp. 63–167. Pergamon, Oxford.

HOROWITZ, M. M. (1975) Herdsmen and husbandmen in Niger: values and strategies. In *Pastoralism in Tropical Africa*, edited by T. MONOD, pp. 387–405. Oxford Univ. Press, London.

LAMPREY, H. (1976) *Survey of Desertification in Kordofan Province, Sudan*. UNEP, Nairobi.

LE HOUÉROU, H. N. (1974) Deterioration of the ecological equilibrium in the arid zones of North Africa. *Agron. Res. organiz. Volcanic Centre*, Beit Dagan, Israel, *Spec. Pub.* **39**, 54–57.

MEIGS, P. (1953) World distribution of arid and semi-arid nomo-climates. In *Reviews of Research on Arid Zone Hydrology*, pp. 203–204. UNESCO, Paris.

MONOD, T. (1975) Introduction. In *Pastoralism in Tropical Africa*, pp. 2–183. Oxford Univ. Press, London.

MOTT, G. O. (1960) Grazing pressure and measurement of pasture production. *Proc. 8th Int. Grassland Congr. Reading, England*, pp. 606–611.

PRATT, D. J. and GWYNNE, M. D. (1977) *Rangeland Ecology and Management in East Africa*. Univ. London Press, London.

RAPP, A. (1974) A review of desertization in Africa, water, vegetation and man. *Sec. for Internat. Ecol.*, Sweden, 77 pp.

RAYNAUT, C. (1977) Lessons of a crisis. In *Drought in Africa*, 2, edited by D. DALBY, R. J. H. CHURCH and FATIMA BEZZAZ, pp. 3–16. Internat. Afr. Inst. London.

SANDFORD, S. (1976) *Pastoralism under Pressure*. Overseas Devel. Inst. London, Review No. 2.

SEDES (1975) *Éléments de statistique pour une analyse du développement socioèconomique dans les six pays du Sahel.* 266 pp.

STEBBING, E. P. (1953) *The Creeping Desert in the Sudan and elsewhere in Africa.* Sudan Govt. Khartoum.

STREET, F. A. and GROVE, A. J. (1979) Global maps of lake-level fluctuations since 30,000 B.P. *Quaternary Research*, **12**, 83–118.

TOUPET, C. (1975) Le nomade, conservateur de la nature? L'example de la Mauritanie centrale. In *Pastoralism in Tropical Africa*, edited by T. MONOD, pp. 455–467. Oxford Univ. Press, London.

UNCOD (1977a) *Case Study on Desertification. The Eghazer and Azawak Region, Niger.* UN Desertification Conference, Nairobi, A. Conf. 74/14.

UNCOD (1977b) *Case Study on Desertification: Oglat Merteba Region, Tunisia.* UN Desertification Conference, Nairobi, A. Conf. 74/12.

WARREN, A. (1970) Dune trends and their implications in the central Sudan. *Zietschr. für Geomorphologie*, Suppl. 10, pp. 154–180.

WARREN, A. and MAIZELS, J. K. (1977) Ecological change and desertification, UNCOD document. Also in *Desertification, its Causes and Consequences*, pp. 169–260. UN Desertification Secretariat, Pergamon, Oxford.

WILLIAMS, O. B. (1969) Studies in the ecology of the Riverline Plain, V. Plant density response of species in a Danthonia caespitosa grassland to 16 years of grazing by merino sheep. *Aust. J. Bot.* **17**, 255–268.

Index

KEY ENVIRONMENTS

Other Titles in the Series
